ANNALS OF THE NEW YORK ACADEMY OF SCIENCES

Volume 988

EDITORIAL STAFF

Director, Publishing and New Media
SARAH GREENE

Managing Editor
JUSTINE CULLINAN

Associate Editor
LINDA HOTCHKISS MEHTA

The New York Academy of Sciences
2 East 63rd Street
New York, New York 10021

THE NEW YORK ACADEMY OF SCIENCES
(Founded in 1817)

BOARD OF GOVERNORS, September 2002 – September 2003

TORSTEN N. WIESEL, *Chairman of the Board*
JOHN F. NIBLACK, *Vice Chairman*
JOHN T. MORGAN, *Treasurer*
ELLIS RUBINSTEIN, *Chief Executive Officer* [ex officio]

Honorary Life Governors
WILLIAM T. GOLDEN   JOSHUA LEDERBERG

Governors

ELEANOR BAUM	KAREN E. BURKE	PRAVEEN CHAUDHARI
R. BRIAN FERGUSON	GERALD D. FISCHBACH	RONALD L. GRAHAM
MARNIE IMHOFF	JACQUELINE LEO	BRUCE McEWEN
PAUL MARKS	RONAY MENSCHEL	SANDRA PANEM
PETER RINGROSE	LEE G. VANCE	DEBORAH WILEY

HELENE L. KAPLAN, *Counsel* [ex officio]

CHEMICAL EXPLANATION

CHARACTERISTICS, DEVELOPMENT, AUTONOMY

ANNALS OF THE NEW YORK ACADEMY OF SCIENCES
Volume 988

CHEMICAL EXPLANATION

CHARACTERISTICS, DEVELOPMENT, AUTONOMY

Edited by Joseph E. Earley, Sr.

The New York Academy of Sciences
New York, New York
2003

Copyright © 2003 by the New York Academy of Sciences. All rights reserved. Under the provisions of the United States Copyright Act of 1976, individual readers of the Annals are permitted to make fair use of the material in them for teaching or research. Permission is granted to quote from the Annals provided that the customary acknowledgment is made of the source. Material in the Annals may be republished only by permission of the Academy. Address inquiries to the Permissions Department (permissions@nyas.org) at the New York Academy of Sciences.

Copying fees: For each copy of an article made beyond the free copying permitted under Section 107 or 108 of the 1976 Copyright Act, a fee should be paid through the Copyright Clearance Center, Inc., 222 Rosewood Drive, Danvers, MA 01923 (www.copyright.com).

♾ The paper used in this publication meets the minimum requirements of the American National Standard for Information Sciences—Permanence of Paper for Printed Library Materials, ANSI Z39.48-1984.

Library of Congress Cataloging-in-Publication Data

Summer Symposium on the Philosophy of Chemistry and Biochemistry (6th : 2002 : Georgetown University)
  Chemical explanation: characteristics, development, autonomy / edited by Joseph E. Earley, Sr.
    p. ; cm. -- (Annals of the New York Academy of Sciences ; v. 988)
  "This volume is a result of a conference entitled Sixth Summer Symposium on the Philosophy of Chemistry and Biochemistry, organized by the International Society for the Philosophy of Chemistry and held on August 04–08, 2002, at Georgetown University in Washington, D.C."—Contents p.
  Includes bibliographical references and index.
    ISBN 1-57331-456-0 (cloth : alk. paper) -- ISBN 1-57331-457-9 (paper : alk. paper)
    1. Chemistry—Philosophy—Congresses. 2. Biochemistry—Philosophy—Congresses. 3. Philosophy—Congresses. [DNLM: 1. Biochemistry—Congresses. 2. Chemistry—Congresses. 3. Philosophy—Congresses. QU 4 S955c 2003]   I. Earley, Joseph E. II. International Society for the Phiosophy of Chemistry. III. Title. IV. Series.
  Q11.N5 vol. 988
  [QD6]
  500 s--dc21
  [540/   2003002236

GYAT/PCP
Printed in the United States of America
ISBN 1-57331-456-0 (cloth)
ISBN 1-57331-457-9 (paper)
ISSN 0077-8923

ANNALS OF THE NEW YORK ACADEMY OF SCIENCES

Volume 988
May 2003

CHEMICAL EXPLANATION

CHARACTERISTICS, DEVELOPMENT, AUTONOMY

Editor
JOSEPH E. EARLEY, SR.

This volume is the result of the **Sixth Summer Symposium on the Philosophy of Chemistry and Biochemistry**, organized by the International Society for the Philosophy of Chemistry and Biochemistry and held at Georgetown University in Washington, DC on August 4–8, 2002.

CONTENTS

Foreword. *By* ROM HARRÉ ix

Introduction. *By* JOSEPH E. EARLEY, SR. xi

Part I. General Considerations

Structural Explanation in Chemistry and Its Evolving Forms. *By* ROM HARRÉ . 1

Explaining Explanation in Chemistry. *By* GRANT FISHER 16

On Explanatory Practice and Disciplinary Identity. *By* ANDREA I. WOODY ... 22

The *Ignis Fatuus* of Reduction and Unification: Back to the Rough Ground.
   By J. VAN BRAKEL 30

Autonomy, Explanation, and Theoretical Values: Physicists and Chemists on
   Molecular Quantum Mechanics. *By* ROBIN FINDLAY HENDRY 44

Natural Kinds, Explanation, and Essentialism in Chemistry.
   By REIN VIHALEMM 59

The Primary Properties? *By* S.H. VOLLMER 71

Varieties of Properties: An Alternative Distinction among Qualities.
   By JOSEPH E. EARLEY, SR. 80

Property Reduction in Chemistry: Some Lessons.
By G. K. VEMULAPALLI   90

Chemical Substances and Intensive Properties. *By* PAUL NEEDHAM   99

Paradoxes of Measurement. *By* PATRICK A. HEELAN   114

Part II. Specific Applications

Chemical Kinetics and Dynamics. *By* ILYA PRIGOGINE   128

Reaction Mechanisms and Chemical Explanation. *By* GIUSEPPE DEL RE   133

Explanation in Organic Chemistry. *By* WILLIAM GOODWIN   141

On the Ordinality of Causes in Complex Autocatalytic Systems.
By ROBERT E. ULANOWICZ   154

Chirality and Handedness: The Ruch "Shoe–Potato" Dichotomy in the
Right–Left Classification Problem. *By* R. BRUCE KING   158

How Harmful Is the First Law? *By* GEORG JOB AND TIMM LANKAU   171

Physical Explanation of the Periodic Table. *By* V. N. OSTROVSKY   182

Gauge Theory and Chemical Structure. *By* JAMES MATTINGLY   193

Symmetry in Basic Physical Laws. *By* ELMAR KÜHL AND TIMM LANKAU   203

Part III. Representation and Instrumentation

The Metaphorical Foundations of Chemical Explanation.
By THEODORE L. BROWN   209

"Causes" in Chemical Explanations. *By* JANET D. STEMWEDEL   217

John Dalton and the Aesthetics of Molecular Representation.
By TAMI I. SPECTOR   227

Writing as Thinking. *By* JEFFREY KOVAC   233

Statement Analysis in Chemistry. *By* CLAUS JACOB   239

Beyond the Dimensionality of Visualization in Chemistry.
By ANDRZEJ BUREWICZ AND NIKODEM MIRANOWICZ   244

Justifying Instrumental Techniques of Analytical Chemistry.
By DANIEL ROTHBART AND LADISLAV KOHOUT   250

Negotiated Identities of Chemical Instrumentation: The Case of Nuclear
Magnetic Resonance Spectroscopy, 1956–1969. *By* JODY A. ROBERTS . .   257

Part IV. Development and Societal Impact

Chemical versus Biological Explanations: Interdisciplinarity and
Reductionism in 19th Century Life Sciences. *By* JOACHIM SCHUMMER . .   269

Richard Rufus's Theory of Mixture: A Medieval Explanation of Chemical
Combination. *By* MICHAEL WEISBERG AND REGA WOOD   282

The Lavoisier–Kirwan Debate and Approaches to the Evaluation of Theories.
By MICHAEL AKEROYD   293

Social Background of the Discovery and the Reception of the Periodic Law of the Elements: Recognizing the Contribution of Dmitri Ivanovich Mendeleev and Julius Lothar Meyer. *By* MANSANORI KAJI 302

How G. N. Lewis Reset the Terms of the Dialogue between Chemistry and Physics. *By* PAUL A. BOGAARD 307

The Superiority of "Chemical Thinking" for Understanding Free Human Society According to Hegel. *By* MARK R. NOWACKI AND WILFRIED VER EECKE 313

An Organic Framework for a Philosophical Appreciation of Chemical Phenomena. *By* RICHARD K. KHURI 322

Chemistry Beyond Positivism. *By* WERNER W. BRANDT 335

Chemical Self-Organization, Complexification, and Process Metaphysics. *By* JAMES F. SALMON 345

The Philosophy of Kitaro Nishida and Current Concepts of the Origin of Life. *By* KO HOJO 353

Constructivism, Relativism, and Chemical Education. *By* ERIC SCERRI 359

The National Science Foundation and the Philosophy of Chemistry. *By* BRUCE E. SEELY 370

Index of Contributors 377

Assistance was received from several Georgetown University sources:

- **GEORGETOWN COLLEGE**
- **THE GRADUATE SCHOOL**
- **THE SCHOOL OF SUMMER AND CONTINUING EDUCATION**
- **THE DEPARTMENTS OF PHILOSOPHY AND OF CHEMISTRY**
- **THE CENTER FOR SCIENCE AND RELIGION**
- **OFFICE OF THE PRESIDENT**

The New York Academy of Sciences believes it has a responsibility to provide an open forum for discussion of scientific questions. The positions taken by the participants in the reported conferences are their own and not necessarily those of the Academy. The Academy has no intent to influence legislation by providing such forums.

Foreword

ROM HARRÉ

Department of Psychology, Georgetown University, Washington, DC 20057, USA

That there is a philosophy of chemistry at all presupposes that the nature and methods of the chemical sciences cannot be redescribed exhaustively in terms of concepts drawn from physics and its philosophy. There are a number of dimensions along which the autonomy of chemistry can be discussed. Here are some. Do the explanations of chemical phenomena eventually modulate into and terminate in explanations in which only concepts from physics are invoked? Are chemical kinds and species reducible to complex arrangements of physical kinds and species? Is quantum chemistry just a branch of quantum physics? At least one important role for the philosophy of chemistry is the critical exploration of these dimensions.

Philosophers of science interested in chemistry can take an apologetic stance, manning the defenses for a last-ditch stand against salvos of reductionist arguments. On the other hand, they can address the issue of autonomy positively, by discussing the ways in which the science of chemistry is pursued and drawing out the presuppositions on which the rationality of these ways depends. They will be critically examining various specifications of its characteristic subject matter, methods of enquiry and types and standards of explanation. If, at the end of the day, these seem to share features that are unlike those of adjoining sciences, the issue of the reduction of chemistry to something else has been settled, on the hoof, so to say. Even if the world consists of only one kind of ultimate stuff, the ways of investigating its various modes and manifestations are highly various. Each discloses aspects of material reality that are obscured when viewed in some other way. We might think of the array of the natural sciences as like the repertoire of stains with which a microbiologist can color a cell. Each stain reveals an aspect of the anatomy of the cell invisible when another stain is employed. One brings out the mitochondria, another the nucleus, another the cell wall, and so on. What does the chemical approach to material nature reveal that could be revealed in no other way?

Readers of the papers collected in this volume will surely notice the upbeat way in which the authors discuss ontological, methodological, and historical aspects of the science of chemistry. The science of chemistry exists as a distinct institution. Those who call themselves chemists pursue their work in a certain way. What are its characteristics? For example, the role of visualizability in the methodology and pedagogy of chemistry has profound ontological significance. For the physicist, the very idea of *visualizing* quantum states or field potentials makes no sense. It follows that,

Address for correspondence: Department of Psychology, Georgetown University, Washington, DC 20057.
 harre@georgetown.edu

in chemistry, space and time play a role that is quite distinct from that which they play in physics. This observation emerges from a study of chemistry as a practice. It is not hauled into view to block an aggressively reductionist takeover bid from a philosopher imbued to excess with the spirit of the absolute unity of science. The unique way that the objects of chemical research fulfill the philosophical criteria for the identification of distinct substances illustrates the same feature of these discussions in another way. Concepts like "natural kind," "shape," "intensive property," "feature," and so on are made use of by philosophers in their efforts to understand a variety of scientific disciplines. Yet, in investigating the materials to which chemists address their research efforts, those concepts take on a specially "chemical" character.

In these and other ways, work in the philosophy of chemistry displays chemistry as an autonomous science, owing a great deal to geology, mathematics, and physics, and inspired by problems in biology—but neither reduced to or reducing the sciences with which it is so intimately associated.

Introduction

JOSEPH E. EARLEY, SR.

Department of Chemistry, Georgetown University, Washington, DC 20057, USA

The current state of philosophical understanding of scientific explanation has been described as "an embarrassment for the philosophy of science."[1] A previously existing consensus on a general model[2] of explanation has collapsed.[3] Explanatory practices of scientific disciplines as diverse as archaeology[4] and econometrics[5] are now being examined closely—in the spirit of Wittgenstein's injunction, "don't think, but look!"[6]

Chemistry is an ancient discipline—and is also currently among the most productive and populous of scientific fields, with intimate connections to every area of human life and culture. Chemistry and the emerging discipline of the philosophy of chemistry both have much to contribute to ongoing philosophical discussion concerning the nature of scientific explanation.[7]

The volume—based on papers presented at the Sixth Summer Symposium of the International Society for the Philosophy of Chemistry, held at Georgetown University in August, 2002—is organized into four sections of roughly equal size.

- The first section considers questions of central philosophic importance: the nature and development of "structural" explanation (especially characteristic of chemistry), relationships of chemical explanation to other sorts of discourse (nomological explanation, reduction, supervenience, and emergence), and disciplinary identity. Also considered are aspects of "natural kinds," qualities, and properties—and the relevance of certain concepts of the "phenomenology" of E. Husserl to philosophy of chemistry.

- The second part of the volume deals more directly with specific scientific problem areas. A main focus is mechanistic explanation in chemistry (see also notes 8 and 9). Autocatalysis, self-organization, chirality, thermodynamics, gauge theory and symmetry breaking are also discussed.

- Papers in the third part of the book generally deal with representation and instrumentation. Topics include: how chemists use causal language, metaphor and representation; the fundamental significance of instrumental techniques in chemistry; and the complex process of instrument development.

- In addition to its importance for philosophy of science, chemistry and chemical patterns of thought have major significance for many parts of culture. The final segment of this collection of papers concerns aspects of the historical development of chemical thought and practice, and also the interrelationship of chemical ways of thinking and wider social concerns. One historical study deals with a medieval contribution to the problem, posed by Aristotle, of how chemical composition (*mixis*) is distinguished from mere aggregation. Others concern the chemical revolution of the eighteenth century, the social reception

Mendeleev's periodic table, and relationships between chemical and biological explanations in the nineteenth century. The mutual impact of chemical thought and wider cultural concerns are explored in terms the relationship between Hegel's notion of "chemism" and the functioning of free societies. Ideas of several widely influential philosophers (Kitaro Nishida of Japan, the philosopher-chemist Michael Polanyi, and the Jesuit paleontologist, Pierre Teilhard de Chardin). Deleterious impacts of "social constructivist" ideas on chemical education and opportunities for the support of research in philosophy of chemistry by NSF are also considered.

Readers will notice many contrasts among the papers in this volume—philosophical styles range widely: analytic, constructivist, continental, empiricist, instrumentalist, physicalist, processive, pragmaticist, structuralist approaches are all represented. But many connections among these essays can also readily be recognized. One unifying notion is a general confidence that the ancient tradition of chemical thinking remains vigorous, and autonomous[10]—it can contribute in important ways to philosophical dialogue, both now and in the future (*pace*, Isabelle![11]).

As editor, I would like to express sincere thanks to all who contributed in any way to the production of this volume, and to the international conference on which it was based. Special thanks are due to the Local Committee (Rom Harré, James Mattingly, and Daniel Rothbart) all the authors represented here, and to many officers, faculty, students and staff of Georgetown University who helped in the project. In particular, I want to express appreciation to the following: the Chairs of the Philosophy and Chemistry Departments (Wayne Davis and Miklos Kertesz—succeeded by Richard Bates); the Deans of Georgetown College (Jane Damen McAuliffe), the Graduate School (David Lightfoot), and the School of Summer and Continuing Education (Michael Collins); the Director of the Georgetown Center for Religion and Science (John Haught); the Provost (James O'Donnell),[12] and the President (John DeGioia)—all of them contributed significantly and generously to the success of the symposium.

NOTES AND REFERENCES

1. NEWTON-SMITH, W. 2000. Explanation. *In* A Companion to the Philosophy of Science. W. Newton-Smith, Ed.: 131. Blackwell. Oxford, UK.
2. HEMPEL, C. 1965. Aspects of Scientific Explanation. Free Press. New York.
3. SALMON, W. 1998. Causality and Explanation. Oxford University Press. Oxford, UK.
4. WYLIE, A. 2002. Thinking from Things: Essays in the Philosophy of Archaeology. University of California Press. Berkeley, CA.
5. CARTWRIGHT, N. 1989. Nature's Capacities and Their Measurement. Clarendon Press. Oxford, UK.
6. WITTGENSTEIN, L. 1973. Philosophical Investigations *(section 66, line 8). Translated by* G.E.M. Anscombe, 3rd ed. Macmillan. New York. p. 31e.
7. While this volume was in press, another significant book on chemical explanation was published: ZIELENAKA-LIS, EWA. 2003. Filozofiezne Koncepcje Wyjapnienia Naukowego a Wspóczesna Chemia. Adam Mickiewicz University Press. Poznan, Poland.
8. MACHAMER, P., LINDLEY DARDEN & CARL CRAVER. 2000. Thinking about mechanisms. Philosophy of Science **67**: 1–25.
9. GLENNAN, S. 2002. Rethinking mechanistic explanation. Philos. Sci. **69**: S342–S353.

10. JANICH, P. & NIKOLAOS PSARROS, Eds. 1998. The Autonomy of Chemistry: 3rd Erlenmeyer-Colloquy for the Philosophy of Chemistry. Würzburg, Königshausen & Neumann.
11. BENSAUDE-VINCENT, B. & ISABELLE STENGERS. 1996. A History of Chemistry. *Translated from the French by* Deborah van Dam. Harvard University Press, Cambridge, MA. The concluding chapter of this excellent history (that chapter probably mainly written by Stengers, a chemist and philosopher) seems rather less sanguine about prospects for the future of the chemical tradition.
12. The International Society for the Philosophy of Chemistry (ISPC) (for further information, see http://www.georgetown.edu/earley/ISPC.html) is a recently founded and relatively impecunious organization. There is no president of the ISPC. In August 2002, Provost O'Donnell was also serving a President of the American Philological Society. He was kind enough officially to open the ISPC Summer Symposium, thereby exemplifying the potential of "executive outsourcing."

Structural Explanation in Chemistry and Its Evolving Forms

ROM HARRÉ

Department of Psychology, Georgetown University, Washington, DC 20057, USA

ABSTRACT: Historically, there has been a close relationship between the concepts of structure and of shape. Is the relationship a matter of fact or is it conceptual? Exploring this question with examples enables us to distinguish several senses of structure. Clearly, spatial shape is not conceptually related to constitutive structure, since a molecule may consist only of an ordered set of potentialities. The concept of structure in chemistry developed in the nineteenth century from an ambiguous but essentially taxonomic use in type theory to a realist hypothesis as to the spatial distribution of atoms in the molecule. This development required a defense of the Daltonian view of elements as constituted by minute corpuscularian atoms. Once a realist concept of structure had been established, the issue of the nature of structural explanation could then be addressed. Template explanation formats as well as jigsaw formats can be found in chemistry. The problem of halting a regress of structural explanations leads to a sketch of a way in which physics can serve as a heterogeneous foundation for chemistry without reductionist implications.

KEYWORDS: structure; atom; molecule; type; taxonomy; realism; explanation; description; regress

PRELIMINARIES

Two seemingly closely related concepts are often invoked in analyzing the nature of material entitites: namely, "structure" and "shape." It is taken for granted that, if a certain structure of atoms can be achieved, then, *ceteris paribus*, a certain shape can be created. This principle is very important in biochemistry and chemical pharmacology. In an important discussion of the structure/shape relation, Woolley[1] explicitly treats molecular structure and molecular shape as synonyms; that is, for him the concepts are internally related.[a] I want to begin this analysis of structural explanations and their presuppositions by querying and subsequently breaking the seeming necessity of the relation between "having a structure" and "being of a certain shape." This move will allow me to introduce some varieties of the concept of struc-

---

Address for correspondence: Department of Psychology, Georgetown University, Washington, DC 20057.
harre@georgetown.edu

[a]Woolley's skepticism about the use of the concept of shape in the context of molecular ontology derives from his skepticism about the "classical molecular model of atoms joined by bonds" (Ref. 1, p. 1073). I am grateful to an anonymous reviewer for drawing my attention to this paper, and to the comments on it by Weininger.

ture that will prove helpful in working out the ways that structural explanations can be developed in chemistry.

From a philosophical point of view, it will allow us to give a reasoned verdict on whether the "structure-to-shape" relation is symmetrical, that is, entails a "shape-to-structure" relation of equal strength. It will, *inter alia,* enable us to decide whether the relation between shape and structure is necessary or contingent. If the latter, it may be the result of a dateable discovery.

By the "shape of something," I will mean its distinctive geometrical form as manifested in its extension in physical space. Such adjectives as "spherical," "square," and so on spring to mind. There are shapes without names, however, and there is a condition of shapelessness, a kind of mush, and of polymorphism, and so on. Shape has to do with the configurations of the boundaries and surfaces of material beings.

The structure of something has at least two main meanings. There is "constitutive structure," by which I will mean the set of material entities that, standing in certain relations, are constitutive of an entity. One might contrast "mereological structure," the set of abstract entities, say the four triangles, the four vertices, and so on that are, in a certain sense, constitutive of a tetrahedron, when standing in certain relations.[b]

I want to play off the contrast between *shape* and *structure* against the contrast between *chemical element* and *chemical atom* in trying to get clear about the nature and varieties of structural explanations in chemistry. It is fairly clear that there is no necessary relation between the concepts of *chemical element* and *chemical atom.* Distinctive chemical elements, as displayed in bite-sized bits of stuff, might not have been constituted of distinctive types of atoms, one atom type to each element type. They might have been constituted of stable clusters of qualities as Lamarck maintained. They might have been constituted by stable proportions of the Hot, the Cold, the Wet, and the Dry. They might have been constituted as distinctive clusters of members of a universal atom type.

The same logical point arises *vis à vis* the atom/element relation as for that between the shape/structure relation. Is it a necessary or a contingent truth that blobs of elementary "stuffs" are clusters of atoms that are all members of the same atomic type? Was this the result of an empirical discovery? On the other hand, is it a semantic rule, fixing the way the concepts of *element* and *atom* are to be used?

If *shape* and *constitutive structure* are tightly related, then it seems that this requires that chemical elements are, if isolated in a pure state, congeries of distinctive chemical atoms.

THE LOGICAL RELATIONS BETWEEN SHAPE AND CONSTITUTIVE STRUCTURE

We can have distinctions of shape without any corresponding distinctions of constitutive structure. For example, the insensible corpuscles of the material world, as conceived by Locke and Boyle, differed in shape. Some were spherical, some pointed, and so on. However, they were not made of parts. They were "little Particles of several sizes and shapes."[2] They served as the constitutive parts of observable material things by virtue of their bulk, figure (shape), texture (arrangement in structures),

[b]Thanks are due to Robin Hendry for pointing out this use of the concept of structure.

and motion and accounted for some of their sensory qualities. For many chemists of the seventeenth century, the shapes of corpuscles explained the manifest qualities of things and substances. For example, though it sounds naïve to us today, Boyle suggested that "... particles of Nitre ... may come to have parts ... the sharpness of the sides and points may fit them to stab and cut, and perhaps sear the nervous and membranous parts of the organ of taste."[3] The shapes of the corpuscles were unanalyzable variants of the primary quality *form* and not susceptible of further explanation in terms of the concepts of natural philosophy. In particular, they were not the geometrical forms of elementary constituents arranged in quasi-invariant structures. Textural properties also had perceptible effects. For instance, Boyle[4] says, "Not upon the Account of the Predominancy of this or that Principle in them, but upon that of their texture, and especially the Disposition of their superficial parts" do they cause sensations of color.

The distinctions in the shapes of fundamental corpuscles are distinctions in mereological rather than constitutive structure, a mathematical not a material property. While crystallography of the period of Abbé Huy was the science of mereological structure, corpuscularian chemistry uses both constitutive structure (texture) and mereological structure (basic shape) as explanatory concepts. Clearly, for Boyle, Locke, and their contemporaries, these were logically independent.

We can have distinctions of constitutive structure without any corresponding manifest differences in shape. The reaction products of reagents with some given stuff may not be detached constituents, but the realization of potentials, with respect to the stuff in question. The stuff can be thought of as a structure of possibilities, none of which has a preexistent precursor as a distinctive material constituent of the four humors are at the back of the ancient elements—earth, air, fire, and water. The proportion of the wet, the cold, the hot, and the dry may be invariant under all sorts of transformations, we might well want to say that the elements of Earth are the humors, cold and dry.[c] However, they are not present in the way that earth and water are constituents of potter's clay.

This is not just some odd and outmoded medieval fancy. This is more or less what a Bohrian would say about the relation between electrons as particles offered as constitutive of atoms. Atoms may afford electrons under certain experimental circumstances. We ought not take for granted, based on that matter of fact, that atoms are constituted of electrons, among other bits and pieces. That an experimental setup affords electrons does not entail that the material stuffs involved in the experiment have electrons as constituents.

It seems to me that there were some who took the type theory of chemical structure to be a Bohrian account of chemical metaphysics. The repertoire or "structure of possibilities" that limited the obtainable reaction products was not the expression of a set of distinct material precursors such as radicals and elementary atoms. Butlerov's complaint that type theorists offered contradictory type formula for the same stuff only makes sense if the sources of the distinct reaction products were potentialities. Sticking to a "repertoire of possibilities" interpretation enabled one to stand on the sidelines of the atomic debates as an agnostic. In a recent discussion of the

---

[c]The exposition of Richard Rufus on Aristotle's theory of mixtures (Weisberg and Wood, this volume) shows the Oxford Master taking a remarkably Bohrian line on the status of the humors as constituents of stuff.

structure/shape relation, Weininger revives a Bohrian account of "molecular structure" as a product of interactions. "Molecular structure results from an *interaction* between a molecule and its environment in a way for which *ab initio* theory cannot account."[5]

Putting together the two lines of analysis, we have cases of shape without material constituent structure and material constituent structure without shape. The current way that these concepts are treated, as if they were internally or necessarily related, is a basic principle of engineering chemistry, pharmacology, and so on. It is well to remind ourselves that the relation is contingent. In the bulk of this paper, I want to explore the philosophical question of how that contingency came to be treated as a necessity.

THE ADVENT OF STRUCTURAL FORMULAS: UPS AND DOWNS OF THE ELEMENT-TO-ATOM PRESUPPOSITION

Here is the orthodox story. In the late eighteenth and early nineteenth centuries, a flurry of research projects resulted in the extraction and isolation of many elementary substances, believed to be the fundamental constituents of the material stuffs of the natural world. For example, Lavoisier demonstrated that metal ores were compounds of the metal with oxygen by analytical methods that first separated the metal and gaseous oxygen, and then recreated the calx or ore, by recombining the constituents. Humphrey Davy used electrolysis to isolate sodium metal from molten soda by passing a powerful electric current through the liquid. He used the same method to extract other metallic elements from their "calxes."

Like many at this time, Davy believed that material substances were made up of myriads of separable corpuscles, of distinct species. He also believed that the "glue" that held the constituents of compounds together was electrical. This suggested to him that a sufficiently powerful electric current should tear them apart, atom from atom.

We might formulate the dominant presupposition behind this story as the Dalton/Davy principle: that material substances are made up of myriads of separable corpuscles of distinct species, each chemical element being defined by a distinct species of chemical atom.[6]

However, about the same time as Davy was heroically ripping molecules apart into elementary shreds, the question of the legitimacy of combining the concept of an element with the atomic hypothesis came to prominence. Were stable and consistent ratios of equivalent weights of the elementary substances combined chemically, the outward and macroscopic manifestation of inward clustering of electrically tied microscopic atoms of types exactly parallel to the types of the observable elements?

The history of this debate and the confusions that led up to the Kahlsruhe Conference and Cannizaro's memoir are very well known. But what of the work to clarify the philosophical presuppositions of this issue in relation to the meaning of structural formulae?[d] The upshot of the meeting was the clarification of the relation between equivalent and atomic weights. But it is well to remind ourselves that this did not entail a belief in the reality of chemical atoms.

---

[d] I am setting aside any discussion of the criteria for "element-hood." This is a deep philosophical issue. I shall assume that the "no further analysis" criterion is in place.

WHAT ARE STRUCTURES, STRUCTURES OF?

The thesis that atoms are the constituents of elements and compounds manifested as observable chunks of stuff seems an obvious precursor and presupposition of the idea there were a small number of definite patterns of the constituent atoms of material substance. If compound molecules were clusters of elementary atoms, surely the way the atoms are arranged would be relevant to their chemistry.

There seem to be two possibilities given this presupposition. Two manifestly different substances might be composed of the elements in the same proportions, but with the constituent atoms arranged in different patterns. Alternatively, the same pattern would underlie different manifest substances by virtue of atoms of different types occupying the same sites in the structure. Chemical types should be generic forms or patterns of structural patterns of atomic constituents of the constitutive elements.

This was what the theory of chemical types ought to have been to maintain consistency with the Dalton/Davy principle by which the concepts of "element" and "atom" were tightly linked. This linkage would have been a Wittgensteinian "grammatical rule," since there was no empirical display of the atomic constituents of a sample of an element. In fact, it was used and promoted as taxonomy.

ANTI-ATOMISM IN THE MID-NINETEENTH CENTURY

The fact that there were popular nonconstitutive interpretations of types as nothing but convenient taxonomic categories makes sense only in the light of the rampant anti-atomism of the 1850s and 60s. This entailed the rejection of the Dalton-Davy principle to be replaced by a variety of alternatives.[e]

The dominant figure in British chemistry was Sir Benjamin Brodie, professor chemistry at Oxford. His current obscurity is inversely proportional to his contemporary fame. He was the first to apply Boolean algebra to a science, in his *Calculus of Chemical Operations*. The main principle of Brodie's chemical ontology was the reality of weight-changing operations on units of space. The operation of oxygenating space resulted in that space being eight times the relative weight of a unit of hydrogenated space. There were no chemical atoms, only sequences of weight-changing operations. There were no chemical elements, only numerically invariant units of weight.

In this view the very best that a theory of chemical types might be is a chemical taxonomy, using notional "atomic" formulae. The presentation of the classification of ethyl alcohol as a member of the water type in the following formula should entail no ontological consequences whatever.

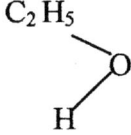

---

[e] At a meeting of the Royal Chemical Society in the 1860s, the proposition that the concept of "chemical atom" does not denote a real constituent of nature was passed with only one dissenting voice, that of Williamson.

ORIGINS OF THE THEORY OF CHEMICAL TYPES

Let us turn to a brief historical sketch of the development of the Theory of Types in order to site the philosophical issue of contingency/necessity of the two principles, shape to structure and element to atom, more precisely. In the 1830s, Laurent realized "that chlorine, this substance so different from hydrogen, may in certain circumstances, come to take the place of this and play its role, without changing the arrangement of the atoms of the compound in which it enters."[7]

Here we have both the structural hypothesis and the hint of a technique by which structures could be experimentally explored. However, something else was needed. Laurent had mentioned only the atoms of elements as substitutions. What if there were groups of atoms, radicals that behaved like the atoms of elements and could be substituted for the hydrogen atoms in the molecules that exemplified the types?

In 1837 Leibig had suggested that ammonia, represented by the formula NH_3, expressing the proportions of the elements in the compound, might also represent a general pattern that other compounds also exemplified, particularly certain organic compounds. In 1849, Hoffman (1818–1892) had used Laurent's principle of substitution to explore compounds of the "ammonia" type. In these cases organic radicals, stable groups of atoms, played the same role as chlorine had played in Laurent's speculation. Hoffman[8] represented the results of his studies of the ethyl derivatives of aniline in the following scheme:

$$\begin{matrix} C_{12}H_5 \searrow & & C_{12}H_5 \searrow & & C_{12}H_5 \searrow \\ H \rightarrow N & & C_4H_5 \rightarrow N & & C_4H_5 \rightarrow N \\ H \nearrow & & H \nearrow & & C_4H_5 \nearrow \end{matrix}$$

The "ammonia type" is represented by the following formula:

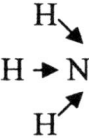

$$\begin{matrix} H \searrow \\ H \rightarrow N \\ H \nearrow \end{matrix}$$

The substitutions are complex but stable groups of atoms, for example, C_4H_5 behaves as if it were a chemical entity like an atom of hydrogen.

Because it was known that water was a compound of two elements, one part of oxygen to two parts of hydrogen by volume, both being gaseous as products of the decomposition of water, the "water type" was an obvious next step. Thinking in atomic terms suggested that the water molecule consisted of an atom of oxygen combined, somehow, with two atoms of hydrogen, our familiar H_2O. However, to think this way was to tacitly accept the necessity of the "element-to-atom" Dalton/Davy principle, which was at that time, as I have remarked, unusual.

Williamson (1824–1904) followed the fashion with the "water type." His presentation of the type principle was bold. He wrote,[9]

I believe that throughout inorganic chemistry, and for the best known organic compounds, one simple type will be found sufficient; it is that of water, represented as containing 2 atoms of hydrogen to 1 of oxygen, that is:

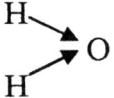

He used this to make sense of compounds in which the "ethyl" radical appeared.

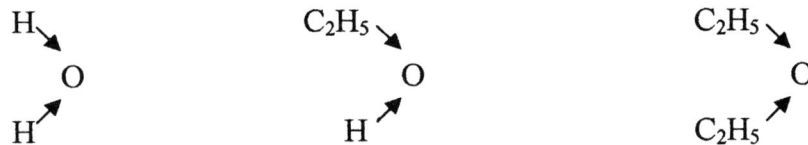

Were these patterns representations of real spatial arrangements, as well as the proportions of the constituent elements? Williamson certainly thought so, though he was in a tiny minority in Britain.

In France, the advocates of type chemistry were subjected to both abuse and professional persecution. Laurent, who preferred scientific freedom to professional preferment,[10] was shuffled off to Bordeaux, where he amused himself by writing an opera. Nevertheless, Laurent found a supporter and later a collaborator in the aggressive and talented person of C. F. Gerhardt (1816–1856). Gerhardt's final scheme involved four types: water, hydrochloric acid (HCl), ammonia, and hydrogen (HH). Gerhardt answered the "ontological" question unhesitatingly. These patterns were heuristic, guides to practice. They were not pictures of molecules. To make the exegesis of the views of these authors even more puzzling, Laurent had presented a three-dimensional pyramidal model of molecular shape in his dissertation of 1837. Gerhardts' point was expressed with great clarity by F. A. Kekulé.[11] The man, who in a later and massive change of heart invented the benzene ring, said this: "The types' (rational formulae) are only reaction formulae, not constitutional formulae, and they do not express the relative positions of the atoms in the compound."

FROM HEURISTIC TYPES TO REAL STRUCTURES

Philosophers use the term *reify* to refer to the treatment of an abstract entity as a thing, a constituent of the world of material reality. In the mid-nineteenth century, the type formulae were gradually superceded by structural formulae, abstract taxonomic categories were reified and treated as real material structures. Convenient representations of molecular structures in terms of their reaction products began to be reified and to be treated as pictures of arrangements of atoms in space. They came to be seen as more than heuristic devices to guide research. For this move to make sense, however, elements had to be replaced by atoms as constituents of *molecules* of compounds. The idea of real spatial structure as a prime attribute of the chemical molecule served to resolve all sorts of puzzles. It was the final step in the exploitation of the distinction between atoms, the smallest units that could enter into compounds, and molecules, the smallest units in which compounds could exist. For example, there was a very large number of hydrocarbons, compounds of carbon and hydrogen only. By the 1860s it had become commonplace to think of some carbon compounds as based on chains of carbon atoms and others on rings. A hexagon consisting of a closed chain of six carbon atoms proved to be a particularly useful image. Since hypotheses that these were the structures of the molecules under study were the result of inferences from observable chemical behavior, it was possible to interpret them as no more than helpful visual aids. It was another step altogether to take these structures to be genuine maps of real molecular terrains, shapes in physical space.

The story of the transformation of structural diagrams from heuristic device representing taxonomic categories to iconic model involves one of those unpleasant episodes in the checkered history of science in which small-minded nationalism deprived a pair of original thinkers of due credit. It seems clear that A. M. Butlerov, then in Kazan, was the first to enunciate a realist reading of structural formulae *and* to provide the necessary supporting principles for the break with a purely taxonomic interpretation of structural formulae. The story of the sources of Butlerov's realist reading of structural formulae is complicated. Butlerov spent his study leave of 1857 in Heidelberg. He became friendly with the Scottish chemist, A. S. Couper, who was visiting at the same time. Couper had developed a full-scale structural theory of carbon compounds about this time that was strongly realist. The publication of Couper's paper was entrusted to a colleague of Kekulé's. However, it was delayed until after that of Kekulé in 1858, in which very similar proposals were to be found.[f] Butlerov's paper, which provided the necessary reasoned foundations for realism in chemistry, appeared in 1861. Finally, we must credit the first clear description of the tetrahedral geometry of carbon atoms, a necessary foundation for treating types as real structures, to Butlerov in 1862.

The type theorists had felt free to write several different formulae for the one substance, each representing the potentiality for some particular set of products of different chemical operations on the compound in question. For example, as Brock[12] puts it concerning the classification of alcohol in the water type, "oxidation of alcohol to acetic acid justified Liebig's acetyl radical, C_4H_3; its esterification also justified Liebig's ethyl radical; while its conversion to chloral justified Dumas' etherin." Nye[13] describes the pre-Butlerov situation as follows: "Neither the chemical theory of the atom nor the chemical theory of the benzene ring corresponded to a single explanatory device, but rather to overlapping, partially incommensurate explanations of chemical behavior."[g]

In realists' terms these would not count as explanations at all. Butlerov realized that if a diagram was to be interpreted as an iconic model of the spatial arrangement of the atoms in the molecule then each substance could have one and only one formula. Each distinct diagrammatic formula must represent the constituent structure of just one molecule. In general, then, he argued that "the chemical nature of a compound molecule depends on the nature and quantity of its elementary constituents and its chemical structure."[14] It must, therefore, be capable of giving rise to different reactions products under different circumstances.

What did Butlerov mean by "chemical nature" and "chemical structure"? The phrase "mechanical structure" was also in use. Were chemical structure and mechanical structure names for the same feature of molecular reality? According to Parting-

---

[f]Couper was deeply upset by all this, and very soon abandoned chemistry altogether.
[g]I owe notice of this well-organized book to an anonymous reviewer.

ton,[15] in a paper of 1862, Butlerov had claimed his formulae represented "la position topographique des atomes." This sounds very much as if he had elided the concepts of chemical structure, that is, proportion of constituents, with mechanical structure, that is, the arrangement in space of real atomic constituents. Yet, Turkevich[16] remarks that in 1862 Butlerov dismissed Kekulé's hypothesis "that spatial positions of the atoms, in addition to their nature, determined the properties of organic compounds." At the same time, he says that Butlerov supported the interpretation of structural formulae as expressing "chemical structure" rather than "physical structure," apparently opting for their formal rather than their spatial significance. I am not convinced that this is consistent with the "topographique" quotation above.[h] The exact dating of the various contributions to the transformation of type theory is crucial to adjudicating the priority issue. It seems that credit should go to Couper for the original insight and to Butlerov for the provision of the clinching arguments for a realist reading of structural formulae. (For a very clear and detailed account, see Brock.[12])

The historical niceties are not of central importance to this paper, though I would like to get them right and the ambiguities resolved. From a philosophical point of view, the point is the distinction between a taxonomic use of structural presentations and a realist one. It is obvious that these are not incompatible. Nevertheless, the antirealism of the mid-nineteenth century can be felt in the caution with which the realist interpretation was expressed by many authors. Only Williamson seems to have held firmly to the strict interpretation of the Dalton/Davy principle throughout the period in question.

By 1858, Kekulé had abandoned his view that structural formulae were only summaries of chemical reactions. In later years, he told the story of how the idea of real molecular structures had come to him in 1854 while dozing in a London bus. I cannot resist quoting it, though I am sure it is very well know to everyone who glances at this paper.

> I fell into a reverie. The atoms were gamboling before my eyes. I had always seen them in motion, these small beings, but I had never succeeded in discerning the nature of their motion. Now, however, I saw how, frequently, two smaller atoms united to form a pair, how a larger one embraced two smaller ones.... But look! What was that? One of the snakes [chains of atoms] has seized hold of his own tail, and the form whirled mockingly before my eyes.'[17]

The German chemical establishment seemed to have refused to acknowledge Butlerov's priority in presenting an unambiguous version of the structural theory of atoms in the molecule, really arranged in space as the formulae presented them in just the same way as Couper's contribution was sidelined. As some of the Russians saw it, the German chemical establishment not only ignored Butlerov's claims to priority, but at the same time embarked on a campaign of denigration of the work of the Russian chemists who had been inspired by Butlerov. The underlying offence, in the eyes of the Germans, it has been suggested, was the impertinence of a provincial in proposing a radical theoretical concept owed, they thought, to no less a person than Kekulé himself. Unlike the volatile Couper, Butlerov was famous for his easygoing nature and his tact. His only known response to the German claims to priority is a

---

[h]I owe to M. Kaji the observation that Butlerov did not attend the Karlsruhe conference.

paragraph in his report to Kazan University about a conference he had attended in Germany. Here is what he said:

> One feature of these German congresses is particularly striking to us foreigners, so strange that it cannot be passed over in silence; it is an aspiration to assert their own nationality at every opportunity.[18]

His pupil, V. V. Markovnikov (1838–1904), was not so reticent. He published an article on the history of the structural hypothesis and set out on a counterattack. However, this had little immediate result in setting the historical record straight. True, Kekulé had at first seen the move from type theory to more specifically structural formulae as merely practical. He had once asserted that "It is clear that one cannot determine the positions of atoms in a specific compound by a study of metamorphoses," that is, a study of the results of substitutions of hydrogen atoms by organic radicals. In whatever way that problem was to be resolved, he nevertheless became increasingly sure that structural formulae were pictures of the spatial arrangements of atoms in the molecule.

By mid century the realist interpretations of structural formulae were well established, though the seminal role of Butlerov's insights was scarcely mentioned. However, could a more direct method of discovering the spatial arrangement of atoms be devised than inferences from the analysis of products of reactions? Could Kekulé's methodological question be answered?

Almost one hundred years were to pass before this question was indisputably answered in the affirmative. Inspired by Bernal, and making use of the techniques pioneered by von Laue and the Braggs, father and son,[19] Dorothy Hodgkin showed by example how the structures of even the most complex molecules could be ascertained.[20] X-ray crystallography makes use of the effects of the structures or lattices of atomic constituents on the diffraction patterns of electromagnetic radiation. These effects are physical, not chemical products of the compounds in question.

If constituent structure is nothing other than the arrangement of atoms in some stable spatial configuration, then necessarily constituent structure entails shape. In order for that principle to be sustained as a Wittgensteinian grammatical rule, and hence to be noncontingent, however, the Dalton/Davy principle that links distinguishable elements to distinctive atoms must be presupposed. It was this principle that was at a discount in the mid-nineteenth century.[21]

INTERIM SUMMARY

The following three concepts of structure seem to be in play:

(1) Chemical structure: this has had several different meanings in the history, which I have briefly sketched: (a) "Proportion of elements" in a compound. For anti-atomists this "structure" would have been expressed in percentages, such as those I read in the ancient chemical textbooks in my school telephone room when, during the Second World War, I was waiting to place a call home. So much carbon, so much nitrogen, so much oxygen, and so on. (b) "Range of reaction products." This seems to have been what the anti-atomist type theorists understood by it.

(2) "*La position topographique*": this could make sense only to those who took the Dalton/Davy attitude to the nature of elements. It could mean nothing else but the arrangement of atomic constituents of molecules in space. If this is what Butlerov really held to be the proper structural concept, then his writings should not be interpreted using either of the above meanings of "chemical structure."

(3) "Mechanical structure": this seems to refer to a Newtonian atomism, and must surely be read as a particular case of "*positions topographiques des atomes*." As I was reminded by Joachim Schummer, there was a discussion in the mid-nineteenth century as to whether "chemical structure" was synonymous with "mechanical structure." Since the concept of mechanical structure is the more differentiated of the two, it suggests that there was no general agreement on what chemical structure ought to be taken to mean.

What was the driving force that moved chemists from a quasi-Lamarckian view of compounds as sets of possibilities, given a powerful mathematical rendering by the ingenious Brodie, to a full-blown constitutive atomic structure metaphysics, which had been sketched by Dalton and Davy and then lost sight of? Nye[22] suggests that Dumas' generalization of the concept of structure from the proposals of such biologists as Sainte-Hilaire for a universal ground plan for all animals languished, since it "presented few useful strategies in the chemical laboratory." Though, no doubt, this was a factor, the prevailing anti-atomism surely counted against making realist assumptions about structured clusters of atoms as the real essences of distinctive chemical compounds.

PATTERNS OF STRUCTURAL EXPLANATION

By the time we have reached the mid-twentieth century, analytical methods, such as those deployed with such effect by Dorothy Hodgkin, are taken unproblematically to bring to light the structures of molecular aggregates of atoms, "*les positions topographiques*." The next step will be to sketch the philosophical aspects of structure and structural explanations.

We need to make clear the sense of explanation at work on this context. When I claim that for the most part type theory analyses were taxonomic rather than explanatory, I am emphasizing one of the uses of the word *explanation*, namely, to denote a description of the causal mechanism or constituent of a causal mechanism that brings about a certain effect or produces a certain substance or product. "Why do cats like to eat mice?" might be answered by saying "Cats are carnivores." Though informally often referred to as an explanation, such a reminder of the category to which a being belongs is not an explanation in the causal sense at all. Rather it serves to assert that this is the nature of cats. It puts a stop to a regress of explanations. Similarly, to offer the type theory to explain Hoffman's substitution experiments, when that theory is used for classificatory purposes, is similarly empty.

Treating type ascriptions as shorthand for lists of possibilities is an intermediate case between taxonomic uses and the strictly explanatory uses invoking a realist interpretation of the formulae. The problem, as Brock[12] points out, is that the total set of possibilities includes incompatibles. Taking a Bohrian stance resolves this paradox, but it is incompatible with the use of the Dalton/Davy metaphysics to interpret the type formulae.

STRUCTURE IN GENERAL

What are structures? We need to tighten up the informal account given at the beginning of this paper. There seem to be three conditions that an aggregate of distinguishable units has to meet to count as a structure:

(1) It must consist of a set of relatively stable elementary beings.
(2) These beings must stand in certain definite relations to one another.
(3) At least some of constitutive relations must be invariant under at least some transformations.

What counts as a structure in any domain will depend on which relations among the constituents are invariant under which transformation. Change the focus of analysis from one type of transformation to another, and one structure fades into the background to be replaced by another.

Chemical structures meet these requirements as follows:

(1) A sample of an element consists of identical molecules, each of which consists of the same catalogue of instances of the same types of "indestructible" atoms.
(2) The constituent atoms of a molecule stand in invariant spatial relations mediated by electromagnetic processes.
(3) The relevant spatial relations are invariant over time, over spatial displacement of the whole molecule, and, in key instances, over substitution of one atom or radical for another.

It is clear then that molecules as conceived by both type theorists and Butlerovian structural realists are structures in the above sense.

STRUCTURAL EXPLANATIONS

How is the existence of such structures to be explained? There are two main forms of structural explanation.

(1) Template explanations: The existence of a certain structure is explained by reference to another structure from which it has been derived by substitution of constituents. This may be by creating an isomorphic copy of the originating structure. In this way, the structure of a blueprint explains the structure of a house. "Why is the staircase there?" "It is shown at that location in the plan." It may be by using the structure as a framework upon which to build something. The metal or wooden "armature" used by sculptors to create a version of the work by sticking clay to the frame is an example. Finally, the original structure may be maintained while elementary parts are replaced by suitable substitutes. The famous case of the much-darned silk stockings that eventually became entirely worsted as the silk threads were replaced by woolen yarn is an example of this kind of explanation. Why does

the house have such and such a structure? Why does the statue have such and such a structure? Why do these woolen garments have such and such a structure? In each case the explanation format is the same: because the structure of the entity from which they were derived was isomorphic to that structure. In chemistry, the structure of a protein molecule is derived from the structure of a segment of DNA by the intermediation of an isomorphic molecule of messenger RNA.

(2) Jigsaw explanations: The existence of a certain structure is explained by reference to the structural properties of the constituent parts. A jigsaw puzzle can go together only in one way, as each piece has a distinctive shape and fits only with its proper neighbours.

(3) Hybrid explanations: Sometimes the structure of a material thing can be subject to both kinds of explanation. The structure of a car is what it is because parts are structured and can go together in only one way. However, the structure of the car is also isomorphic to the structure of a designer's blueprint. The roof the new court of the British Museum is an even stronger example of the convergence of both explanatory formats. It is isomorphic to a designer's blueprint. However, I believe that each piece of glass in the roof is different from every other, and so can be fitted together in only one way.

All three structural explanation formats are poised as moments in potential regresses. How do we explain the structure of the engineer's blueprint or the sculptor's armature? Template or jigsaw? How do we explain the structure of the parts of the jigsaw? Jigsaw or template?

THE EXPLANATION OF CHEMICAL STRUCTURES

The history of the "structural" idea in chemistry presents, at first sight, two neatly differentiated cases.

Type-theory is a case of the template explanatory format. How we account for the formula of ethyl alcohol as $C_2 H_5 O H$? The answer draws on the water type

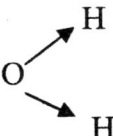

One of the hydrogens has been substituted by the ethyl radical. Because the type theory was initially offered only as a taxonomic device, the second-order question, "What is the explanation of the type structure?" did not then arise.

Butlerovian structuralism, of which the Hodgkin structures are indirect descendents, surely exemplifies the jigsaw format. The Russian, British, and German originators of strong structuralism all took the type formula to be a picture of the spatial layout of the atoms in the molecule and to be the consequence of the differentiated structural layout of what shortly thereafter came to be called "valence bonds." Hydrocarbons and benzene and its derivates shared this explanatory format. Unlike the taxonomic "structures" of the early type theory, the Butlerovian structures opened up the regress I have already drawn attention to. What explains the structural differences between the distinctive orientation of valence bonds in each kind of atom? For that too, we have had to wait until the twentieth century and the application by Coulson and others of quantum mechanics to the understanding of the chemical bond.

A COMPLICATION

The metaphysics of structure involves another presupposition so far unexamined. In setting out the constituents of a structure, I did not consider the question of the invariance of the elementary components of that structure. The specification included the invariance of at least some of the constituent relations under relevant transformations. However, there is also the question of the invariance of the elementary components, for example, Daltonian atoms as developed by the use of the Bohr model. The type of an atom is defined by the number and array of the subatomic particles presumed in the model. In the simple model, however, atoms gain, share, and lose electrons while maintaining their type status. We use the phrase "hydrogen *ion*" for H^+, for example. Whether atom or ion, it is of the hydrogen type. Qualitative identity criteria get rather messy in the application of the Bohr version of the Daltonian atom.

There is another dimension to this complexity. In metaphysics, we distinguish internal from external relations. The former are relations that are partly constitutive of the type of the beings so related. If internal relations change, so does the type of the beings so related. It seems obvious that, in the classical model, electron sharing is an internal relation. The only external relations admitted in the physical sciences are spatio-temporal, though it is not easy to find a set of structure-constituting relations that are spatio-temporal and nothing else. Even scaling is not a wholly external relation, since there are physical quantities that are not invariant under all scale transformations.

There is a working compromise evident in contemporary structural chemistry. The picture of the Hodgkin icon model of the penicillin molecule crystallizes into a standard molecular formula with neat and bounded letters and lines.

ICONIC THINKING

This topic can be looked at in yet another way. Recently, the question of the role of visual imagery and iconic thinking has come to the fore again in philosophy of science. The two modes of structural explanation fit rather well with the distinction between descriptive models and explanatory models. Descriptive models are no more than convenient visual or imaginative devices for representing distinctions that can be made in detail in other and more direct ways. The story of the theory of chemical types seems to relate to a transition from a heuristic to a realist explanatory role for just such a collection of models. Explanatory models are linked to scientific realism. Though they may serve merely heuristic purposes, the main value of such models is to lead from the realm of actual observation to that of possible observation and beyond, through representations of entities, structures, and processes we are currently unable to observe. The history of Butlerovian structuralism in chemistry through to Hodgkin's work is a splendid example of the role of a structural realism in motivating a very long-running research program.

THE QUESTION OF TERMINATION

How would a regress of structural explanations terminate, if it terminated at all? In principle, it seems to me, whether the explanatory regress invokes a sequence of templates, a sequence of jigsaw units, or is a template/jigsaw hybrid, there is no logically necessary point of termination. The same question is posed by regresses of explanations in terms of causal powers. If the causal powers of beings at one level are explicated by reference to the mode of organization of lower level beings that themselves have causal powers, such a regress can terminate only in a level of beings the causal powers of which are exhaustive of their natures. In the case of chemical regresses to which the Dalton/Davy Principle is applied, the spatial orientations of valence bonds, atom by atom, defines ground zero for chemistry. Unless elements are really atomic, structure need not entail shape.

REFERENCES

1. WOOLLEY, J. 1978. Must a molecule have a shape? J. Am. Chem. Soc. **100**: 1073–1078.
2. BOYLE, R. 1661. The Sceptical Chymist. Crooke. London. Proposition I.
3. BOYLE, R. 1675. The Mechanical Production of Tastes. Davis. London. p. 10.
4. BOYLE, Sceptical Chymist, 327.
5. WEININGER, S.J. 1984. The molecular structure conundrum. J. Chem. Educ. **61**: 939–944, (p. 939).
6. KNIGHT, D.M. 1970. Atoms and Elements. Hutchison. London.
7. Quoted in PARTINGTON, J.R. 1964. A History of Chemistry. Macmillan. London. p. 392.
8. HOFFMAN, A.W. 1849. Compte Rendue **xxix**: 182.
9. WILLIAMSON, A.W. 1851. The Chemical Gazette **ix**: 334.
10. NOVITSKI, M. 1992. Auguste Laurent and the Prehistory of Chemistry. Harwood. Chur.
11. Quoted in HARTLEY, H. 1971. Studies in the History of Chemistry. Clarendon Press. Oxford, UK. p. 214.
12. BROCK, W.H. 1992. Fontana History of Chemistry. Chapters 6 and 7. Fontana. London.
13. NYE, M. J. 1993. From Chemical Philosophy to Theoretical Chemistry. University of California Press. Berkeley, CA.
14. BUTLEROV, A.M. 1861. Z. Chem. 549–560.
15. PARTINGTON, History of Chemistry, p. 550.
16. TURKEVICH, J. 1963 (1965). Soviet Men of Science. Van Nostrand. Princeton, NJ.
17. Quoted in PARTINGTON, History of Chemistry, p. 537.
18. LEICESTER, H.M. 1966. "Controversies on Structure." *In* American Chemical Society: History of Chemistry, Vol. 61. American Chemical Society. Washington, DC.
19. BRAGG, W.H. & W.L. BRAGG. 1915. X Rays and Crystal Structure. Routledge. London.
20. FERRY, G. 1998. Dorothy Hodgkin: A Life. Granta Books. London.
21. ROCKE, A.J. 1984. Chemical Atomism in the Nineteenth Century. Ohio University Press. Columbus, OH.
22. NYE, Chemical Philosophy, p. 71.

Explaining Explanation in Chemistry

GRANT FISHER

Division of History and Philosophy of Science, School of Philosophy, The University of Leeds, Leeds LS2 9JT, UK

ABSTRACT: The notion that explanations are arguments, central to the covering law account, persists in developments of accounts of explanation in general philosophy of science. I argue that these developments fail to capture chemical explanation—motivating the explanatory autonomy of chemistry requires the investigation of alternative accounts. Ubiquitous use of models in chemistry is indicative of their importance to explanation. I argue that contemporary accounts of models that emphasize their partial autonomy from fundamental physical laws provide a basis for investigating explanation in chemistry. Furthermore, I sketch the importance of the semantic view of theories in articulating model autonomy, and thus why the semantic view might fulfill a significant role in explicating the explanatory autonomy of chemistry.

KEYWORDS: covering law account; unification; models; model autonomy; semantic view of theories

My aim in this short paper is to suggest that the explanatory autonomy of chemistry should be considered indicative of a need for accounts of explanation that diverge from traditional approaches in philosophy of science. Perhaps the most familiar view of explanation is Hempel's seminal deductive-nomological or "covering law" account. An event or phenomenon is explained if it can be deductively subsumed under sets of laws and initial conditions. The significance and power of this account (and the logically weaker inductive-statistical approach) lies partly in its familiarity, since explanations are *arguments*. In spite of the considerable critical attention it has received, Hempel's model still provides the kernel for subsequent accounts of explanation. I argue that explanation in chemistry cannot be accommodated by developments of the covering law account. Moreover, in view of the recent growth in literature regarding models in the philosophy of chemistry, I argue that any account of explanation in chemistry must make models central and that we can learn much from the contemporary debate in philosophy of science concerning the autonomy of models. In this respect, I argue that the semantic view of theories provides a potentially important perspective for an analysis of explanation in chemistry.

According to Hempel and Oppenheim, explanations must satisfy necessary "conditions of adequacy."[1] For example, the explanandum must follow deductively from

---

Address for correspondence: Grant Fisher, Division of History and Philosophy of Science, School of Philosophy, The University of Leeds, Leeds LS2 9JT, U.K. Voice: +44 0113 274 6456.
phlgaf@leeds.ac.uk

the explanans; the explanans must contain at least one general law; and the explanans sentences must be true and not merely confirmed. (A further condition is that the explanans must have empirical content.) The following example, drawn by Brody from chemistry, illustrates what is a satisfactory explanation according to the covering law account yet lacks explanatory power:

(1) sodium normally combines with bromine in a ratio of one-to-one;
(2) everything that normally combines with bromine in a ratio of one-to-one normally combines with chlorine in a ratio of one-to-one;
(3) therefore, sodium normally combines with chlorine in a one-to-one ratio.[2]

Brody argues that the combination of sodium and chlorine cannot be explained by the combination of sodium and bromine, not only because the latter does not tell us the cause of the combination, but also because there should be a statement in the explanans that attributes an essential property to sodium and chlorine.[a] The required essential property is atomic structure, and this motivates a preference for an explanation in terms of the theory of the chemical bond. The justification for evoking essential properties and the justification for distinguishing causes from the role played by essential properties in explanation are not matters that I shall consider here. The issue I wish to focus on is relatively straightforward. The combination of sodium and bromine does not explain the combination of sodium and chlorine, irrespective of whether it lacks essential attribution. The combination of sodium and chlorine is, according to Brody, an event.[3] But why does another event, the combination of sodium and bromine, appear in the explanans? Furthermore, whereas the second statement (2) might appear law-like, it is not of sufficient generality to warrant its place in the explanans, and tends to make Brody's example appear trivial.

The point is that Brody's example is a poor contender for an explanation in chemistry since it does not necessarily fulfill the conditions of the covering law account. While it might not be difficult to provide a more robust example to motivate the notion that explanations in chemistry are arguments, the most difficult problem is satisfying the requirement that the explanans sentences be true. As is well known, nomic necessity is contentious in chemistry. For example, Scerri and McIntyre argue for a pluralistic view of laws.[4] Although there are exceptions to the periodic law, it is nonetheless "law-like" from the perspective of chemistry. Similarly, Christie argues that chemical laws need not be exact to be law-like.[5] But if the condition of the truth of the explanans sentences is not met in chemistry, this must challenge the very notion that explanations are arguments. It should be noted that absence of nomic necessity is not a matter that distinguishes the nature of explanation in chemistry from

---

[a]Causation is invoked in order to address the notorious "symmetry" problem associated with the covering law account. Bromberger points out that from Newton's laws and statements concerning the length of a piece of string with a bob, we can deduce the period of the bob's oscillation, thus explaining why the bob has a certain period of oscillation. But from the same laws and from the period of oscillation, we can deduce the length of the string and bob. To say that we have explained why the string and bob have a certain length from the appropriate laws and a statement of the period of the bob's oscillation doesn't seem causally right. While an explanation of the period of the bob's oscillation *does* depend on the length of the string, an explanation of the length of string *does not* depend on the period of oscillation. The length would be what it is regardless of whether the bob was oscillating. Cf. Bromberger, S. 1966. Why-Questions. *In* Mind and Cosmos. R.G. Colodny. Ed.: 86–111. University of Pittsburgh Press. Pittsburgh.

that of physics. Cartwright has famously attacked the covering law account and the privileged "true" status of fundamental laws in physics.[6] Cartwright argues that physical laws are subject to *ceteris paribus* conditions and truth or "descriptive accuracy" is reserved only for phenomenological laws, and this is partly intended to motivate the rejection of the covering law account.[b]

Kitcher provides another account of explanation that employs an example from chemistry.[7] He attempts to preserve the "deductivism" inherent in the covering law account but argues that the aim of scientific explanation is *unification*. Put simply, sets of scientific arguments (rather than unrelated individual arguments) are employed in explanatory contexts. Unification is construed as the derivation of as many conclusions as possible from as few "patterns" of derivation (rather than as few premises as is possible).[8] For example, Kitcher argues that the post-Daltonian explanation of why substances form compounds in certain ratios was achieved according to a simple argument pattern that was first extended by the introduction of the notion of valence. This allowed for the derivation of conclusions concerning the possible formulas for compounds and a deeper understanding of the relationships between weights of compounds.[9] Next came the introduction of premises characterizing the atom as having a shell-like structure and principles concerning ionic and covalent bonding. This allowed for derivations of possible compound formulas and for an understanding of the seemingly arbitrary valence rules, and the derivation of atomic weight relationships was effected by introducing further premises concerning atomic constitution in terms of protons, neutrons, and electrons.[10] Ultimately, the derivations associated with the shell model can be embedded within the quantum mechanical treatment of the atom and bond-formation. While Kitcher remarks that these derivations are mathematically tractable only in elementary cases (which is indicative of the time of writing), further extension of the explanatory derivations is possible in principle.[11]

The embedding of the derivations generated by the shell model and the principles governing bond formation within quantum mechanics constitutes an "explanation extension." Kitcher argues that his account is distinct from reductive explanation since the extending theory is an "explanation extension" even if some of the concepts of the extended theory cannot be formulated in quantum mechanical terms.[12] This might appear to tend towards a notion of approximate "reduction," or rather an intermediate position between the realism of chemical models and reduction to quantum mechanics as advocated by Scerri.[13] By contemporary standards, Kitcher's notion of explanation extension appears to be often realized, since quantum mechanical calculations are tractable in more complex cases, subject to approximation. (The nature and status of approximation in semi-empirical and *ab initio* methods respectively is an issue that I will not pursue here.) Yet Kitcher does not appear to offer an account that accommodates approximate reduction. According to Kitcher, *all* explanation is deductive (including probabilistic explanation),[14,c] and if explanatory pow-

---

[b]Recently, Cartwright *et al.* have rejected not only the truth of theoretical laws but also deductivism itself while emphasising the autonomy of phenomenological models. Cf. Cartwright, N., T. Shomar & M. Suárez. 1995. The Tool Box of Science—Tools for the Building of Models with a Superconductivity Example. Poznan Stud. Philos. Sci. Hum. **44:** 137–149.

[c]The motivation for Kitcher's deductivism is the need to compare the relative unifying power of different sets of arguments, and this would be difficult if nondeductive arguments were also included. (See Ref. 7, p. 448.)

er is construed in terms of unification, this much is consistent with reductive explanation. Nagel, for example, famously employed the covering law account as a basis for reductive explanations.[15] He attempted this while allowing for the preservation of some of the terms of the reduced theory that were related to terms of the reducing theory by correspondence rules. Whether explanations are arguments or patterns of arguments, the result appears to be the same. There is no motivation for the explanatory *relevance* of autonomous chemical concepts such as "chemists' orbitals." (I borrow the phrase "chemists' orbitals" from Scerri, Ref. 13, p. 52.)

Kitcher provides a particular defence of what he calls "deductive chauvinism" concerning probabilistic explanations with reference to the difficulties associated with deriving deductive explanations of macroscopic events from quantum mechanics. One way of deriving such explanations is to idealize the phenomena. Here, Kitcher allies himself, up to a point, with Cartwright's "simulacrum" account of explanation.[16] In order that we might fit an equation to some phenomenon, we must first prepare a description of it, and this will amount to providing a sanitized representation of that phenomenon. This much might be consistent with Kitcher, but Cartwright's focus for the representation and explanation of phenomena to which equations can be fitted are *models*, not the laws themselves. Models incorporate idealizations not realized in concrete systems, yet some of the properties of the model are shared by the real systems. Hence, the model can fulfill an explanatory role, since it acts as a "simulacrum" of the system modelled.[17]

The use of models in chemistry has received some considerable articulation in recent years.*f* In general philosophy of science, contemporary debates emphasize the autonomy of models from fundamental theory. This autonomy arises because there are many models that are simply not deduced from our high-level theories. For example, Morrison has developed the notion of "mediating models" that are partially autonomous from fundamental theory and phenomena in terms of construction and function.[18] These models are "partially" autonomous because, while they incorporate *some* theoretical structure in their construction, they are not deductively subsumed under fundamental laws. Therefore, partial autonomy is of direct relevance to the autonomy of explanation in chemistry. Briefly, approximate reduction can be interpreted as the incorporation of quantum mechanical structure in models that enables approximate solutions to Schrödinger's equation. One problem that arises, however, is that the partial relations between fundamental theories and models in explanation cannot be accommodated by the logical relation between explanans and explanandum.

The semantic view of theories provides a means for understanding model autonomy and a potential basis for articulating the autonomy of explanation in chemistry. First, theories are presented as families of extra-linguistic structures or models. According to the so-called "received view," theories are formal-linguistic entities, and models are seldom more than superfluous interpretations of a theory.[19] It is therefore not clear how models have any explanatory role at all, but denying such a role would seriously misrepresent chemical practice. Second, although the relations between the models of a theory have been traditionally articulated within the semantic view

---

*f*For example, see the special issues on models in chemistry in Hyle **5(2)**, November 1999; Hyle **6(1)**, March 2000. Further references are provided in van Brakel, J. 2000. Philosophy of Chemistry. Leuven University Press. Leuven, Belgium. p. 154, n. 382.

in terms of inter-structural isomorphisms, some (for example, French and Ladyman[20]) favor *partial* isomorphisms. The point here is that if models are related by isomorphisms, there is no means for capturing the partial autonomy essential to model autonomy and explanation in chemistry because isomorphism implies structural identity. In fact, Morrison eschews the semantic view partly because it implies that all models are intimately related to high-level theory. But if the relations between models are expressed as partial isomorphisms between partial structures, the kind of independence of model construction and function that Morrison advocates can be accommodated.

The above is merely a sketch of the importance of the concept of model autonomy for explanation in chemistry and the potential that the semantic view might provide in precisely articulating what such autonomy amounts to. However, a thorough account of explanation is itself yet forthcoming in structuralist terms. I can only suggest here that one means for outlining how a semantic view of explanation might proceed is to associate the explanatory function of models with how they represent. The association between explanation and representation is important to both Cartwright's simulacrum account of explanation and Morrison's notion of mediating models. I believe that the ubiquitous use of "visual" models that characterizes chemistry demands an account of explanation with an extra-linguistic basis. Consequently, the semantic view is central. However, this is a contentious issue. For example, Hendry criticizes the semantic approach as inadequate in accounting for the varieties of media, including the importance of language, in representation.[21] Hendry favors an "interactionist" view of scientific theories that emphasizes the importance of both language and extra-linguistic models.[22] This debate should provide an important focus for investigating the nature of explanation in chemistry, and both the semantic view and interactionist account require further articulation regarding how explanations are to be understood.

ACKNOWLEDGMENTS

I would like to thank Steven French and four anonymous referees for helpful suggestions for improving this paper. Some of the research that contributes to this paper was undertaken with financial support from the Arts and Humanities Research Board.

REFERENCES

1. HEMPEL, C.G. & P. OPPENHEIM. 1948. Studies in the logic of explanation. *Reprinted in* Theories of Explanation. 1988. J.C. Pitt, Ed.: 9–46. Oxford University Press. Oxford, UK
2. BRODY, B.A. 1972. Towards an Aristotelian theory of scientific explanation. Philos. Sci. **39**: 20–31.
3. Ibid., p. 24.
4. SCERRI, E.R. & L. MACINTYRE. 1997. The case for the philosophy of chemistry. Synthese **111**: 213–232.
5. CHRISTIE, M. 1994. Philosophers versus chemists concerning "laws of nature." Stud. Hist. Philos. Sci. **25(4)**: 613–629.
6. CARTWRIGHT, N. 1983. How the Laws of Physics Lie. Clarendon Press. Oxford, UK.

7. KITCHER, P. 1989. Explanatory unification. *In* Scientific Explanation: Minnesota Studies in the Philosophy of Science. Vol. XIII. P. Kitcher & W. Salmon, Eds.: 410–505. University of Minnesota Press. Minneapolis.
8. Ibid., pp. 432–434.
9. Ibid., p. 446.
10. Ibid.
11. Ibid. 447.
12. Ibid.
13. SCERRI, E.R. 2000. Realism, reduction, and the "intermediate position." *In* Of Minds and Molecules: New Philosophical Perspectives on Chemistry. N. Bhushan & S. Rosenfeld, Eds.: 51–72. Oxford University Press. Oxford, UK.
14. KITCHER, Explanatory unification, p. 448.
15. NAGEL, N. 1974. Issues in the logic of reductive explanations. *Reprinted in* Philosophy of Science: The Central Issues. 1998. M. Curd & J.A. Cover, Eds.: 905–921. Norton. New York.
16. KITCHER, Explanatory unification, p. 453.
17. CARTWRIGHT, How the Laws, p. 152.
18. MORRISON, M. 1999. Models as autonomous agents. *In* Models as Mediators: Perspectives on Natural and Social Science. M. Morgan & M. Morrison, Eds.: 38–65. Cambridge University Press. Cambridge, UK.
19. BRAITHWAITE, R.B. 1962. Models in the empirical sciences. *In* Logic, Methodology and Philosophy of Science: Proceedings of the 1960 International Congress. E. Nagel, P. Suppes & A. Tarski, Eds.: 224–231. Stanford University Press. Stanford, CA.
20. FRENCH, S. & J. LADYMAN. 1999. Reinflating the semantic approach. Int. Stud. Philos. Sci. **13(2)**: 103–121.
21. HENDRY, R.F. 2001. Mathematics, representation and molecular structure. *In* Tools and Modes of Representation in the Laboratory Sciences. U. Klein, Ed.: 221–236. Kluwer Academic Publishers. Dordrecht.
22. HENDRY, R.F. 1999. Theories and models: the interactive view. *In* Visual Representations and Interpretations. R. Paton & I. Neilson, Eds.: 121–130. Springer-Verlag. London.

On Explanatory Practice and Disciplinary Identity

ANDREA I. WOODY

Department of Philosophy, University of Washington, Seattle, Washington 98195, USA

> ABSTRACT: This essay explores connections between explanatory discourse and disciplinary identity. While explanation is frequently assumed to be one of the central aims of the sciences, I contend that if we want to appreciate explanation's role within scientific practice, we need to look elsewhere. Explanations encode the aims and values of particular scientific communities, telling practitioners what they should want to know about the world and how they should reason to get there. By inculcating patterns of reasoning consistent with these aims and values, explanatory discourse is crucial for the formation of cohesive scientific disciplines. Yet because such discourse is frequently grounded in exemplars, the implicit nature of exemplar-based training must be confronted. I argue that exemplars are an appropriate vehicle for communicating skill-based knowledge and methodological norms and, additionally, that exemplars are a practical means of grappling with the indeterminacy of theory application.
>
> KEYWORDS: scientific explanation; exemplar; disciplinarity; Kuhn; scientific practice

INTRODUCTION

It seems undeniable that the modern sciences aim, among other things, at explanation. Indeed, according to Ernest Nagel, a considerably stronger claim can be defended: "It is the desire for explanations which are at once systematic and controllable by factual evidence that generates science; and it is the organization and classification of knowledge on the basis of explanatory principles that is the distinctive goal of the sciences."[1] Thus, many have assumed that uncovering the common elements of scientific explanations, features presumably shared widely across distinct disciplines, will help to illuminate the essence of science itself. In the following discussion, I contend that while explanation is crucial to all sciences, its function is often misconstrued. Seeing how this is the case will clarify what we mean by "explanation" at the same time that it reveals the close connections between explanatory discourse and disciplinary identity.

Address for correspondence: Department of Philosophy, University of Washington, Box 353350, Seattle, WA 98195. Voice: 206-685-2663; fax: 206-685-8740.
a.woody@u.washington.edu

EXPLANATIONS

Explanations grace the pages of scientific textbooks. Teachers frequently provide explanations to their students. Experts may explain things to close colleagues, to other scientists, or sometimes, to the general public. So conceived, explanations are things embedded in language; they are straightforwardly propositional. Yet much language communication obviously is not explanatory. How then do we differentiate explanations from other statements? Often, an explanation serves as a response to some explanation-seeking question.[2] We can judge the explanatory status of such statements according to whether they legitimately satisfy the demands of the original question. Such satisfaction, in turn, might be judged by acceptance of the explanation by the intended audience or, alternatively, evaluated by the relationship between the content of the explanatory response and the content of the original question.

Precisely because the first mode of judgment can seem dangerously subjective, philosophical analysis has sanctioned the second. Statements should be accepted as explanatory if, and only if, the right sort of relationship exists between the *explanans* and the *explanandum*. The analytic challenge becomes proper characterization of this relationship. Some researchers[3,4] have hoped for a purely logical, syntactic characterization, while others[5] have embraced one or another set of semantic constraints based on metaphysical presuppositions. In practice, however, judgments of explanatory adequacy rely almost exclusively on the first mode of judgment. We accept claims as explanatory based on the effect they have on an audience.

But not simply any audience will suffice. Neither the nodding approval of a small child nor the wide-eyed curiosity of a classroom of engineering students would indicate the adequacy of my explanation of high conductivity in copper wire. If, on the other hand, a gathering of solid-state chemists approved my explanation, I could justifiably accept their judgment. This would be the case even if the explanation were so sketchy that no possibility existed of applying logical analysis of the sort just mentioned. On the surface, then, explanations are more directly linked to human communication, which happens to be language-based, than to statements as logical units of content, severely abstracted from practice.

Moreover, much explanatory discourse lacks conspicuous reference to any explicitly posed question. There is no obvious back and forth, for instance, to the explanatory content of many textbooks. In fact, there seem to be two distinct contexts for explanatory discourse. In some circumstances, genuine questions are clearly posed. Such questions are grounded in the assumption that an explanation surely exists, and any potential response is judged by a prior conception of the requirements for explanation. In other circumstances, in place of explicit questions we find descriptions aimed at conveying what sort of information *should* be judged explanatory. This discourse implicitly establishes the standards endorsed by a given community in judging explanatory adequacy. In this respect, textbooks function as instruction manuals as well as repositories of content. This second context obviously provides foundation for the first. Not only is it the source of explanatory norms assumed in the first context, but it communicates these norms to novice practitioners. Because the community will judge novel explanations based on their acceptance by practitioners, it is essential these folks be well-trained, reliable judges of explanatory norms.

We might anticipate a third context of explanatory discourse in which practitioners grapple more directly, and generally, with what constitutes explanatory power.

At least in communities with well-established disciplinary character, such dialogue seldom occurs. The discussions so pervasive among philosophers of science are rare among scientists.

EXEMPLARS

That experts must train novices to recognize adequate explanations is not surprising.[a] It is harder, though, to make sense of how this training is achieved. Contrary to what one might assume from my characterization of textbooks as "instruction manuals," inexperienced practitioners are rarely, if ever, given "rules" for explanation. Instead, they are instructed obliquely through examples (which often occupy a majority of textbook space and lecture time). Problem sets, those mainstays of modern pedagogy, are largely lessons in recognizing conditions for explanatory adequacy and transferring them to new cases. Here students cultivate analytic skills by mimicking previously exhibited reasoning. Exercises involving Newtonian point masses sliding down an inclined plane, or the one-dimensional particle in its quantum mechanical box, or Punnett square analyses of classical genetics inculcate patterns of thought in the student at the same time as they endorse them.

Exemplars, in other words, display, without explicitly articulating, what a scientific community judges to be explanatory, what model of intelligibility it has chosen to embrace. I recall early in my own graduate training being admonished by a mentor never to use the term "exemplar" in writing; he presumed the term nothing more than a baroque synonym for "example." But such thinking is mistaken. An exemplar is a more self-conscious construct than an example, because it has been imbued with a certain normative status. The same point, from a slightly different perspective, appears in Thomas Kuhn's[6] descriptions of normal science. An exemplar is an example that, through community sanction, we are urged to follow. It has action-guiding force. Point masses and artificially shaped potential wells and Punnett squares take on a certain status when students are required to study them, practice the corresponding analytic techniques, and later display their proper application on examination. Nothing lacking this communal status can be an exemplar. In contrast, something arguably can be an example without being recognized as such. (Exemplification, in this view, is a more straightforwardly two-place representational relation.)

Surprising though it may be, this implicit notion of intelligibility is the stuff out of which coherent scientific communities must be constructed. Seeing this requires us to think more about the social nature of the sciences. At the same time I will explore briefly how the implicit nature of explanatory norms actually addresses a central challenge for all empirical research.

---

[a]It may be instructive, however. If I am correct in asserting that in practice the adequacy of explanations is judged by audience reaction, then it is essential to have audiences that respond reliably. It is also necessary that those seeking explanations be able to identify legitimate audience members. For instance, if we cannot recognize the reliable judgment of a graduate student, his judgments can have no utility for us.

DISCIPLINES

In this day and age there really is no such thing as "science." Rather there are distinct *scientific disciplines*. Let me suggest we take that terminology quite literally. What distinguishes and identifies scientific communities are their unique disciplines, that is, the skills and methods they bring to empirical investigations. But for scientific communities to function as disciplines, the actions and thoughts of individuals must be coordinated to form coherent, cohesive practices.

This coordination must reflect the aims and values of the community. Though it is legitimately tempting to claim that what all sciences share is a general aim to acquire reliable, nontrivial empirical knowledge, we must resist the further temptation to disregard the specific articulation of aims and values that constitutes any particular scientific enterprise. As an idealization generated by ignoring such specificity, "science" can be a convenient fiction. But when treated as descriptive of actual practices, the characterization is anemic and distorting.[b] The sciences comprise a plethora of distinguishable communities. Each must decide what phenomena, and indeed what *aspects* of phenomena, to investigate, what the goals of such investigations should be, what methods to employ, what theoretical structures to consider informative, and how to judge success for each type of activity undertaken. The formidable challenge is to establish practices that are robust while being sensitive to all the variables of the research situation. Communal activities themselves, in other words, must be disciplined—often in highly sophisticated ways.

One of the most fundamental components of this discipline concerns the relationship between phenomena and theory. There must be constraints regarding legitimate theory application, including how exactly a theory is applied and to what specifically it should be applied. But now a pertinent question can be raised: What is a scientific explanation other than an application of theory? In requesting a scientific explanation of some phenomenon, we proclaim "Show me how this phenomenon can be accounted for by the application of a theoretical structure to which I am already committed. The phenomenon will be intelligible to me precisely if this connection to theory is made clear." Note that this gloss on scientific explanation makes natural connection with both nomological and unificationist accounts, in addition to capturing the intuition, mentioned earlier, that explanations concern relations between sets of claims.[3,4,7]

Still, theory application always involves interpretation. Theories do not apply themselves, and they do not come with explicit instructions. Philosophical analysis touches upon this issue most often in the guise of logical problems concerning underdetermination and the role of auxiliary assumptions.[8] In practice, though, the challenges of consistent theory application are met, in large part, by explanatory discourse. By cultivating shared analytic skills and conventions, explanatory discourse is a way of overcoming the indeterminacy of theory application to achieve intellectual coherence within the discipline. Again, Kuhn[9] recognized this element of scientific practice years ago with the phrase "convergent thinking." Through its explanations, a scientific practice communicates choices it has made to undercut the

---

[b]There are methodological consequences to suggesting that what is common to all sciences is highly general and restrictively informative. If we desire to understand how sciences works, we will have to get our hands dirty with the details of individual practices.

ambiguity of theory application. Thus, explanatory discourse is a crucial means of generating and maintaining coherence.

We should not assume, however, that the formation of a coherent scientific practice begins with a clear and determinate set of values, which are self-consciously imposed on the structure of its explanations. Instead, something akin to a feedback mechanism operates. A scientific community adopts a rough set of epistemic and practical aims and begins exploring some domain of inquiry. Certain developments and applications of theory will be successful with respect to these rough aims; others will not. Those that are will influence the evolution of the practice by implicitly reinforcing certain values and undermining the legitimacy of others. When certain representational strategies continue to have great success, their status can be so elevated that they become actual models of intelligibility for the community. When scientists judge a theoretical structure to be highly explanatory, they are either assigning or recognizing such a status.

But what if the models of intelligibility lead us astray? What if they are wrong? On this account, after all, theoretical applications evolve into exemplars by an incremental process of social acceptance. Perhaps the contingencies of a discipline's history produce an exemplar that ultimately misrepresents the phenomena to which it should apply. To pose this question, in this manner, requires jumping back to an unarticulated vision of "science" aiming for "the truth." Scientific disciplines gain control over their fields of inquiry by embracing specific aims and methods in relation to some narrow range of phenomena. An exemplar must be productive relative to these. If not, its status will change. Of course, an exemplar might appear productive and still, ultimately, be misrepresentative. If such errors are strictly irrelevant to the aims of the practice, then arguably they should be ignored. Disciplines can hope to reveal aspects of the truth concerning their empirical domains, but "the truth, the whole truth, and nothing but the truth" is a caricature. Furthermore, errors are often highly informative. We learn from them. The only errors that seem necessarily pernicious are those we have little possibility of recognizing.

History may provide some perspective. We are not tempted to speak harshly about the explanatory successes of Newtonian mechanics. The model of intelligibility provided by its analytic techniques continues to dominate all considerations of terrestrial motion; its pragmatic value has not been diminished by recognition of its shortcomings. (So contrary to what many philosophical accounts suggest, the truth of a theoretical framework does not appear to be a strictly necessary requirement for explanation.) Nor should we be quick to dismiss the explanatory power of the phlogiston theory. For eighteenth century chemistry, phlogiston was a highly explanatory concept because it organized and focused a range of chemical investigations for a budding international community. The diversity of practitioners' training and backgrounds necessitated a unified conceptual structure for the development of disciplinary identity. For a theoretical structure that appears almost perverse to modern eyes, we must appreciate the intellectual benefits of community building: the theory of phlogiston shaped standards that organized researchers so that ultimately they could develop a superior empirical theory.

There is a second worry, raised by the fact that models of intelligibility are communicated through exemplars. Precisely because theory application involves interpretation, would not an explicit set of rules for application more reliably communicate norms and achieve uniformity?[c] If the norms were stated as rules, however, we would confront questions about how to apply the rules. So we simply transfer the application challenge from one set of claims to another. Any further move to provide instructions for the rules sends us down the path of an infinite regress. Exemplar-based training cannot eliminate these issues, but it is designed to grapple with them. Textbook exemplars aim to cultivate communal skills and techniques by a more direct, and yes, implicit, form of communication. Skills are introduced by direct demonstration. Correct theory application is cultivated through mimicry and experience. The original challenge presented by the interpretive gap is not eliminated, but since the exemplars are, by decree, examples of correct application, the challenge is now limited to new cases. On the rule-based system, all cases face the interpretative gap. Thus, exemplar-based training does have a certain bootstrapping quality, and it raises all the epistemic worries typically associated with such support structures. We will return to this issue in Section 5.

My overarching suggestion is that explanatory discourse communicates information that is essential to the formation of cohesive scientific disciplines. Different disciplines (and subdisciplines, for that matter) embrace different models of intelligibility, which in turn make different forms of investigation attractive and feasible. Explanations encode the aims and values of particular scientific communities, telling practitioners what they should want to know about the world and how they should reason to get there. More specifically, exemplars communicate norms of theory application, the lack of which would undermine consistent practice. The epistemological worries raised earlier are legitimate, but they are worries faced by a practice as a whole; they are not issues for explanation *per se*. Conceivably the coordinated efforts of a community well disciplined in "convergent thinking" are best suited to wrestle with these demons.

PRACTICE RATHER THAN PRODUCT

If the preceding remarks are on track, then the starting point for our discussion was ill-conceived. We began by focusing on explanations, those things produced by scientists in response to certain questions. Subsequently, we discussed how the process of training practitioners through explanatory discourse is a crucial means of establishing discipline. The value of the activity rests in enabling coordinated and cohesive empirical investigations. Yet if this is so, it is the activity of explaining that is important and productive, rather than the resulting explanations themselves. Consequently, the stress on linguistic analysis has been misplaced. We care not about the explanans, as product, but about the role that producing it has for the practice generally. The Latin construction is telling; it is "explaining" that matters.

Traditional philosophical debates on explanation, it would seem, presuppose both a static propositional conception of explanation and something like the unarticulated (potentially essentialist) vision of "science" mentioned earlier. It is this set-up, I believe, that has dictated a particular analytic project: identifying necessary conditions for individual explanations. But no semblance of consensus has emerged.[2–5,7] Each approach nicely captures some widely accepted explanations and excludes others.

---

[c]The "correspondence rules" of logical empiricism were envisioned to function in just this manner.

As soon as we acknowledge the diversity of viable scientific practices, each with its own specific aims, and shift our attention to explaining as an activity, all reasons to expect unified structure in the explanations themselves evaporate.

More importantly, these same traditional analyses fail to provide a satisfying account of why explanatory power is considered to be a general epistemic virtue. On many accounts, explanations are just the stories that result from application of a true empirical theory. They are things produced in the aftermath, it would seem, of scientific work. Reorienting ourselves towards explanatory discourse brings the possible epistemic utility of explanation into sharp focus. Admittedly, one still needs to argue for how the social structure supplied by explanatory discourse will be truth conducive, but the rough outline is visible. The social organization of scientific practice facilitates knowledge production. From this perspective, explanation is genuinely valuable to the sciences.

This line of thought concerns explanatory discourse without presupposing what form the discourse will take. Though exemplars typically play a crucial role, is reliance on them necessary? Perhaps we are witnessing simply a historically contingent preference. Perhaps the information could be communicated explicitly. This sounds perfectly sensible, and yet I suspect we have allowed another ticklish conceptual slide. When we speak of "information," it is natural to assume it could be made explicit. But recall that what exemplars display are analytic techniques and their corresponding standards, neither of which is obviously reducible to purely descriptive information. If the techniques are genuinely skills-based, their acquisition may indeed require more than a list of instructions. Recall the stock example of skills-based know-how. You cannot tell someone how to ride a bike. You might show them how you do it, and you can encourage them to get on and give it a try. But instructions alone will be insufficient, perhaps even detrimental. What you wish to share with them is skill, an embodiment of knowledge, rather than information, an abstract representational object of knowledge. Precisely because exemplars are examples, they are an appropriate means of cultivating embodied knowledge. Nor is it easier to conceive the normative content behind various standards being communicated explicitly. Although these brief remarks are far from a principled argument, they do suggest that exemplar-based communication would be difficult to eliminate.

To take stock of our discussion, we return to the opening quotation by Nagel: "It is the desire for explanations which are at once systematic and controllable by factual evidence that generates science; and it is the organization and classification of knowledge on the basis of explanatory principles that is the distinctive goal of the sciences."[1] I cannot agree with what I believe Nagel intends to say here, but if properly translated, this quotation is not far off the mark: "It is the desire for theoretical structures that are systematic and empirically adequate that generates science as we know it; and it is the organization and classification of knowledge made possible by articulated scientific practices, held coherent largely by explanatory discourse, that is the general common goal of the sciences."

CLOSING REMARKS

I have suggested that explanatory discourse is a central mechanism for achieving coherent scientific practices. In essence, explanatory discourse serves to make both the subject matter and the scientific practice itself seem intelligible. While modern western culture idolizes the creative insights of individual scientists, the emergence of coherent, and coordinated, research groups is a better hallmark of the maturity, and genius, of modern science. Scientific practices evolve through time as analytic skills grow ever more focused and powerful in relation to specific epistemic and practical aims. The increasing specialization of the sciences is the articulation and growth of discipline. Explanatory discourse is a principal tool for sculpting this articulation.

REFERENCES

1. NAGEL, E. 1961. The Structure of Science. Harcourt, Brace, and World. New York.
2. VAN FRAASSEN, B. 1980. The Scientific Image. Clarendon Press. Oxford, UK.
3. HEMPEL, C. 1965. Studies in the logic of explanation. *In* Aspects of Scientific Explanation and Other Essays. The Free Press. New York.
4. FRIEDMAN, M. 1974. Explanation and scientific understanding. J. Philos. **71:** 5–19.
5. SALMON, W. 1984. Scientific Explanation and the Causal Structure of the World. Princeton University Press. Princeton, NJ.
6. KUHN, T. 1962. The Structure of Scientific Revolutions. University of Chicago Press. Chicago.
7. KITCHER, P. 1989. Explanatory unification and the causal structure of the world. *In* P. Kitcher & W. Salmon, Eds. Scientific Explanation. University of Minnesota Press. Minneapolis.
8. DUHEM, P. 1954. The Aim and Structure of Physical Theory. Princeton University Press. Princeton, NJ.
9. KUHN, T. 1977. The essential tension. *In* The Essential Tension: Selected Studies in Scientific Tradition and Change. University of Chicago Press. Chicago.

The *Ignis Fatuus*[a] of Reduction and Unification

Back to the Rough Ground

J. VAN BRAKEL

Higher Institute of Philosophy, Catholic University of Leuven, 3000 Leuven, Belgium

ABSTRACT: Given the enormous variety of possible intertheoretical relations, the proliferation of definitions of reduction, supervenience, emergence, unification, and so on, as well as the fact that empirical studies have provided support for almost any metaphysical option or, alternatively, have shown rather conclusively that empirically the case is inconclusive, I suggest a moratorium on the use of words such as "reduction," "supervenience," and "unification" and to go back to the rough ground and give perspicuous renderings of the practice of chemistry. What is needed are very detailed case studies and further discussions about them, instead of bickering about whether chemistry can be reduced to physics, supervenes on it, can be unified with it, and similar "metaphysical" concerns.

KEYWORDS: philosophy of chemistry; reduction; supervenience; emergence; unification; basis sets; EPR correlations; quantum chemistry

INTRODUCTION

Over the past few years there have been a considerable number of publications and conference presentations in philosophy of chemistry on the issue of reduction; usually, but not exclusively, focusing on the question whether chemistry can be reduced to physics.[1–8] The main source of the idea that chemistry can be reduced to physics might be traced back to one remark of one physicist, though a famous one, namely, Dirac[9]: "The underlying laws necessary for the mathematical theory of a large part of physics and the whole of chemistry are thus completely known, and the difficulty is only that exact applications of these laws lead to equations that are too complicated to be soluble."

In this paper my aim is not to argue for or against the latter thesis; nor is it to argue for or against a particular concept of reduction (or one of its congeners such as supervenience). Instead I will try to give a perspicuous rendering of the conceptual intricacies of "the debate on reduction" and how these intricacies relate to the history and current state of the art in the philosophy of science. I choose the (Wittgensteinian)[10] expression "perspicuous rendering" and not, say, "conceptual analysis," because the latter already presupposes that "reduction" is one thing and it is the job of

---

Address for correspondence: J. van Brakel, Higher Institute of Philosophy, Catholic University of Leuven, Kardinaal Mercierplein 2, 3000 Leuven, Belgium.
Jaap.vanBrakel@hiw.kuleuven.ac.be

[a]Will-o-the wisp (literal translation, "fleeting flame").

the philosopher to specify the "best" essence of what reduction really is. Instead, I'll avoid the term reduction as much as possible, and I'll introduce a more neutral terminology; that of "intertheoretical relations." Throughout, I will use the letters S and B for the two domains or theories that are in some way related.

I won't argue for or against the thesis "chemistry can be reduced to physics," because, in its generality, the thesis is not meaningful. First, it is unclear how one would delineate and separate chemistry and physics; in particular it is unclear what the meaning of the word *physics* is in this context—I'll come back to that below. Second, it has to be spelled out what it is that is being reduced (theories, concepts, properties, laws, causes, explanations, "everything") and how reduction might differ from related notions such as elimination, integration, supervenience, emergence, and so on, and what *sort* of relation it is (e.g., dependence, identity, causality). I'll come back to that as well. Third, more than two domains or theoretical discourses, not only chemical and physical, need to be distinguished. That is to say, a whole variety of possible intertheoretical relations have to be addressed. In TABLE 1, I have listed a rather long list of examples,[11] not only to emphasise the variety in the examples, but also to draw attention to the great variety of colloquial ways of talking about intertheoretic relations—note that in each entry a different verb is used.

BACON'S IDOLS

Why is it that the debate often focuses on such broad questions as whether one is for or against chemistry being reducible to physics? Because deeply ingrained metaphysical prejudices are hovering in the background, a few examples of which I'll give below (for interesting prejudices among physicists see Dresden[12]). I'm not saying that I can prove these prejudices to be wrong, but I want to stress that these prejudices are upheld by many rather unreflectively, not being aware of the relevant historiography of ideas of which they are part. Both scientists and philosophers use the notion of reduction(ism) for all kinds of different rhetorical purposes and with numerous meanings—meanings they are often unclear about themselves.

Success Proves Any Story You Tell about It

A very general prejudice is that because science is successful, it proves whatever your favourite story about science is. If you can give at least one plausible example from the history of science or can make plausible that there might be such an example, you are allowed to generalize across all of science. Your view of science being proven, you start looking around and, lo and behold, you find confirmations of your view everywhere. Any problematic case you might be confronted with can be moved to "future research." For example, there is no denying that the quantitative approach effected via quantum chemistry has made a considerable contribution to chemistry in a variety of different areas. But nothing follows from that about the relevant intertheoretical relations involved.

A specific case of this prejudice is that many scientists and philosophers believe that there are numerous success stories for the reductionistic program. Partly, this is because simple slogans stick (see below), but there is a more subtle point. As Wimsatt points out,[13] often a heuristic principle succeeds, in part, by transforming a prob-

TABLE 1. Making the question "Can chemistry be reduced to physics?" more concrete and varied

- Can chemical thermodynamics be reduced to statistical mechanics?
- Can the notion of chemical bond be defined in purely quantum mechanical discourse?
- Is "being water" *nothing but* "being H_2O"? Are all properties of water realized by the properties of H_2O?
- What underpins the concept of boiling point at the most fundamental level of description?
- What is the quantum-mechanical correlate of thermodynamic entropy?
- Is heat and mass transfer in chemical engineering fully covered by "fundamental" phenomenological equations such as Fourier's law for heat conduction and Fick's law for molecular diffusion?
- Which more fundamental theoretical notions underlie the Onsager relations?
- Are the nonlinear differential equations of chemical kinetics grounded in linear quantum mechanics?
- How does the Pauli principle fit into quantum mechanics?
- Does the periodic table supervene on the properties of elements, the latter understood in terms of nuclei and orbitals?
- How do the concepts of orbital and hybridization figure at different levels of description (if any)?
- How is chemical reactivity brought about in terms of the four fundamental physical forces?
- Can we calculate all possible structures and stereoisomers for all possible isomers from a given compositional formula?
- Are the properties of sodium chloride "resultant" or "emergent" relative to the properties of sodium and chlorine?
- Can we give a statistical mechanical description of the three-phase line between one solid and two immiscible liquids?
- Can we give a precise account of the textbook wisdom that a neutron consists of three quarks?

lem into a different but related problem (which is easier to solve). Some *part* of the complex behavior is picked out by the available experimental techniques and considered worthy of further investigation, clarification, and thereby amplification. If this works very effectively, there will be a tendency to identify the new problem as the old one. In particular, if there are many small adjustments, a substantial change in paradigm may result. This *ex post facto* reification is central to the high opinion many have of reductionistic methodologies.

The Craving for Simple Slogans

People, philosophers and scientists not excluded, like simple slogans. That's why the notorious comment of Dirac had such an impact.[14] The physicist Gell-Mann gives the proper response to Dirac's one-liner[15]:

Of course [Dirac] was exaggerating.... In principle, a theoretical physicist using QED [quantum electrodynamics] can calculate the behaviour of any chemical system *in which the detailed internal structure of atomic nuclei is not important* .... In order to derive chemical properties from fundamental physical theory, it is necessary, so to speak, to ask chemical questions of the theory.... The laws of chemistry can in principle be derived from QED, *provided the additional information describing suitable chemical conditions is fed into the equations*; moreover, those conditions are special—they do not hold throughout the universe. [italics added]

But I haven't found anybody yet who has quoted Gell-Mann in full, though there are publications in which Gell-Mann is simply quoted as saying "chemistry is in principle derivable from elementary particle physics," thus propagating what I'll call "Dirac's myth." Note, by the way, that the proviso Gell-Mann adds is very general. It won't go away if physics is unified, or computers get bigger.

The Shadow of Kant

That mathematics is the bench mark of "real" science is exemplified by physics, and is an idea upheld by many chemists. But where does this idea come from? Why is its truth taken to be self-evident? This prejudice goes back to Kant's view that science is "proper" to the extent that mathematics is applied within it. According to Kant, chemistry is rational (because it uses logical reasoning), though not a *proper* science, because it misses the basis of mathematics and the synthetic *a priori*. Chemistry is an *uneigentliche Wissenschaft*—an "improper" science. Though the metaphysics and the synthetic *a priori* was later dropped, the idea that proper science is something that uses mathematics has been prevalent to the present day. To put it crudely, the logical positivists, following Hume, threw out metaphysics, replacing it with logic to line it up side by side with mathematics as the metaphysical foundation of all proper science. The traces of Kant's view can be found throughout the history and philosophy of chemistry.[16]

The Politics of Unified Science

Reduction is often associated with the idea of the unity or unification of science. To build a science is to build a unified system. Moreover, scientific legitimacy is granted by tracing a discipline's pedigree back to the "patriarchal" science of physics (via reduction). The idea of unity or unification has much to do with metaphysics and theology, but very little with being a scientific hypothesis subject to critical scrutiny. Here is a bit of historiography, drawing on the work of Galison.[17]

The pre-war unity of science notion of the Vienna Circle was primarily a commitment to the *methodological* unity of science. Ontologically or metaphysically speaking, it was a mixed bag of views. Carnap wanted to provide an *Aufbau* of the world, securing objectivity by way of the relation between its elements. Neurath's idea of unity was very different. He didn't aim at expressing all science in the language of physics, but at the creation of a heterogeneous jargon. He didn't want a pyramid of knowledge but an orchestration of different instruments. Today Neurath might be grouped with those advocating a disunifed science.[18,19]

The post-war unity of science movement differs strikingly from the pre-war concern of the *Wiener Kreis*. In 1947 Frank convinced the Rockefeller Foundation to give a grant to the Unity of Science Movement, whose members included Carnap, Morris, Frank, Reichenbach, the physicist Van Vleck, the psychologist S.S. Stevens,

Norbert Wiener, and many other celebrities. At stake was the fate of the world: The Unity of Science Movement would combat both right-wing organismic metaphysics (including Thomism) and left-wing dialectical physicalism (i.e. communism). The metaphysical commitment to the unity of nature became an often-repeated creed; few scientists disagreed. In the fall of 1952, Frank and his colleagues started work on a basic operational dictionary, which was to include three hundred basic concepts, including mass, matter, message, noise, order, congruence, utility, market, capital, degree of assent, id, love, gestalt, God, and damnation. Whatever one thinks about the politics, I think anybody who favors the unity of science should reflect on the movement's idea of 300 basic concepts. One wonders why the Rockefeller Foundation took this old boys' network metaphysics seriously.

REDUCTION AND ITS CONGENERS

Though among chemists and philosophers of chemistry there is considerable interest in the issue of reduction, actually reduction talk is a rather outdated discourse, except in the context of specific research paradigms such as structuralism.[20] All arguments against reduction that I know of, in publications by theoretical chemists and philosophers of chemistry, direct their attacks to some variant of the classical notion of reduction. In TABLE 2 I've given illustrations of what I take to be the classical view of reduction.[21] Note again, that the different wordings emphasize different aspects of the reduction relation—as in TABLE 1 a different verb is used in each entry. Because this is by now a degenerating research program (in the sense of Lakatos), I won't say much about it, except for stressing that the issues of bridge laws and approximations have never been resolved satisfactorily. In particular, the issue of initial and boundary conditions has been neglected almost completely, although it more or less automatically undermines any notion of reduction. This neglect is one of the reasons for the success of the notorious quote of Dirac. It is also noteworthy that the defenders of unification and reduction all share a belief in the unproblematic character of approximation procedures, in contrast to their opponents.

Due to the failure of reductionistic programs in the philosophy of mind and philosophy of biology, reduction talk has been replaced by talk of supervenience in mainstream analytic philosophy.[22] It is notoriously difficult to give a satisfying definition of supervenience; one of the more concise formulations is as follows: "a predicate p (of discourse S) is supervenient on a set of predicates B, if and only if p does not distinguish any entities that cannot be distinguished by B." The intended difference with reduction will be obvious. The supervenience relation is the generalized successor of the bridge laws, limiting the relation to a token-token, instead of a type-type relationship. The hope was that supervenience relations would have all the advantages of reduction and none of its disadvantages. That is to say: it would offer at the same time ontological reduction (i.e. physicalism) *and* conceptual autonomy for different disciplines or theories, making room for a potentially highly pluralistic, if not promiscuous, realism. But still everything happening would be constrained by the fact that "really," deep down, the only things and causes there are, are physical.

After the initial hype about supervenience being *the* answer if one favors a form of nonreductive physicalism (i.e., ontological reductionism plus conceptual autonomy), it was realized that few if any goods were delivered, whereas alternative defi-

TABLE 2. Alternative "informal" characterizations of "classical" reduction (B and S refer to the reducing and reduced theory or domain, respectively)

- True S statements are translatable into B statements.
- Every true S statement is logically equivalent to a B statement.
- The concepts (properties, natural kinds) necessary for the description of S can be redefined in an extensionally equivalent way by the concepts of B.
- The B-discourse can be used to express the same facts as the S-discourse.
- The laws governing S can be derived (approximately) from those of B.
- There is isomorphism between entity and predicate terms of the two theories S and B.
- For every event or process in S there is some mechanism belonging to B that (causally) explains the event or process.
- S-behavior and S-properties are (mechanistically) explicable in terms of B-properties and their interactions.
- The S-properties can be predicted using mathematical equations governing the B-domain.
- The S-entities are constituents of the same elementary substrates with the same elementary interactions as the B-entities.
- The S-ontology can be replaced by the B-ontology.
- The reduced theory can be replaced by the reducing theory.

nitions of supervenience proliferated—in the past few years interest has shifted somewhat to emergence. This notion goes back to the 1920s, being the predecessor of the notions of reduction and supervenience. In the 1920s, Broad,[23] using chemical examples such as the "emergence" of potassium chloride, suggested that microstructures have an explanatory role, but the emergent properties cannot be *defined* in terms of (deduced from) microstructural properties. Whether Broad's views were actually much different from current definitions of supervenience is a moot point. As in the case of supervenience, many definitions of emergence have been proposed, but all contemporary proposals combine emergentism with physicalism.[24] As yet, as far as I know, the notions of supervenience and emergence have only been touched upon in passing in the philosophy of chemistry.[7,8,25]

There are many other intertheoretic relations that have been discussed in the literature apart from reduction, supervenience, and emergence; for example: determination, unification, empirical equivalence, specialization, explanatory integration, and intertheoretic approximation. Perhaps the closest cousin of reduction is the idea of unification, already mentioned above. Unification, just like other proposed intertheoretical relations, cannot be uniquely characterized. In a detailed examination of three physical and two biological case studies of unification, Morrison has found a bewildering variety of uses of the words "unity" and "unification."[26] Different degrees of reduction, synthesis, and integration yield distinct ways in which phenomena can be united under a common theoretical framework. Not all forms of theory unification depend on isomorphism or reduction. Often unification is at odds with explanation. Hence, in its generality, unification cannot be linked to truth. In particular, unification is not sufficient for theory acceptance. It is one of the many epistemic virtues that can play a role in theory acceptance, but it is not a determining factor.[27] Even within the framework of the localized settings of specific theories different kinds of unity emerge.

Furthermore, it should be noted that in cases of successful unification in physics there is always a rather abstract mathematical framework that plays an essential role. The more general structure will accommodate more phenomena. Hence, the unified theory seems to score well in terms of empirical support. But its generality allows specific features of the relevant systems/phenomena to be ignored. The empirical support is limited to quantitative predictions of carefully specified types of laboratory experiments.

INTER-THEORETICAL RELATIONS AND PHYSICALISM

The ordinary language expression that perhaps most often exemplifies the ambiguity in referring to intertheoretical relations is the use of the verb *to underlie* and its cognates in scientific practice. There are many different ways of understanding that B underlies S (cf. TABLES 1 and 2). For example one might say about the relation between an S- and B-discourse or domain that B *fixes* S; that S is *brought about by* or *realized in* B phenomena; that S properties are possessed *in virtue of* B properties; that S is *grounded in* B; that B *underpins* S; that S is *determined by* or *dependent on* B; that S *boils down to* or *comes to nothing more* than B; that B *controls* S; that the *constitution of* S is a function of the *constitution of* B; that B *causes* S; that S *is the same* as B (but at a different level of description); that S should be *understood or explained in terms of* B. In unreflective scientific language use, such expressions seem to make no clear distinction between things being identical; being somehow related; one thing constituting, causing, controlling, or determining the other; one thing making some sort of contribution to another; one thing providing an approximate estimate of another; one thing being, *ceteris paribus*, statistically relevant for another; and so on. I suggest the vagueness exemplifies a kind of covering up in order to avoid properly addressing the question of intertheoretical relations.[22]

Most recent literature on supervenience and emergentism is committed to nonreductive physicalism. The latter seems to have three commitments:

1. All intertheoretical relations rest on an ultimate physical base: "deep down" everything is physical (i.e., has some sort of dependence of S on B).

2. Intertheoretical relations must be explained in terms of things indiscernible from the B-view being also indiscernible from the S-view (i.e., co-variance between S and B).

3. Each S-discourse should be autonomous at the conceptual and explanatory level (i.e. nonreducibility).

Contemplating the intuitive meaning of "dependence," "co-variance," and "nonreducibility" suggests that the task set is not an easy one (if coherent at all). The intended dependency or determination relation is asymmetrical, whereas covariance is an entailment or necessitation relation, which is nonsymmetrical. Hence, there is a tension between the intended asymmetry of the intertheoretical relation and the required autonomy of the S-discourse.

I suggest the following metaphor to make some sense of this tension or dilemma of wanting S to be both dependent on B as well as being autonomous (i.e., independent of B). Recent work on supervenience and emergence relations is best seen as a sophisticated version of the isomorphism thesis or the mirror of nature paradigm. The picture is sophisticated because more than one mirror is invoked and each mirror is two-sided, mirroring the autonomous pictures of (part of) the world into one another. One mirror—the "ideal" physical one—however, mirrors reality as it *really* is. Therefore, all other mirrors, eventually, overlay (part of) the ideal mirror. That is to say, intertheoretical relations are meant to be asymmetrical in a deep sense: the "ideal" physical picture is more real than any overlying picture. One could perhaps say that, on the one hand, the overlying mirrors have somehow emerged, but, on the other, they picture mere appearances, without cosmic significance. The purpose of work on intertheoretical relations can then be seen as aiming to show that the content of the overlying mirrors is not completely redundant (epiphenomenal), although the ideal physical mirror is assumed to fix the world completely. In such a view, presumably, chemistry would supervene on the ideal mirror of physics but not be part of the ideal mirror (though philosophers of science often lump physics and chemistry together as constituting the base domain on which everything else supervenes; a recent example is Rosenberg[28]).

In all this, it is an open question what is to be included in the ideal mirror of physics. Physicalism is the ontological position according to which the physical facts determine all the facts. However, to make some sense of this, we need to presuppose some antecedent understanding of what is meant by "physical." That is to say, it is not clear how to differentiate "ideal physics" (physics basically as we know it, but better) and "IDEAL PHYSICS" (whatever we will have at the end of inquiry—it might include reference to souls, ESP, or something as yet unknown).[8,29–31] Owing to this problem, we lack any general criterion of "physical object, property, or law," framed independently of existing physical theory. There is also the problem of chance entering into the most fundamental physical picture of what there is.

Finally, there is a big difference between the possibly correct claim that physics has something to say about everything in the world of the (physical) space-time manifold and the claim that physics has *everything* to say about every thing—leaving aside the question how to relate ordinary life to "really" living in the ten-dimensional space of final physics. The explanations that physicists provide are no doubt broad as well as deep, but their breadth has more to do with their touching the outer reaches of the universe than with their covering the same ground as more ordinary explanations. Moreover, even in physical terms the "universality" or "completeness" of quantum mechanics (or the Theory of Everything) is less obvious than is generally assumed.[12,18,32] If one looks at details, it is not clear at all in what way quantum mechanics would cover *on its own* the behavior of chemical molecules or any macroscopic physical behavior. Often it seems that the intertheoretic relation looks more like a macro reduction instead of a micro reduction: for example, a quantum chemical explanation tends to go downwards, that is, *from* molecular structure *to* the motions of the parts *to* the quantum chemical models *to* the pure mathematics of quantum mechanics.[8] There is nothing new to the idea that macro phenomena can cause micro events: The field of a magnet alters the velocities of particles in its vicinity. The Gibbs' potentials of thermodynamic systems cause the system's constituent molecules to bring the system into equilibrium.

Instead of an absolute commitment to the program of unified science grounded in IDEAL PHYSICS, a more modest claim would be that neither the claim of unity nor that of disunity is adequate for providing a descriptive, explanatory, or even metaphysical account of nature; yet in certain contexts and for certain purposes, either may be sufficient to capture the phenomena at hand. The phenomena at hand may or may not be amenable to a unified treatment. Alternatively, one could say that the modest claim takes an agnostic stance with respect to the metaphysical view. Instead it would stress going back to the rough ground, to which I now turn.

MORATORIUM ON METAPHYSICS: BACK TO THE ROUGH GROUND—PERSPICUOUS RENDERING

Prima facie there is a large family of intertheoretical relations with differences and similarities cropping up and disappearing without having a common core. And I suggest that the most plausible future situation (including the "real" metaphysical structure of the world) is that this variety is not likely to go away. One might think that questions of reduction or supervenience are simply a matter of empirical investigation. For example, Scerri and McIntyre say[25]: "The question of the supervenience of chemistry on physics would seem to depend precisely on the empirical facts and the conclusions which they support and not from more general philosophical musing about chemistry and physics."

But, unfortunately, if we drop the philosophical musing, the metaphysical presuppositions will enter via the back door. Given the strong influence of socio-historical bias (cf. section on Bacon's idols), one should be extra weary of metaphysical prejudices hidden in one's "what is obvious." Even if one prefers physics as one's role model, there are plausible ways of understanding modern physics that support dropping grand reductive schemes.[12] Already in 1993 Schweber wrote in *Physics Today*[32]:

> These advances [symmetry breaking, the renormalisation group, decoupling] have supported the notion of the existence of objective emergent properties and have challenged the reductionism approach.... Phenomena such as superconductivity are genuine novelties in the universe.... Physics, it could be said, is becoming like chemistry.

Given the enormous variety of possible intertheoretical relations, the proliferation of definitions of reduction, supervenience, emergence, unification, and so on, as well as the fact that empirical studies have provided support for almost any metaphysical option or, alternatively, have shown rather conclusively that empirically the case is inconclusive, I suggest a moratorium on the use of words such as "reduction," "supervenience," and "unification" and to go back to the rough ground and give perspicuous renderings of the practice of chemistry. What is needed are very detailed case studies and further discussions about them, instead of bickering about whether chemistry can be reduced to physics, supervenes on it, can be unified with it, and similar "metaphysical" concerns. Let me illustrate what I mean by going back to the rough ground by giving an example for which I take the exchange between the philosopher (and chemist) Scerri and the theoretical chemist Schaefer in *Chemistry in Britain* in 1992.

Example from the Rough Ground: Basis Sets

Here are the facts (as I have selected them of course). Scerri had pointed out that in the work of Schaefer et al. on the bond angle of methylene, extrapolations are made from calculations performed on other molecules; hence it was not truly *ab initio*.[33,34] Schaefer replied rather indignantly[35]:

> Those who have read the 1970 paper will join me in wondering what it is about the work [on the methylene molecule] that was not *ab initio*. The only remote possibilities that come to mind are the use of a) basis sets designed to give lowest possible energies for the carbon and hydrogen atoms, and b) configuration interaction methods proven to give accurate bond distances for very small diatomic molecules.

The disagreement between Scerri and Schaefer is about the theoretical work of Bender and Schaefer in the 1970s, who predicted a bond angle for methylene of about 135°, convincing experimentalists that the methylene molecule was bent. In an article of 1986, reviewing the earlier work, Schaefer claims this as a big success showing that, as he says, now theory is a "full partner with experiment." But Schaefer concludes the success story on methylene by saying[36]: "Fortunately, the equilibrium bond angle of triplet methylene is no longer a matter of debate. A *definitive experiment* [showed that] the angle is $133.8° \pm 0.1°$." [italics added]

What should we think of this? I would say that being a great success for theory and becoming a "full partner with experiment," don't imply that the bond angle of methylene was derived *solely* from quantum mechanical first principles. More generally, I would say that the example shows that if theoretical models get better, it will happen more and more often that the predictions of these models can challenge results from experimental chemistry or make predictions before experimental measurements are available, thus becoming a "full partner with experiment." If this is so, it shouldn't surprise if *ab initio* methods are sometimes more precise and reliable than experiments with small molecules, especially for unstable systems. But the reliability of the choice of basis set and configuration interaction models that makes this success possible draws support from trying out the models on other molecules through trial and error guided by chemical "experience" and experimental data.[37]

But I'm now talking at the level of slogans. An example of going back to the rough ground would consist of working out this example in great detail. What we need is very detailed accounts of what is happening when basis sets and configuration interaction models are chosen. To what extent do the use of "basis sets" and "configuration methods" introduce background assumptions, models, and intuitions that are not part of quantum mechanical principles? If the basis sets have been chosen for the particular system (i.e., they are not imported from other calculations), what sort of considerations govern that choice (physical or chemical intuition? "experience"?).

Though this going back to the rough ground requires close interaction between philosophy of chemistry and chemistry, I wouldn't use the expression "cooperation." A contestatory relation might be more productive. After all, neither philosophers of chemistry nor theoretical chemists form a unified whole. My own experience is that theoretical chemists often have a low opinion of the philosophical and even scientific acumen of many of their colleagues. There is therefore room for another type of going back to the rough ground as well, which is to carry out anthropological and historiographic studies of the life world of chemists; for example, how does hybridization talk or views on orbitals[38,39] function in chemical practice?

HOW PHYSICALISM ELIMINATES ITSELF

Having urged all philosophers of chemistry to carry out more case studies that will give insight into the variety of intertheoretical relations, I can't resist adding some more general comments. I will argue that physicalism eliminates itself and that we do need a more disunified view of science to accommodate the variety of intertheoretical relations.

For a start it is instructive to look at Churchland's eliminative materialism in the philosophy of mind. According to Churchland, folk psychological kinds such as beliefs and desires are not natural kinds; hence, in the end, they have to be eliminated, to be superseded by neurobiological natural kinds.[40] At best they are merely practical kinds, makeshift and fleeting, only useful in particular local environments—irrelevant from a cosmic point of view. The only true natural kinds are the kind of things that figure in the most fundamental laws of nature. Not only beliefs and desires, but also natural kinds such as water and gold are merely *practical* kinds: the specification of what water or gold is, is but one configuration in an infinite range of combinations of more primitive elements. A scientific way of looking at the world, as advocated by Churchland, shows there's *nothing to see*. Churchland's eliminativism starts out as a form of reductive scientific realism. But he pushes the priority of physicalism so far that it eliminates itself: *nothing* remains, except for the properties of quarks or that sort of thing. But why stop there? Why not take quantum (electro/chromo) mechanics seriously?

The holism of quantum mechanics gives *prima facie* evidence that, even accepting the universality and autonomy of physics, there is nothing to favor one level of description as more basic than another, because according to the most fundamental physical theory *nothing* can be said about the world we know and even if something could be said, it would refer to highly esoteric events far away in space and time or created in laboratories carefully shielded from the rest of the world. It would not have anything to say about ordinary events without inserting some common human prejudices about macroscopic objects (including measurement equipment).

In standard quantum chemical approaches, a molecule is taken to be a set of nuclei and electrons, but in the physicist's description electrons and nuclei are correlated by Einstein-Podolsky-Rosen correlations; hence, neither electrons nor nuclei exist as individual objects. A consequence of the holism of quantum mechanics is that an object can only be defined in terms of its relations to its environment.[41–45] In quantum theory an atom or molecule in a stationary state has no extension in space or time, so that it makes no sense to talk about the size or shape of an atom or molecule in such a state. That molecules do not occur with a definite shape according to quantum mechanics, quite in contrast to what is believed in chemistry, was already noted by Hund in 1927: isomeric molecules are described by the *same* Hamiltonian[46]: "The fact that the right-handed or left-handed configuration of a molecule is not a quantum state might appear to contradict the existence of optical isomers."

Physical systems are automatically coupled to their environment and are never closed in a strict sense. When and where to suppress the EPR-correlations is not something that can be derived from quantum mechanics—more precisely: the only truly *ab initio* approach would not allow *any* EPR correlation to be suppressed. But it is the decisions when to suppress them that, as it were, abstract objects out of the quantum mechanical formalism. Any system (including a measurement apparatus) is in an entangled state with the environment. Hence the notion of a closed system is always an approximation. Quantum mechanics describes the universe, in principle, as one whole. To separate out objects from this whole (including electrons, molecules, and apples) requires a justification that lies outside the principles of quantum mechanics.

Moving on to more fundamental physical theories such as quantum field theory will not change the picture. In quantum field theory, entities that are treated as physical systems in quantum mechanics (such as electrons) are conceived as properties of quantum fields. Again, because of entanglement, only the whole of all quantum fields is in a pure state. It is only this whole that can serve as a "supervenience" base of everything there is.[47]

NO NEED FOR METAPHYSICAL GLUE

Hence, a thorough-going physicalism leads to a world picture in which we have to wait for the great unification in order to be able to understand what happened during the first second of the universe. But knowledge of such a Theory of Everything will be of little relevance for mundane matters. It won't even help us to understand such ordinary scientific phenomena as turbulence.[32] Finding the metaphysical glue that binds the sciences together has perhaps had occasional local successes, primarily in physics. But at the same time, more and more disparate sciences and new interdisciplinary fields have been claiming their own autonomy. Why not start thinking that the world is more patchy than we had thought?[18,19,48] Perhaps it is not a homogeneous world structured by natural kinds having essences and being related in clear-cut reductive hierarchies. Why wouldn't the world be a very complicated and diverse place governed in different domains by different systems of laws not necessarily related to each other in any systematic or uniform way, by a patchwork of laws—where both the inter-theoretic relations and the laws, kinds, concepts, explanations, and so on proliferate into large families which share no essential core?

As to the alleged reduction of chemistry to physics, why not have physical and chemical and "mixed" descriptions existing side by side, related by a variety of *ceteris paribus* intertheoretical relations, without any universal metaphysical glue, such as supervenience, to bind it all together? If different discourses should be given equal standing in discussions and contestations about the "same subject matter," then we need a different account of the nature of intertheoretical relations than the inherently asymmetrical intuitions of reduction, supervenience, and emergence, an account that I can here indicate only in programmatic terms[48–50]:

1. Intertheoretical relations (bridge laws, supervenience, or emergence relations) are empirical *ceteris paribus* regularities, not metaphysical necessities—the only circumstances in which they might apply strictly are model situations, "completely" isolated from the rest of the universe.

2. Intertheoretical relations are symmetrical, leaving the autonomy of both sides intact (without preventing forms of explanatory interaction or extension *in both directions*, by borrowing, synthesis, criteria of overall coherence, and so on).

3. A strict separation between the two discourses that are connected by intertheoretical relations is usually not possible and primarily serves as a model imposed on a messy world.
4. There is no need to be afraid of causal overdertermination or downward causation.

Perhaps the strength of scientific activities lies, not in formal monolithicity, but in it forming a complex unity of diverse interacting processes of experimentation, theorizing, instrumentation, and the like. Hence, there is no need for grand metaphysical explanation, but perspicuous renderings of the rough ground instead.

ACKNOWLEDGMENT

This research was supported in part by a grant from FWO-Vlaanderen (project number 1.5025.00N).

REFERENCES

1. SCERRI, E.R. 1998. Popper's naturalised approach to the reduction of chemistry. Int. Stud. Philos. Sci. **12:** 33–44.
2. NEEDHAM, P. 1999. Reduction and abduction in chemistry—A response to Scerri. Int. Stud. Philos. Sci. **13:** 169–184.
3. SCERRI, E.R. 1999. Response to Needham. Int. Stud. Philos. Sci. **13:** 185–192.
4. NEEDHAM, P. 2000. Reduction in chemistry—a second response to Scerri. Int. Stud. Philos. Sci. **14:** 317–324.
5. SCERRI, E.R. 2000. Second response to Paul Needham. Int. Stud. Philos. Sci. **14:** 307–315.
6. RAMBERG, P. 2000. Pragmatism, belief, and reduction: stereoformulas and atomic models in early stereochemistry. Hyle **6:** 35–61.
7. VEMULAPALLI, G.K. & H. BYERLY. 1999. Remnants of reductionism. Found. Chem. **1:** 17–41.
8. HENDRY, R.F. 1999. Molecular models and the question of physicalism. Hyle **5:** 117–134.
9. DIRAC, P.A.M. 1929. Quantum mechanics of many-electron systems. Proc. R. Soc. London **A123:** 714–733.
10. WITTGENSTEIN, L. 1972. Philosophical Investigations. Blackwell. Oxford. §122.
11. VAN BRAKEL, J. 2000. Chapters 4 and 5. *In* Philosophy of Chemistry: Between the Manifest and the Scientific Image. Leuven University Press. Leuven, Belgium.
12. DRESDEN, M. 1998. The Klopsteg memorial lecture: fundamentality and numerical scales—diversity and the structure of physics. Am. J. Phys. **66:** 468–482.
13. WIMSATT, W.C. 2000. Emergence as non-aggregativity and the biases of reductionisms. Found. Sci. **5:** 269–297.
14. VAN BRAKEL, Pilosophy of Chemistry, §5.1.
15. GELL-MANN, M. 1994. The Quark and the Jaguar: Adventures in the Simple and the Complex. Freeman. New York.
16. VAN BRAKEL, J. 2003. Kant's legacy for the philosophy of chemistry. To appear in a volume of the Boston Studies in the Philosophy of Science. E. Scerri, D. Baird, and L. McIntyre, Eds. Kluwer. Dordrecht.
17. GALISON, P. 2001. The Americanization of unity. *In* Science in Culture. P. Galison, S.R. Graubard & E. Mendelsohn, Eds.: 45–71. Transaction Publishers. New Brunswick, NJ.
18. CARTWRIGHT, N. 1999. The Dappled World: A Study of the Boundaries of Science. Cambridge University Press. Cambridge, UK.

19. DUPRÉ, J. 1993. The Disorder of Things: Metaphysical Foundations of the Disunity of Science. Harvard University Press. Cambridge, MA.
20. BALZER, W., D.A. PEARCE & H.-J. SCHMIDT, Eds. 1984. Reduction in Science: Structure, Examples, Philosophical Problems. Reidel. Boston.
21. VAN BRAKEL, Pilosophy of Chemistry, §2.3.
22. VAN BRAKEL, J. 1996. Interdiscourse or supervenience relations: the priority of the manifest image. Synthese **106**: 253–297
23. BROAD, C.D. 1925. The Mind and its Place in Nature. Kegan Paul, Trench, Trübner & Co. London.
24. VAN BRAKEL, Pilosophy of Chemistry, §2.4.
25. SCERRI, E.R. & L. MCINTYRE. 1997. The case for the philosophy of chemistry. Synthese **111**: 213–232.
26. MORRISON, M. 2000. Unifying Scientific Theories: Physical Concepts and Mathematical Structures. Cambridge University Press. Cambridge, UK.
27. VAN BRAKEL, J. 1999. Epistemische deugden en hun verantwoording. Tijdschr. Filosofie **60**: 243–268.
28. ROSENBERG, A. 2001 Reductionism in a historical science. Philos. Sci. **68**: 135–163.
29. CRANE, T. & D.H. MELLOR. 1990. There is no question of physicalism. Mind **99**: 185–206.
30. CROOK, S. & C. GILLETT. 2001. Why physics alone cannot define the "physical": materialism, metaphysics, and the formulation of physicalism. Can. J. Philos. **31**: 333–360.
31. SPURRETT, D. 2001. What physical properties are. Pacific Philos. Q. **82**: 201–225.
32. SCHWEBER, S.S. 1993. Physics, community and the crisis in physical theory. Physics Today, November: 34–40.
33. SCERRI, E.R. 1992. Quantum chemistry truth. Chem. Brit. **28**: 326.
34. SCERRI, E.R. 1992. Quantum extrapolation. Chem. Brit. **28**: 781.
35. SCHAEFER, H.F. 1992. Quantum dispute. Chem. Brit. **28**: 604.
36. SCHAEFER III, H.F. 1986. Methylene: a paradigm for computational quantum chemistry. Science **231**: 1100–1107.
37. VAN BRAKEL, Pilosophy of Chemistry, §5.4.
38. Ibid., Block 5.9.
39. SCERRRI, E.R. 2001. The recently claimed observation of atomic orbitals and some related philosophical issues. Philos. Sci. Proc. **68**: S76–S88.
40. CHURCHLAND, P.M. 1989. A Neurocomputational Perspective. The MIT Press. Cambridge, MA.
41. AMANN, A. 1991. Chirality: a superselection rule generated by the molecular environment? J. Math. Chem. **6**: 1–15.
42. AMANN, A. 1993. The gestalt problem in quantum theory: generation of molecular shape by the environment. Synthese **97**: 125–156.
43. PRIMAS, H. 1983. Chemistry, Quantum Mechanics and Reductionism: Perspectives in Theoretical Chemistry. Springer. Berlin.
44. WOOLLEY, R.G. 1991. Quantum chemistry beyond the Born-Oppenheimer approximation. J. Mol. Struct. Theochem. **230**: 17–46.
45. STAMATESCU, L.-O. 1996. Stochastic collapse models. *In* Decoherence and the Appearance of a Classical World in Quantum Theory. D. Giulini *et al.*, Eds.: 249–255. Springer. Berlin.
46. JOOS, E. 1996. Decoherence through interaction with the environment. *In* Decoherence and the Appearance of a Classical World in Quantum Theory. D. Giulini *et al.*, Eds.: 92. Springer. Berlin.
47. ESFELD, M. 1999. Physicalism and ontological holism. Metaphilosophy **30**: 319–337.
48. VAN BRAKEL, J. 2001. The world: an unruly mess. Found. Chem. **3**: 251–262.
49. VAN BRAKEL, J. 1999. Supervenience and anomalous monism. Dialectica **53**: 3–25.
50. VAN BRAKEL, J. 2002. Chemistry and anomalous monism. To appear in the Proceedings of the 2000 ISPC Symposium, E. Zielonacka-Lis, D. Sobczynska, and P. Zeidler, Eds.

Autonomy, Explanation, and Theoretical Values

Physicists and Chemists on Molecular Quantum Mechanics

ROBIN FINDLAY HENDRY

Department of Philosophy, University of Durham, Durham DH1 3HN, UK

> ABSTRACT: The emergence of quantum chemistry in the early twentieth century was an international as well as an interdisciplinary affair, involving dialogue between physicists and chemists in Germany, the United States, and Britain. Historians of science have recently documented both the causes and effects of this internationalism and interdisciplinarity. Chemists and physicists involved in the development of quantum chemistry in its first few decades tended to argue for opposing views on acceptable standards of explanation in their field, although the debate did not divide along disciplinary lines. The purpose of this paper is to investigate these different positions, through the methodological reflections of John Clarke Slater, Linus Pauling, and Charles Coulson. Slater tended to argue for quantum-mechanical rigor and the application of fundamental principles as the values guiding models of molecular bonding. Although they were on different sides of the debate between the valence-bond and molecular-orbital approaches, Pauling and Coulson both emphasized the recovery of traditional chemical explanations and systematic explanatory power within chemistry.
>
> KEYWORDS: quantum chemistry; explanation; Slater; Pauling; Coulson

INTRODUCTION

In 1929, not long after the birth of quantum mechanics, P.A.M. Dirac made a comment whose confidence is the starting point of many arguments in the philosophy of chemistry[1]:

> The underlying physical laws necessary for the mathematical theory of a large part of physics and the whole of chemistry are thus completely known, and the difficulty is only that the exact application of these laws leads to equations much too complicated to be soluble.

Address for correspondence: Robin Findlay Hendry, Department of Philosophy, University of Durham, 50 Old Elvet, Durham DH1 3HN, UK.

r.f.hendry@durham.ac.uk

In the less often quoted remarks that follow, Dirac emphasizes that detailed application will require approximate methods:

> It therefore becomes desirable that approximate methods of applying quantum mechanics should be developed, which can lead to an explanation of the main features of complex atomic systems without too much computation.

Now Dirac's call can be answered with different kinds of activity. On the one hand, the problem situation—the intractability of exact quantum mechanics—suggests that the novelty will be essentially mathematical, in teasing out structure that is in some sense "already there" in the exact equations. On the other hand, the phrase "approximate methods of applying quantum mechanics" might also suggest some *descriptive* novelty. Perhaps new, more tractable kinds of quantum-mechanical model will be required to understand the chemical and spectroscopic behavior of polyatomic molecules in the terms of that theory. There may be no guarantee that that behavior can be treated by the same methods the pioneers of quantum mechanics used to solve the Schrödinger equation for the hydrogen atom. On this latter view, molecular quantum mechanics need not be the reductionist enterprise it is often seen as.

The early history of quantum chemistry is complicated by the fact that, from the 1930s, there were *two* approximation schemes that gave workable results for molecules. Both were built on physical insight as well as mathematical theorems. Both were conceived in Europe in the late 1920s, and elaborated in the United States in the early 1930s. The Heitler-London-Pauling-Slater valence-bond method modeled wavefunctions for molecules as superpositions of states corresponding to classically bonded structures. The resulting quantum-mechanical states were "resonance hybrids" of the canonical structures from which they were formed. The Hund-Mulliken molecular-orbital approach built up delocalized molecular orbitals from available atomic orbitals. In the ensuing debate about the relative merits of the two schemes, appeal was commonly made to more general epistemic principles and values. Two different sets of such principles and values are discernible, two different positions on how approximate models should be assessed. I will call them the views from physics and chemistry, even though their adoption is not a straightforwardly disciplinary matter. Chemists can, in other contexts, be found appealing to the "downward-directed" theoretical values of rigor and generality, while physicists often cite such "upward-directed" virtues as detailed explanatory power with respect to individual cases. The labels are apt, however, because within the local context of quantum chemistry, the issues concern autonomy, authority, and disciplinary affiliation. Should molecular quantum mechanics be regarded as a descriptively complete import from physics, an instance of the quantum mechanics of systems of charged particles? Or may quantum mechanics be adapted as required to the explanatory demands of chemistry, tailored case by case to the recovery of known structures? Note that the opposition between these two sets of theoretical values is not the same as that between the valence-bond and molecular-orbital approaches. Pauling and Coulson each appeal to (what in this case are) the chemists' values of explanatory power within chemistry. Slater, as we shall see, is not directly engaged in the debate between the two rival approximation schemes, for according to the view from physics there should, properly speaking, be no such debate.

THE VIEW FROM PHYSICS

It is perhaps natural for theoretical physicists to see the application of physical theories to chemical systems as an imperialist process in which a lawless domain that borders physics is brought to order: so natural, in fact, that the hubris of some physicists predated the emergence of quantum mechanics. In a presentation of "old" quantum-theoretic atomic models published well before the advent of quantum mechanics, Max Born commented that[2]

> When we contemplate the path by which we have come we realize that we have not penetrated far into the vast territory of chemistry; yet we have travelled far enough to see before us in the distance the passes which must be traversed before physics can impose her laws upon her sister science.

Born's "we" presumably invokes a community of physicists, but the founders of quantum mechanics—Niels Bohr, Werner Heisenberg, Wolfgang Pauli, Erwin Schrödinger, and Arnold Sommerfeld—did not themselves develop detailed applications of quantum mechanics to molecules. That work was done instead by what Mary Jo Nye has called the "mathematically trained, but chemically oriented scientists"[3] whom they inspired. Schrödinger excepted, the founders of quantum mechanics were all heavily involved, in the twilight years of the old quantum theory, with the development of detailed models of atoms to account for their spectroscopic behavior. Once the new quantum mechanics was on the table, most of its founders were, according to Nye,[4] less interested in its detailed application to molecules than they were in its (disputed) interpretation, and the allied developments of relativistic quantum mechanics and nuclear and solid-state physics. As Joseph Hirschfelder later put it[5]:

> At this point, the theoretical physicists left the chemist to wallow around with their messy molecules while they resumed their search for new fundamental laws of nature.

Now it would be natural to expect those who played only a distant part in the development of quantum chemistry to take a loftily ambitious view of what quantum mechanics could explain in chemistry. But John Clarke Slater, one of the architects of the valence-bond approach, was similarly convinced of the primacy of fundamental physical principles.

After completing a Ph.D. at Harvard under P.W. Bridgman, Slater obtained a Sheldon traveling fellowship to work with Bohr in Copenhagen in the academic year 1923–24.[6] His time there was not a success, however.[7] Slater had been developing some ideas on the relationship between photons and the electromagnetic field. Although convinced of the reality of photons himself, he was browbeaten into co-authoring a theory that rejected them (the short-lived Bohr-Kramers-Slater theory).[8] On his return from Europe, Slater made, during the late 1920s and early 1930s, a slew of important contributions to the quantum mechanical theory of many-electron systems, and of molecular structure and bonding.[9] Although they constructed the same model of directed valence, Slater's style and motivation were quite different from Linus Pauling's. Buhm Soon Park argues that Slater's papers on hybridization were an attempt to "lay out the procedure for solving the Schrödinger equation for molecules in general, rather than carry out computations for actual examples."[10] Moreover, Slater sought general theoretical justification for his approximations, rather than computational tractability and physical plausibility in the case at hand.

Slater professed a pragmatism influenced by Bridgman and was consequently uninterested in, even hostile to, discussion of the interpretation of quantum mechanics, or any "philosophical consequences" the theory might have.[11] Like other American scientists of the time, he preferred to get on with applying the theory.[12] In some tension with this pragmatism, however, were the reductionist philosophical views he also professed, and the methodological outlook they inspired. In 1933, Slater complained[13] that "The study of molecular structure has been too much based on particular models, rather than on general principles," the "general principles" being those of physics. Moreover, as he argued in an undelivered lecture in written in 1937, physics is[14]

> ...the simplest science, and the one which most nearly gets us back to first causes. Its laws...its fundamental particles, seem as basic in the scheme of things as scientists have succeeded in getting. Next, now hardly distinguishable from it, comes chemistry.

It was not that he was against approximate or semi-empirical methods *per se*. In an article for the *Bulletin of the American Mathematical Society* he set out a legitimate role for semi-empirical methods[15]:

> (A) General theory has important uses even in the absence of exact solutions for special cases; for the general theory tells us the type of functional behavior we can expect, in terms of certain undetermined constants, and in our practical work we have determined these experimentally, in cases where we could only estimate them crudely from numerical calculation.

That legitimate role involves the continuation of the mathematical development and elucidation of intractable theory by other means. But Slater did (in 1950) express particular worries about the chemists' misuse of approximate methods in quantum mechanics, in whose development he had played so large a part[16]:

> [T]he theory of chemical valence was taken up by the chemists, who rather naturally did not have enough fundamental insight into the nature of wave mechanics to make the contributions to the theory which were really needed to put it in a satisfactory form, but instead proceeded on the one hand to a great extension of its mathematical formalism, on the other to stretching the theory far beyond what was justified by its crude nature in making comparisons with experiment, with the result that most of the results obtained by the chemists since that time are suspect by the physicists.

While the chemists argued over the relative merits of the valence-bond and molecular-orbital theories, and sought to defend and extend their explanatory uses, in his textbook *An Introduction to Chemical Physics* (1939), Slater presented them as two approximation methods, differing not in the "fundamentals," but only in the "analytical steps used." A proper understanding of the forces between atoms and molecules should, he thought, appeal to "fundamental principles, without reference to exact methods of calculation."[17] Slater's ecumenism on the issue of valence-bond versus molecular-orbital methods looks to be a further expression of his reductionism.[18] The two schemes should, in his view, be regarded not as competing theories of molecular bonding, but as logically dispensable (though not practically dispensable) aids to understanding the content of quantum mechanics with regard to the behaviour of molecules.

Slater saw himself as upholding standards of rigor that others engaged in applying quantum mechanics to molecules failed to observe. Towards the end of his life, he set out his own positive views on what these standards should be[19]:

> I was never satisfied with any theory that didn't come out of the fundamental principles of quantum mechanics perfectly straight forwardly, and I still stick to that point of

view. Whereas a great many people were willing to make what they would call models—put in something that would look plausible, out of which you could get formulas that would describe the facts, but where you could not derive this model from fundamental theory—I felt here we had a fundamental theory and a challenge to be able to explain everything we see around us in terms of that fundamental theory. So I have always discarded anything that didn't fit into that.

With due caution about putting words into the mouths of past scientists, there is a consistent view on the aim of approximate theorizing to be found here, which I will try to flesh out in the rest of the section. The key to understanding the view is how an aim—subsumptive unification under physical principles—dictates a set of norms for approximate theorizing, or what one might call *theoretical values*.

The first and most obvious element of this view, familiar to philosophers, is that quantum-mechanical explanation should be a subsumptive process in which laws, facts, and tendencies associated with molecular structure and bonding are shown to be special cases of the quantum mechanics of systems of charged masses. The "fundamental principles" of quantum mechanics, plus the special assumptions and constants required for its application to systems of charged particles, are sufficient to determine a canonical quantum-mechanical treatment for each system of interest to chemistry; that is, for every atom and molecule. The explanatory power of the theory with respect to particular molecules must be delimited precisely by what can be derived from this "exact" treatment. (Note that the word "exact" appears in scare quotes because of the idealization that masses in the system interact only electrostatically.)

Of course, the intractability of quantum-mechanical equations makes their explanatory content somewhat inscrutable, but the ideal of subsumptive explanation can be upheld even in idealized and approximate treatments, so long as there is warranted satisfaction that whatever has been calculated using idealized or approximate quantum-mechanical equations reflects the exact equations. In short, idealized or approximate "model" treatments are legitimate in so far as they are proxies for the intractable exact treatments, which are themselves legitimate applications of the "fundamental principles" of the theory.[20] That failure of autonomy dictates certain restrictions on the proper uses of such treatments in explanations. Firstly, a feature of the model's structure can legitimately be appealed to in explanation only if it is shared by the exact treatment (otherwise, the explanation is not really quantum mechanical). Hence the approximate or idealized model should have a clear mathematical relationship to the exact treatment, for if the empirical value of some molecular constant is compared to one calculated from a model treatment, we need to know that the model calculation gives us something close to the "exact" value. For instance, the energy of the system in the approximate model might equate to the sum of an initial portion of an expansion series whose sum provably tends to the "exact" energy in the infinite limit. Where a model treatment is based on idealizing assumptions, conclusions derived in the model can be expected to hold only in those systems in which those assumptions hold. Lastly, it is important not to misinterpret features of an approximate model that will disappear when the limit of the exact equations is approached (and so are not shared by the exact treatment). On this construal, Slater's complaint about the chemists' models is not an objection to approximation, which is inevitable. It is more that artefactual features of the model, features that are not provably shared by the "exact" quantum mechanics, are cited in the chemists' explanations. Arbitrary features of schemes of approximation have no place in a proper (physical) understanding of chemical bonding. The chemists' mistake was not in using approximate methods, but in their *mis*using them.

THE VIEW FROM CHEMISTRY

Although quantum chemistry began with Heitler and London's 1927 treatment of the hydrogen molecule,[21] new subdisciplines are shaped by textbooks as well as journal papers. In this section I will therefore concentrate on Linus Pauling and Charles Alfred Coulson, the authors of two influential textbooks of quantum chemistry (*The Nature of the Chemical Bond* [1939] and *Valence* [1952]), associated with the valence-bond and molecular orbital approaches, respectively. Each book went through multiple editions and was read by successive generations of students studying quantum chemistry.

Pauling

Pauling's rise to prominence was closely associated with the importation of the quantum-mechanical concept of resonance into chemistry. In early 1926, he traveled to Europe on a Guggenheim fellowship to study in the main centers of the new quantum mechanics, including Munich, Copenhagen, Göttingen, and Zurich.[22] On his return to California he published, in 1928, a presentation to a chemical audience of the methods used by Heitler and London on the hydrogen molecule, and a brief note relating them to G.N. Lewis's theory of bonding, with a suggestive sketch of how they might be extended to account for the tetrahedral structure of methane. It was a few years before Pauling was able to fulfil this promissory note: he was heavily engaged also with X-ray crystallography at this time, determining chemical structures and developing what he later called a deep "chemical intuition" about what would and would not work, physically, at the molecular level.[23] In a groundbreaking series of seven papers starting in 1931, he developed a quantum-mechanical theory of bonding in polyatomic molecules, consisting of a series of rules that were only partially justified by quantum mechanics, because they were based on detailed calculations for diatomic molecules.[24]

When Pauling first published on molecular quantum mechanics in the late 1920s, fresh from his travels in Europe, he emphasized the continuity of the new methods with the physical ideas in which they originated[25]: "[I]t has become evident that the factors mainly responsible for chemical valence are the Pauli exclusion principle and the Heisenberg-Dirac resonance phenomena."

By 1956, in contrast, he emphasized the relative autonomy of the chemical applications of resonance[26]:

> The theory of resonance was well on its way toward formulation before quantum mechanics was discovered.... It is true that the idea of resonance energy was then provided by quantum mechanics...but the theory of resonance has gone far beyond the region of application in which any precise quantum mechanical calculations have been made, and its great extension has been almost entirely empirical with only the valuable and effective guidance of fundamental quantum principles.

The two quoted passages may not be strictly incompatible, but there is certainly a change in emphasis. Why? In a pioneering paper on many-body quantum mechanics, Heisenberg had represented the two-electron states of the helium atom as superpositions of degenerate states, each corresponding to products of one-electron states. The term "resonance" arose from the formal analogy between quantum-mechanical degeneracy and the interactions of coupled oscillators. Heitler and London had treated the hydrogen molecule in the same way, with a superposition of two equivalent states in which the electrons are assigned to the atomic orbital of first one hydrogen atom, then the other. Pauling's contribution was then to tie these superposed states to the static electronic structures of G.N. Lewis.[27] In the process, the relatively inscrutable quantum-mechanical notion of resonance became easily understandable even to those chemists who were not familiar with the mathematical details of the theory. The great product of Pauling's approach was *The Nature of the Chemical Bond*, a textbook of quantum chemistry, published in 1939, addressed to the non-mathematical reader, rich in qualitative rules and explanations that, once again, were only partially justified by derivations from quantum mechanics.[28]

Not that Pauling's efforts to accommodate chemical intuition were without their critics among the founders of quantum chemistry. For Robert Mulliken, Pauling was a mere "showman" whose deceptively simple explanations were "crude" though "popular."[29] This is not simply an architect of the molecular orbital methods impugning the rival valence-bond method: the worry was that Pauling's valence-bond explanations were too quick, and threatened to set the new discipline of quantum chemistry back by appealing to what were essentially arbitrary features of an approximate mathematical scheme. The explanatory status of resonance was questioned even by proponents of the valence-bond approach. Thus G.W. Wheland (an advocate and architect of the application of resonance ideas in organic chemistry) argued against Pauling that resonance was a "man-made" artifact of a particular scheme of approximate methods[30]:

> [R]esonance is not an intrinsic property of a molecule that is described as a resonance hybrid, but is instead something deliberately added by the chemist or physicist who is talking about the molecule. In anthropomorphic terms, I might say that the molecule does not know about the resonance in the same sense in which it knows about its weight, energy, size, shape, and other properties that have what I might call real physical significance. Similarly... a hybrid molecule does not know its total energy is divided between bond energy and resonance energy. Even the double bond in ethylene seems to me less "man-made" than the resonance in benzene. The statement that the ethylene contains a double bond can be regarded as an indirect and approximate description of such real properties as interatomic distance, force constant, charge distribution, chemical reactivity and the like; on the other hand, the statement that benzene is a hybrid of the two Kekulé structures does not describe the properties of the molecule so much as the mental processes of the person who makes the statement.

But Pauling did not agree that bonds and resonance could be separated in this way, and therefore felt that observing the theoretical niceties of the "fundamental" theory may have a heavy price: drastic loss of explanatory power[31]:

> [Y]ou mention that if the quantum mechanical problem could be solved rigorously the idea of resonance would not arise. I think that we might also say that if the quantum mechanical problem could be solved rigorously the idea of the double bond would not arise.

For Pauling, the double bond and benzene's resonance were intimately connected: if one was "man-made" then both were. Pauling also appealed to this linkage between resonance and classical structure in later defenses of the concept of resonance. In the third (1960) edition of *The Nature of the Chemical Bond*, he incorporated a chapter entitled "Resonance and its Significance for Chemistry," in which he emphasized its continuity with the classical notions of tautomerism. According to Pauling there was "no sharp distinction": the difference was a matter of the degree of the stability of the individual valence-bond structures.[32]

In a retrospective article entitled "Fifty Years of Progress in Structural Chemistry and Molecular Biology" published in 1970, Pauling once again defended resonance by means of its connection to classical structural theory[33]:

> We may say that the cyclohexene molecule is a system that can be shown by experiment to be resolvable into six carbon nuclei, ten hydrogen nuclei, and forty-six electrons, and it can be shown to have certain other structural properties, such as values 133 pm, 154 pm, and so on, for the average distances between nuclei in the molecule in its normal state; but it is not resolvable by any experimental technique into one carbon–carbon double bond, five carbon–carbon single bonds, and ten carbon–hydrogen single bonds; these bonds are theoretical constructs, idealizations, which have aided chemists during the past one hundred years in developing the convenient and extremely valuable classical structure theory of organic chemistry. The theory of resonance constitutes an extension of this theory. It is based upon the use of the same idealizations—the bonds between atoms—as used in classical structure theory, with the important extension that in describing the benzene molecule two arrangements of these bonds are used, rather than only one.

Classical structures in general, and double bonds in particular, are extremely well supported by a wide variety of chemical and physical evidence. Resonance theory is merely an extension of this, and is independently supported by its explanatory successes as a "new semi-empirical structural principle" that is "compatible with quantum mechanics and supported by agreement with the facts obtained by experiment."[34] Note that Pauling requires only consistency of the applications with the fundamental theory, and appeals to *both* physics *and* classical chemistry (and experiment, of course) for justification. Slater (see above) requires that applications "come out of the fundamental principles of quantum mechanics," and allows justificatory appeal only to physical principles. These contrasts in style and outlook between Slater and Pauling extended both to their ground-breaking research papers[35] and their textbooks.[36]

Whereas in Pauling's approach, consistency with classical bonding is built in, on the physicists' view, the proper application of quantum mechanical methods may well turn out to be highly revisionary of standard chemical explanations. For Pauling, resonance, as the basis of bonding, was a direct development of classical theories of valence and partially independent of quantum mechanics. To reject resonance as man-made was to reject the chemical bond. Without a ready and "rigorous" quantum-mechanical explanation of what chemists (and spectroscopists) explain by reference to double bonds, revisionary chemical physics would face explanatory loss. Perhaps predictably, an extreme version of this revisionism is voiced by a leading practitioner of the "fundamental" science, physics: Max Born thought that physicists would need to calculate the binding energies of molecules to "establish the cases in which the valence theory of the chemist is reliable."[37]

Coulson

Coulson was the first Cambridge research student of J.E. Lennard-Jones, the first occupant of the first Chair in Theoretical Chemistry in the United Kingdom. After receiving his Ph.D. in 1936 for a thesis covering the electronic structures of H_3^+ and methane, a Prize Fellowship at Trinity College allowed Coulson time to pursue laboratory work on the effects of radiation on bacteria and also a molecular-orbital treatment of fractional bond orders in arenes and conjugated polyenes, complementing a valence-bond treatment published by Pauling and others in 1935. A subsequent senior lectureship at Dundee lasted through most of the war (for religious reasons, Coulson was a conscientious objector), after which he was appointed to a fellowship at the Physical Chemistry Laboratory in Oxford, and then a chair in theoretical physics at King's College London, in 1947.[38] His prominence was secured in 1952, with his appointment as Rouse Ball Professor of Mathematics at Oxford, and the first edition of his influential textbook *Valence*.

As an advocate for and contributor to the molecular orbital theory, Coulson could be expected to be critical of the valence-bond approach, and he did once call "resonance" a "dirty word."[39] He argued a number of times that resonance is not a real phenomenon, for neither resonance structures nor the oscillation between them are really exhibited by molecules.[40]

In 1947, however, Coulson wrote a broadly sympathetic popular account of the explanatory applications of resonance, though his concluding remarks were more sceptical[41]:

> One last question arises: is resonance a *real* phenomenon? The answer is, quite definitely, no. For consider benzene with its two Kekulé structures. We cannot say that the molecule has either one or the other structure, or even that it oscillates between them. There is, in fact, simply the C–C aromatic bond. Putting it in mathematical terms, there is just one full, complete, and proper solution of the Schrödinger wave equation that describes the motions of the electrons. Resonance is merely a way of dissecting this solution: or, indeed, since the full solution is too complicated to work out in detail, resonance is one way—and that not the only way—of describing the approximate solution. It is a "calculus," if by calculus we mean a method of calculation; but it has no physical reality. It has grown up because chemists have become so used to the idea of localized electron-pair bonds that they are loth to abandon it, and prefer to speak of a superposition of definite structures, each of which contains familiar single or double bonds and can be easily visualized.

Coulson goes on to admit, however, that resonance is very useful in "correlating, explaining, and even predicting an astonishingly large and varied body of chemical experience." Like Pauling, Coulson links resonance to the classical bond, but the linkage does not automatically save resonance. He agrees with Pauling that the chemical bond is a theoretical notion[42]:

> Sometimes it seems to me that a bond between two atoms has become so real, so tangible, so friendly that I can almost see it. And then I awake with a little shock: for a chemical bond is not a real thing: it does not exist: no-one has ever seen it, no-one ever can. It is a figment of our own imagination.

On an uncharitable interpretation, these doubts seem unsophisticated, an unmotivated and nonspecific rejection of the unobservable. That kind of skepticism would equally call electrons, atoms, and even molecules into question, unless "seeing" is interpreted broadly so as to include detection. (I leave aside the nice question of whether molecular orbitals would fare any better than bonds on this construal, and whether bonds themselves are really undetectable.) However, his skepticism about the bond is not only a matter of detectability or observability, but also one of idealization, as he argued much later, in 1970[43]:

> From its very nature a bond is a statement about two electrons, so that if the behaviour of these two electrons is significantly dependent upon, or correlated with, other electrons, our idea of a bond separate from, and independent of, other bonds must be modified. In the beautiful density diagrams of today the simple bond has got lost.

Moreover, Coulson was less sanguine than Pauling about the long-term role of the bond in chemical theory, wondering sometimes whether chemistry would outgrow the notion, and require "something bigger."[44] In *Valence* he added that "it is an approximation to suppose that the pairing together of any two electrons in atoms A and B is sufficiently unique to render all other possible pairings irrelevant."[45] There is a certain amount of uncharitable interpretation here too: why should claims about bonds be construed as claims about electron pairs that are *entirely* separate and independent of other electrons, rather than as claims about the relatively independent atomic centers they connect? In any case, he also admitted that "it is equally an approximation to suppose, as in the m.o. theory, that there are m.o.s at all."[46] The physical realization of the chemical bond, or the lack of it, is not a subject to pursue here, however.

Valence promoted the molecular orbital approach, to the extent that Gavroglu and Simoes[47] have called it "The Book that Mulliken Never Managed to Write." The book displayed, in terms amenable to the chemist, the explanatory wares of the molecular orbital approach, especially within organic chemistry where the valence-bond approach had hitherto been regarded as more successful.[48] Coulson was generally even-handed, even ecumenical, in the book, however. He included chapters on the valence-bond method, whereas Pauling had not done the same for molecular-orbital theory in *The Nature of the Chemical Bond*. He also incorporated comments from Pauling into the second and later editions.[49] Like Slater, he also included comparisons of the two methods,[50] noting that in fully sophisticated applications of the two methods, they approached equivalence. Unlike Slater, however, his ecumenism did not seem to undermine the explanatory autonomy of the two schemes. Both, he admitted, introduced gross approximations, compounded by "limitations in the amount of calculation that is worthwhile," and so their value lay in "qualitative understanding, and not quantitative calculation."[51] Despite their equivalence in the limit of fully sophisticated calculations, practical applications would frequently diverge, but "where both theories predict similar conclusions, there is considerable ground for believing that these conclusions are correct."[52]

At the beginning of his presentation of valence-bond methods in Chapter V of the book he remarks[53]:

> It is not unfair to say that…in practically the whole of theoretical chemistry, the form in which the mathematics is cast is suggested, almost inevitably, by experimental results. This is not surprising when we recognize how impossible is any exact solution of the wave equation for a molecule. Our approximations to an exact solution ought to reflect the ideas, intuitions and conclusions of the experimental chemist.

By "experimental results" Coulson presumably means experimentally calibrated theoretical constants associated with molecular structure: the turbulent history of molecular structure in the nineteenth century surely qualifies as "theoretical" any experimental results presented in terms that presuppose it. In comparing the valence-bond and molecular-orbital treatments of bond polarity, he argues that although the molecular orbital presentation is "more natural and conceptually the simpler,…the structures used in the v.b. description do correspond to pictures long familiar, in classical form, to experimental chemistry." Moreover, "this link with the older and more conventional language is of considerable value."[54]

Note the difference between Coulson's remarks about approximations "reflecting the ideas, intuitions and conclusions of the experimental chemist," and the caution one finds in the "view from physics" about misinterpreting arbitrary elements of a scheme of approximation. In the "view from physics" the explanatory power of a model treatment should depend on its particular mathematical form only in so far as this is shared by the exact treatment. In Chapter IX of *Valence*, Coulson compares valence-bond and molecular-orbital treatments of conjugated polyenes and arenes. He concludes that although they are far from quantitatively reliable, they provide "general outlines" which[55]

> enable us to understand a very large part of the field of organic chemistry, and to organize it and correlate its different sections: they provide us with at very least a qualitative understanding of the essential processes at work. Taken as a whole, both the v.b. and m.o. approximations seem about equally good; any theoretical conclusion cannot be regarded as substantiated unless it is predicted by both.

On one measure, Coulson's requirements of approximate models are weaker than Slater's. Slater demanded that a quantum mechanical of molecules "come out of the fundamental principles of quantum mechanics perfectly straight forwardly" (see above). Coulson required rather that results for molecules within either the valence-bond and molecular-orbital approaches be corroborated by corresponding results within the other framework. However, he also adds a criterion that is independent of physics: explanatory and predictive efficacy in systematizing chemical knowledge.

At a conference on molecular quantum mechanics in Boulder, Colorado in 1960, Coulson commented in an after-dinner speech on the trends he had discerned in the conference papers.[56] He worried that a split among theoretical chemists would result from the advance of computational quantum chemistry which, he noted, was thought by some quantum chemists to be "so remote from the normal natural conventional concepts of chemistry, such as bonds, orbitals, and overlapping hybrids, as to carry the work itself out of the sphere of real quantum chemistry."[57] Foreseeable electronic computers offered "effectively exact" solutions to the Schrödinger equation, but although *ab initio* approaches were ostensibly more accurate, he argued, they faced a practical limit of about 20 electrons and were difficult to interpret in terms of traditional chemical concepts. In contrast, chemists working on the "posterior" approaches sought the opposing epistemic virtues of visualizability and ease of interpretation in terms of traditional chemical concepts, rather than accuracy or rigor for its own sake. For this group[58]

> [t]he role of quantum chemistry is to understand these concepts and show what are the essential features in chemical behavior. These people are anxious to be told why, when one vinyl group is added to a conjugated chain, the uv absorption usually shifts to the red; they are not concerned with calculating this shift to the nearest angstrom; all they want is that it should be possible to calculate the shift sufficiently accurately that they can be sure that they really do possess the fundamental reason for the shift. Or, to take another example, they want to know why the HF bond is so strong, when the FF bond is so weak. They are content to let spectroscopists or physical chemists make the measurements; they expect from the quantum mechanician that he will explain *why* the difference exists. But any explanation *why* must be given in terms which are regarded as adequate or suitable. So the explanation must not be that the computer shows that $D(\text{H–F}) \gg D(\text{F–F})$, since this is not an explanation at all, but merely a confirmation of experiment. Any acceptable explanation must be in terms of repulsions between nonbonding electrons, dispersion forces between the atomic cores, hybridization and ionic character.

These explanatory concepts are characteristic of chemistry as a discipline, and its relations to other disciplines[59]:

For chemistry itself operates at a particular level of depth. At that depth certain concepts have significance and—if the word may be allowed—reality. To go deeper than this is to be led to physics and elaborate calculation. To go less deep is to be in a field akin to biology.

Now Coulson was a scion of Cambridge mathematics, confident of the central role of advanced mathematical tools in solving physical problems, and many of his efforts certainly were directed towards developing mathematical methods for quantum chemistry.[60] As Simoes and Gavroglu point out, Coulson might be expected to have sympathized with the *ab initio* calculations, but he didn't.[61] Although he acknowledged that *ab initio* and semi-empirical studies may be appropriate for different kinds of theoretical activity, much of his own work was devoted to the recovery within molecular quantum mechanics of such traditional chemical notions as bent bonds and partial valence.[62] Coulson's mathematics was designed to *suit* "chemical intuition," not annihilate or revise it.

CONCLUSION

There must be many reasons why Coulson and Pauling advocated applications of quantum-mechanical methods within chemistry that were designed to extend and deepen chemical explanations, rather than change them radically. Both men were seen, and saw themselves as, contributors to, and advocates for, the explanatory application of quantum mechanics within chemistry. Both wrote textbooks of molecular quantum mechanics aimed at chemistry students. The reception of the methods they advocated could not but depend on how well those methods were adapted to the representational traditions of the discipline—chemistry—whose members they sought to persuade. Organic chemists (to take an example) would have had fewer reasons than physical chemists to accept claims for the explanatory power of quantum mechanics with respect to chemical problems. Leaving aside their interdisciplinary mistrust of hegemonic claims for a "physicist's theory," successful quantum-mechanical treatments of the spin states of the silver atom, say, or the spectrum of hydrogen, would not by themselves show that the theory had anything useful to say about molecules like benzene or naphthalene. Hence informative treatments of complex molecules were high on the theoretical agenda of quantum chemistry. Given the obvious representational shortcomings of exact Schrödinger equations for many-electron systems, an attractive strategy would be to build the known structural features of complex molecules into quantum-mechanical models.

Stephen Brush has analyzed how the relative standing of the valence-bond and molecular-orbital approaches evolved in the first decades of quantum chemistry.[63] Brush argues that in the 1930s, the valence-bond approach was favored by its presentational superiority (and Pauling's visual representation of benzene), as well as the differing personal qualities of Linus Pauling and Robert Mulliken as teachers and explainers. (Brush notes that visual representation is favored also in physics: witness the different receptions accorded to matrix mechanics and wave mechanics.) Moreover, the valence-bond theory "really was closely related to accepted theories of the chemical bond,"[64] especially in the case of benzene, where it echoed Kekulé's oscillating structure. By the 1950s, however, Coulson had produced an account of bonding in benzene within the molecular orbital approach, with a matching visual representation of the molecule (the double-doughnut π-bonds). This, along with the molecular-orbital theory's superior applicability to larger molecules (and later its successful novel predictions in the case of pericyclic reactions) demonstrated, first to theoretical chemists and later to organic chemists, its superiority in accounting for the structure and bonding of organic molecules.[65]

It would be wrong to conclude that the development of nonrevisionary semi-empirical methods was *merely* a pragmatic matter of quantum chemists like Pauling and Coulson convincing sceptical chemical colleagues of the value of their subdiscipline, or making their way in the scientific world first, and in the textbook market second. Nor was it merely a matter of filling a theoretical and explanatory void created by the mathematical intractability of exact treatments. There were positive reasons for adopting a "retentionist strategy" that surely count as epistemic. Firstly, chemical structure theory was as highly confirmed a body of theory as any in existence at the time, and theoretical continuity with that body of theory was consequently as rational as with any other. Direct and indirect evidence, from spectroscopy to reaction dynamics, was accumulating all the time for particular molecular structures, and molecular structure in general. Hence a "retentionist strategy" would have been an independently plausible starting point in molecular quantum mechanics. Secondly, Coulson and Pauling sought, in two different methodological debates (valence-bond versus molecular orbitals; *ab initio* versus semi-empirical), to legitimate the methods they favored by appealing to their success in deepening and systematizing existing structural explanations in chemistry. In one sense this is a familiar strategy in methodological debate: as "an attempt to eliminate the *pressure* of 'first principles' and defend low-level empirically motivated enquiry," Lakatos[66] once called it "defensive positivism." On the other hand, the appeal to explanatory success is a positive response to the particular methodological debates. Consider first the debate between the valence-bond and molecular-orbital methods. By the mid-1930s it was accepted that the two methods were equivalent in the limit of fully sophisticated treatments. But equivalence in the theoretical limit did not dictate that the different ways the valence-bond and molecular-orbital theories simplified many-electron wavefunctions would yield equivalent explanations of the stability of, say, benzene. As Brush points out, partisans of the two approaches cited *chemical* differences between them not only because it was an argument among chemists, but also because there seem to be genuine explanatory differences between the two schemes with respect to chemistry.

Coulson's defence of semi-empirical as against *ab initio* projects in quantum chemistry highlights another feature of the appeal to explanatory power within chemistry. The *ab initio* methods would have won hands down had theoretical rigor (from the point of view of physics) been the only desideratum. The superior tractability of the semi-empirical methods (for systems with more than 20 electrons) would give them only a stop-gap, second-best sort of justification. But in his Boulder talk, he went further and impugned the explanatory power of the high-powered numerical calculations, at least where these are difficult to interpret in structural terms. Coulson identified specific explanatory demands that chemistry makes of quantum mechanics: why the HF bond is so much stronger than that in F_2; why UV absorption in conjugated polyenes shifts to the red with the addition of a vinyl group. He also argued that good answers to these why-questions would need to cite trends, contrasts, and effects that are realized in the physical structure and behavior of molecules. This information, he thought, was exactly what the semi-empirical models represented well. In short, Coulson envisages a quantum chemistry that asks distinct things of physical theory and requires distinct things of the answers to these questions. The style of explanation is far from subsumption, and the explanatory tools required build molecular structure into the very practice of quantum chemistry.[67]

ACKNOWLEDGMENTS

I would like to thank David Knight, Brian Salter-Duke, an anonymous referee, and members of the audience at the ISPC Symposium at Georgetown University for helpful comments on earlier versions of this paper. I would also like to thank Joe Earley for organizing an excellent meeting.

REFERENCES

1. DIRAC, P.A.M. 1929. The quantum mechanics of many-electron systems. Proc. R. Soc. London A **123**: 714–733, p. 714.
2. BORN, M., quoted in NYE, M.J. 1993. From Chemical Philosophy to Theoretical Chemistry. University of California Press. Berkeley, CA, p. 229.
3. NYE, Chemical Philosophy, note 2, p. 227.
4. Ibid., note 2, p. 238–239.
5. HIRSCHFELDER, quoted in NYE, Chemical Philosophy, note 2, p. 239.
6. HODDESON, L. 1990. John Clarke Slater. In Dictionary of Scientific Biography. Vol. 18, Supp, II. F.L. Holmes, Ed.: 832–836. Charles Scribner's Sons. New York.
7. SCHWEBER, S. 1990. The young John Clarke Slater and the development of quantum chemistry. Hist. Stud. Phys. Biol. Sci. **20**: 339–406, p. 350–357.
8. HENDRY, J. 1981. Bohr-Kramers-Slater: a virtual theory of virtual oscillators and its role in the history of quantum mechanics. Centaurus **25**: 189–221.
9. PARK, B.S. 2000. The contexts of simultaneous discovery: Slater, Pauling and the origins of hybridisation. Stud. Hist. Philos. Mod. Phys. **31B**: 451–474.
10. Ibid., note 9, p. 461.
11. SCHWEBER, Young John Clarke Slater, note 7, Section 4.
12. CARTWRIGHT, N. 1987. Philosophical problems of quantum theory: the response of American physicists. In The Probabilistic Revolution Vol. 2. L. Krüger, G. Gigerenzer & M. Morgan, Eds.: 417–435. MIT Press. Cambridge, MA.
13. SCHWEBER, Young John Clarke Slater, note 7, p. 396.
14. SLATER, quoted in SCHWEBER, Young John Clarke Slater, note 7, p. 391.
15. SLATER, J.C. 1946. Physics and the wave equation. Bull. Am. Math. Soc. **52**: 392–400, p. 399.
16. SLATER, quoted in SCHWEBER, Young John Clarke Slater, note 7, pp. 389–390.
17. SLATER, J.C. 1939. Introduction to Chemical Physics. McGraw-Hill. London, p. 368.
18. GAVROGLU, K. & A. SIMOES. 2000. One face or many? The role of textbooks in building the new discipline of quantum chemistry. In Communicating Chemistry: Textbooks and Their Audiences. A. Lundgren & B. Bensaude-Vincent, Eds.: 415–450. Science History Publications. Canton, MA, pp. 427–431.
19. SLATER, quoted in SCHWEBER, Young John Clarke Slater, note 7, p. 396.
20. HENDRY, R.F. 1998. Models and approximations in quantum chemistry. In Idealization in Contemporary Physics: Poznan Studies in the Philosophy of the Sciences and the Humanities. Vol. 63. N. Shanks, Ed.: 23–42. Rodopi. Amsterdam/Atlanta, GA, Section 2.
21. SCHWEBER, Young John Clarke Slater, note 7, p. 398.
22. HAGER, T. Force of Nature: The Life of Linus Pauling. Chapter 5. Simon & Schuster. New York.
23. Ibid., note 22, Chapter 6.

24. PARK, Contexts of simultaneous discovery, note 9, Section 4.
25. PAULING, quoted in NYE, Chemical Philosophy, note 2, p. 241.
26. Ibid., note 2, p. 207.
27. PARK, B.S. 1999. Chemical translators: Pauling, Wheland and their strategies for teaching the theory of resonance. Br. J. Hist. Sci. **32:** 21–46, pp. 24–28.
28. GAVROGLU & SIMOES, One face or many? Note 18.
29. MULLIKEN, quoted in PARK, Chemical translators, note 27, p. 28.
30. WHELAND, letter to Pauling, quoted in A. SIMOES & K. GAVROGLU 2001. Issues in the history of theoretical and quantum chemistry. *In* Chemical Sciences in the Twentieth Century: Bridging Boundaries. C. Reinhardt Ed.: 51–74. Wiley. Weinheim, p. 65.
31. PAULING, letter to Wheland, quoted in PARK, Chemical translators, note 27, p. 43.
32. PAULING, L. 1960. The Nature of the Chemical Bond. Third Edition. Cornell University Press. Ithaca, NY, p. 564.
33. PAULING, L. 1970. Fifty years of progress in structural chemistry and molecular biology. Daedalus **99:** 988–1014, p. 999.
34. PAULING, Fifty years, note 33, p. 1001.
35. PARK, Contexts of simultaneous discovery, note 9.
36. GAVROGLU & SIMOES, One face or many? Note 18.
37. BORN, quoted in NYE, Chemical Philosophy, note 2, p. 229.
38. ALTMANN, S.L. & E.J. BOWEN. 1974. Charles Alfred Coulson. Biogr. Mem. Fellows R. Soc. **20:** 75–134.
39. GAVROGLU & SIMOES, One face or many? Note 18, p. 441.
40. SIMOES & GAVROGLU, Issues, note 30, pp. 67–68.
41. COULSON, C.A. 1947. The meaning of resonance in quantum chemistry. Endeavour **6:** 42–47, p. 47.
42. COULSON, quoted in SIMOES & GAVROGLU, Issues, Note 30, p. 69.
43. Ibid., note 30, p. 69.
44. Ibid.
45. COULSON, C.A. 1961. Valence. Oxford University Press, Oxford, UK, p.146.
46. Ibid., note 45, p. 146.
47. GAVROGLU & SIMOES, One face or many? Note 18, p. 440.
48. Ibid., note 18, p. 441.
49. Ibid., note 18, pp. 442–444.
50. COULSON, Valence, note 45, pp. 146–161 and 274–275.
51. Ibid., note 45, p. 146.
52. Ibid., note 45, p. 158.
53. Ibid., note 45, pp. 113–114.
54. Ibid., note 45, p. 154.
55. Ibid., note 45, p. 275.
56. SIMOES, A. & K. GAVROGLU. 1999. Quantum chemistry *qua* applied mathematics: the contributions of Charles Alfred Coulson. Hist. Stud. Phys. Biol. Sci. **29:** 363–406, pp. 402–403.
57. COULSON, C.A. 1960. Present state of molecular structure calculations. Rev. Mod. Phys. **32:** 170–177, p. 172.
58. Ibid., note 57, p. 173.
59. Ibid., note 57, p. 173.
60. ALTMANN & BOWEN, Charles Alfred Coulson, note 38, pp. 107–110.
61. SIMOES & GAVROGLU, Quantum chemistry, note 56.
62. Ibid., note 56, pp. 380–389; ALTMANN & BOWEN, Charles Alfred Coulson, note 38, pp. 97–105.
63. BRUSH, S.G. 1999. The dynamics of theory change in chemistry: Part 1. The benzene problem 1865–1945; Part 2. Benzene and molecular orbitals, 1945–1980. Stud. Hist. Philos. Sci. **30:** 21–79, 263–302, pp. 51–58.
64. Ibid., note 63, p. 55.
65. Ibid., note 63, p. 291.
66. LAKATOS, I. Newton's effect on scientific standards. *In* The Methodology of Scientific Research Programmes. J. Worrall & G. Currie, Eds.: 193–222. Cambridge University Press. Cambridge, UK, pp. 207–208.
67. NYE, Chemical Philosophy, note 2, Chapters 9 and 10.

Natural Kinds, Explanation, and Essentialism in Chemistry

REIN VIHALEMM

Department of Philosophy, University of Tartu, 50 090 Tartu, Estonia

ABSTRACT: The problem of natural kinds in chemistry is analyzed, proceeding mainly from Rom Harré's and Jaap van Brakel's writings. The problem of natural kinds proves to be different in general philosophy and in philosophy of science. This problem, which originally emerged at the borderline of science and philosophy, belongs to the domain of philosophy of science. Philosophy in general cannot hope that scientific knowledge will help to explain philosophical issues. Chemistry has to be taken very seriously in philosophy of science. Not only do empirical arguments indicate that chemistry should be regarded as a typical science, but it is also relevant for elaborating a theoretical model of science. Chemistry as the science of substances is well suited for the philosophical analysis of the role of the concept of natural kinds in science. The viability of the concept of a natural kind in philosophy of chemistry derives from the explanatory role of chemical natural kinds. The problem of explanation in its turn relates natural kinds to the laws of nature and scientific theories, that is, to classical issues in philosophy of science and to discussions on their (supposed) specificity in chemistry. In philosophy of science, one can demonstrate that there exists a third option between metaphysical realism and internal realism (to use Putnam's terminology). According to this third type of realism our "world versions" (including natural kinds identified by us), but not the world itself, are relative to us.

KEYWORDS: philosophy of chemistry; natural kinds; explanation; laws of nature; essentialism; realism

INTRODUCTION

The topic of natural kinds seems to be quite central in philosophy of chemistry. For instance, Rom Harré even claims in his review of *Philosophy of Chemistry* by Jaap van Brakel that "the philosophical question of the viability of the concept of a natural kind is surely the most important in philosophy of chemistry."[1] On the other hand, the issue of natural kinds, especially that of chemical natural kinds such as water, is very popular among philosophers of mind and language discussing the meaning of natural kind terms. It is well known that these discussions were initiated by H. Putnam[2] and S. Kripke,[3] who made an attempt to renew essentialism. D. H. Mellor,[4] J. van Brakel,[5] J. LaPorte,[6] and others have criticized this new essentialism from the

Address for correspondence: Department of Philosophy, University of Tartu, Lossi 3, 50 090 Tartu, Estonia. Voice: +372 7 375 315; fax: +372 7 375 345.
 Rein.Vihalemm@ut.ee

viewpoint of philosophy of chemistry. Rom Harré has also criticized Putnam's and Kripke's version of the nominal essence/real essence distinction as a theory of meaning for kind terms; however, he says in the aforementioned review that it seems to him "that without the formal pattern *nominal essence/real essence* there would be no chemistry."[7] I would like to discuss some arguments from both sides, that is, arguments presented by those advocating some kind of essentialism and those preferring antiessentialism.

The controversy between essentialism and antiessentialism is related to the problem of realism as well. Rom Harré calls himself a referential realist,[8] whereas Jaap van Brakel classifies himself, perhaps more willingly, as an anomalous monist, but also—recalling that nothing really depends on terminology—as a pragmatic or pluralistic realist or a "realist-with-a-small-r."[9] As for myself, I prefer to call the variety of realism I accept "constructive realism," after Ronald Giere.[10] A later development of Giere's position has been called "perspectival realism"[11] or "scientific perspectivism."[12] Constructive realism or some similar position has also been developed by other philosophers of science who have used different terminology.

I wish to make three points. (1) The most interesting questions in connection with chemical natural kinds are those having to do with their explanatory role. The problem of explanation in its turn relates natural kinds to the laws of nature and scientific theories, that is, to classical issues in philosophy of science and to discussions on their (supposed) specificity in chemistry. (2) The problem of natural kinds proves to be different in general philosophy and in philosophy of science. (3) In philosophy of science, one can demonstrate that there exists a third option between—to use Putnam's terminology—metaphysical realism and internal realism.

THE STATUS OF CHEMISTRY AND THE ROLE OF NATURAL KINDS

Let's start with the question of the viability of the concept of a natural kind in philosophy of chemistry. Why should one regard this question as the most important one only in case of chemistry? Maybe it is equally important, say, in physics as well? Or, on the contrary, maybe it is not relevant in science at all? Indeed, the traditional examples in discussions on natural kinds are those of chemical elements and compounds and biological species. Perhaps this means that one can speak of natural kinds within disciplines that are not sciences proper and which rather belong to the so-called natural history whose character is descriptive-classificatory, not explanatory? So, once again, when one is dealing with issues in philosophy of chemistry, one cannot ignore the traditional problem of the relationship between physics and chemistry. Is chemistry equally relevant for philosophy of science as physics?

In my earlier works[13] I have analyzed the problem of the status of chemistry by comparing chemistry, on the one hand, with physics-like science or constructive-hypothetico-deductive type of knowledge and inquiry; and, on the other hand, by comparing it with natural history or classifying-historico-descriptive type of knowledge and inquiry. I am not going—neither is it possible—to dwell on the arguments and conclusions of these papers here. However, I would like to mention some points concerning my presumptions. I have argued that chemistry, as a science, is not crucially different from physics-like science. But it is important to realize that on the basis of physics it has proved possible to elaborate a theoretical model of science. I have proposed that this idealized physics-like science be called φ-*science*. Chemistry is not merely a physics-like science, but shares important features with the kind of inquiry and knowledge characteristic of natural history. Nevertheless, analyzing chemistry as a science—and this is my intention here—the theoretical model, that is, φ-*science*, should be used. The problem of natural kinds stands differently in natural history and φ-*science*, and it is a difference of principle. And finally, it should be realized that science itself (i.e., φ-*science*) is a conventional, methodological category, which belongs to philosophy of science and is not a natural kind or an empirical object.

Obviously, the concept of a natural kind is important in philosophy of chemistry because the subject of chemistry is particular kinds of matter, that is, substances or chemical kinds. I agree with J. van Brakel that the older definition (used in the 18th and in the first half of the 19th century) of chemistry as a science of substances and their transformations is to be preferred to those of contemporary textbooks saying, for example, that chemistry is the science of transferring electrons between atoms and/or molecules, or something like that.[14] Defining chemistry via the concept of substance, we embrace chemistry as a whole with all its fundamental and applied branches, and not only as a science but also as a prescience, as an applied discipline and technology. Here it should be noted that chemistry as a science is more akin to technology than, for example, theoretical physics, which has become the paradigm of science for philosophy of science. J. van Brakel has insisted, together with other philosophers of chemistry (first and foremost, J. Schummer), that chemistry should be considered a more typical science than physics, since most fields of science are more similar to technology than to theoretical physics. J. Schummer finds that more than 90% of today's fields of science "are more closely related to chemistry than to (the philosophical image of) physics in nearly every aspect. That is why a philosophy of chemistry is in need to achieve a deeper, more realistic understanding of our actual sciences."[15]

In order to clarify the issues concerning the status of chemistry and the role of natural kinds, J. van Brakel also picks up the distinction between the manifest (i.e., the common-sense-human-life-form) and the scientific image of the world which was originally introduced by W. Sellars.[16] I find that, as interpreted by J. van Brakel, this general philosophical issue undoubtedly deserves serious attention in philosophy of chemistry, although R. Harré sees it as a somewhat outdated idea which concedes far too much to hard-line empiricism.[17] J. van Brakel writes:

> But I will use "manifest image" in a different sense than Sellars, avoiding his associations of the manifest image with sense data or phenomenal...data. My use should rather be thought of as akin to the [Wittgenstein's] concept "form(s) of life" [also, e.g., Husserl's "Lebenswelt," Heidegger's "Dasein," Austin's "common sense," Searle's "background"]. That is to say "manifest image" is short for "manifest forms of life, understood interculturally".... In particular, the manifest image in my sense should not be associated with the view that there are basic level categories ("natural kinds") like water and red, which are "manifest" natural categories because of the way the world and our neurophysiology have been fine tuned by evolutionary processes.[18]

It is important to realize that "The manifest/scientific distinction refers to the distinction between what is 'normally' called science and what not. This is a sociological, not an essentialistic definition."[19] The arguments by Sellars (and other philosophers who may use different terminologies) for the primacy of the scientific image proceed usually from the "normal" scientific standpoint. This means that (1) this "normal" scientific standpoint may not be an actual scientific account and (2) even if authentic scientific results are correctly used, there cannot be actual scientific arguments against (or in favor of) the manifest image. Simply the manifest/scientific distinction is not a genuine scientific issue. Neither is it (on the interpretation of van Brakel as it seems to me) a general philosophical issue, but in a sense an issue for philosophy of science. Namely, in the sense that in general philosophy the issue cannot be adequately posed. The importance of van Brakel's approach for philosophy of science generally and for philosophy of chemistry—which is our main concern here—particularly lies in the fact that it enables us to break free from the usual derogatory attitude towards the manifest image and requires a critical assessment of the scientific image. In case the question of priority is raised, the answer has to be, according to van Brakel, that the manifest image has a primacy over the scientific one. Ultimately, the manifest image is the most universal meta-level which, in one way or another, serves as the starting point, and to which one finally always returns when assessing the acceptability of the consequences of any particular conception.

The primacy of the manifest image is quite obvious in the case of chemistry (defined as the science of substances). Still, the question concerning the relationship between the viability of the concept of a natural kind in philosophy of chemistry and the status of chemistry as a science remains. I mean that the need remains for philosophical arguments that would show (1) that chemistry whose subject is substances is *science* proper—not, for example, *natural history*,[20] and (2) that in order to construe chemistry as an autonomous *science* chemical substances should be interpreted as natural kinds. It seems that J. van Brakel does not consider offering a theoretical account or model of science as an essential—or as a possible—task. Sciences are all those disciplines that are science according to the manifest image. So, for example, classical biology undoubtedly qualifies as science, and if anyone should characterize it as natural history, this only indicates that some sciences may be characterized like that, not that biology is really not a science. As mentioned before, van Brakel thinks that, keeping away traditional physics-centered philosophical accounts of science and relying instead on empirical comparative analysis, one should admit that chemistry is an autonomous science and that it is actually a more typical science than physics. On the basis of empirical arguments, therefore, chemistry has to be taken very seriously in philosophy of science. It remains unclear, however, whether chemistry is also relevant for elaborating a theoretical model of science according to which natural history proves to be different from science and hence the problem of natural kinds may stand differently as well.

THE PROBLEM OF NATURAL KINDS IN GENERAL PHILOSOPHY AND IN PHILOSOPHY OF SCIENCE

References to chemistry in philosophical theories need critical analysis from the viewpoint of philosophy of science, whereas some philosophers would seem to agree that scientific knowledge helps to explain philosophical issues. The issue of natural kinds is perhaps the most characteristic one at this point. J. van Brakel has published a whole series of fine critical analyses concerning the favorite thesis of those who theorize about the meaning of natural kind terms—namely, the claim that "water is

H_2O"—and has convincingly shown that these thinkers can win no support for their conclusions from chemistry (1) for the reason that they uncritically use as their starting-point the "normal" scientific image, which needs to be made more precise relying on actual scientific knowledge, and (2) for the reason that, when actual scientific knowledge is adequately used, philosophical questions remain open. "A philosophical theory of natural kinds should be independent of any possible scientific explanation of the structure of matter," van Brakel argues.[21]

Apparently, one must agree with this, since it is metaphysical theories and theories about philosophy of language that van Brakel has in mind here. But is this a valid principle when one is dealing with theories concerning philosophy of science? It appears one may read van Brakel (it was noted earlier that he seems to avoid any theoretical account of science) as saying that the issue of natural kinds falls beyond the scope of philosophy of science. But what about issues which surely belong to philosophy of science—such as scientific explanation, laws of nature, scientific theories? J. van Brakel does not make any clear distinction between philosophy in general and philosophy of science (as some kind of theory of science, which has its own conceptual framework).[22] Among other things, this would explain why R. Harré, who has tried to construct a certain philosophical account of science (a rationale for the natural sciences[23]), which would give justice to theoretical accounts of natural kinds, laws of nature, scientific theory, scientific explanation and so forth, finds in his aforementioned review that van Brakel leaves the questions unanswered or remains skeptical about the possibility of any real answers (e.g., are there any natural kinds?), or that he does not accord the conceptions he seems to accept the significance they deserve (e.g., N. Cartwright's conception of laws of nature).

Now I shall proceed to examine some arguments concerning the controversy, or perhaps just some kind of misunderstanding, between Rom Harré and Jaap van Brakel. Both of them regard chemistry as an autonomous science, but their view of the relationship between this issue and that of the concept of natural kinds is different. First, it is appropriate to examine the arguments to the effect that, in order to understand chemistry *qua autonomous science*, it is essential to have some philosophical account of a natural kind. How does Harré ground his complaint that if, following van Brakel, one decides to "favor the skeptical conclusion on this issue: namely, that there are no natural kinds,"[24] this would mean destroying the status of chemistry as an autonomous science irreducible to physics? On the other hand, concerning van Brakel's critique of essentialism, which is directed towards philosophical theories of natural kinds and which hinders him from seeing any real philosophical content here, will it remain valid if the purely philosophical stance is abandoned and one starts to argue within the conceptual framework of philosophy of science itself?

According to Harré, the concept of a natural kind is important in philosophy of chemistry not because chemistry's subject matter is substances or chemical kinds (considered from this viewpoint, one could say that we are merely dealing here with a specific feature of chemistry), but because the concept of a natural kind plays an important role in the philosophy of science in general. Together with the concepts of natural law, scientific explanation, and theory, it belongs to the central conceptual framework of philosophy of science.[25] Each science has its own system of natural kinds and related concepts on the basis of which it is constituted as an autonomous science.[26] The same should hold true of chemistry if it is in fact an autonomous science. On the other hand, chemistry as the science of substances is well suited for the philosophical analysis of the role of the concept of natural kinds in science.[27]

It should be noted that what is stated above does not apply to biology: although biological examples are often used when analyzing the problem of natural kinds, this does not necessarily mean that one is analyzing the concept of natural kind of science. For instance, Harré claims that "it seems that even in the most liberal reading biology does not really use natural-kind concepts."[28]

As stressed by Harré, the history of this issue in philosophy of science takes us back to Robert Boyle and John Locke, and thereby to the concepts of real and nominal essences of substances, which "became a permanent part of the philosophical armory."[29] This is true in spite of the fact that Kripke's and Putnam's new essentialism, based on these concepts, has been sharply criticized by philosophers of science. On Harré's view, Kripke's and Putnam's account deserves the criticism it has met, but this does not mean that one should drop the distinction of the *nominal essence/ real essence* altogether. This distinction as a formal pattern is crucial in philosophy of science generally; and I would like to stress once more that, according to Harré, there would be no chemistry without this distinction. Essentialism is easily attacked if the argument for it amounts to the following:

> Kinds need essences. Essences are some sort of absolute. There are no absolutes in the field of natural science. So there can be no point in trying to build a theory of kinds on the basis of the concept of "essence." Without that concept kinds are merely convenient human disarticulations of the seamless web of nature. So the idea of "natural kind" is a fancy.[30]

Nevertheless, it is easy to refute this argument, as was also shown by Harré:

> Dispose of the idea that essences must be absolutes and the skeptical argument collapses.... [H]uman decisions to classify natural beings are fixed by fiat in contexts of practice. However, it is a prime aim of natural science to try to find out whether the constitution or nature of those beings, as disclosed by work in a theoretical context, justifies making the divisions into practical kinds where practice has drawn them. "Natural kind" is a concept which can be explicated only within the double framework of practice and theory. Every natural kind is located in both contexts, and cannot be understood by reference to either one alone.[31]

Practical context corresponds, in the formal pattern of *nominal essence/real essence*, to the "nominal essence" and to the "descriptivist" account of the meaning of natural kind terms, whereas theoretical context corresponds to the "real essence" (or to the level of natural laws and scientific explanation) and to the "essentialist" idea that the extensions of kind terms are determined by the essences or natures of the beings. In real cases, however, these contexts are complementary, and there are no absolutes or ultimate explanations, as stressed by Harré.

The argument against essentialism, based on chemistry, which was presented by van Brakel will not "work" in case of Harré's conception of philosophy of science, because, as noted before, it (van Brakel's argument) is not really an argument within a conception of philosophy of science, but is rather meant as a criticism against those who think that scientific knowledge helps to explain philosophical issues. The point is simply that "it is not the task of the scientists to find essences that fit the philosopher's ontological framework, but to give better descriptions of what there is. What there is, is substances and/or natural kinds and compounds thereof."[32] Still, he admits that "as an explanation of the kind of laws scientists discover, it can be instructive to consider the possibility of there being necessary *a posteriori* statements," but these statements need not be expressions of underlying structures as essences of substances. Underlying structures may play an important role in scientific explanations of the structure of matter, but they "do not show the way to the 'true' essence of a particular substance."[33]

To my mind, it is the most important outcome of van Brakel's analysis of the problem of natural kinds that he shows convincingly (as it seems, without having set himself any such goals explicitly) that this problem, which originally emerged at the borderline of science and philosophy, belongs to the domain of philosophy of science. Philosophy in general, or the so-called pure philosophy (i.e., not some kind of special discipline between philosophy and science although, in the present context, this "pure philosophy" means, first of all, philosophy of mind and language) cannot hope that scientific knowledge will help to explain philosophical issues. On the other hand, it is obvious to me that neither can philosophy of science achieve more than just to explain philosophical issues in science. I mean that Harré's reproaches to van Brakel—that he retains a skeptical stance towards issues concerning natural kinds, essentialist realism, causal powers, objective laws, and so forth—are justified only within the framework of philosophy of science. Keeping in mind that van Brakel analyzed these problems in the context of general philosophy, one should perhaps conclude that we may be dealing here with no real controversy, but just some kind of misunderstanding between Rom Harré and Jaap van Brakel.

WHAT KIND OF REALISM IS ACCEPTABLE?

In the context of general philosophy, as it seems to me, Harré's referential realism—as any kind of critical scientific realism,[34] including the aforementioned "constructive realism," "perspectival realism," or "scientific perspectivism"—represents the same line of thinking as van Brakel's anomalous monism, or pragmatic or pluralistic realism, or "realism-with-a-small-r," because one is surely not dealing here (to use Putnam's terminology) with "realism-with-a-big-R" or with metaphysical realism, which is "an impossible attempt to view the world from Nowhere."[35] At the same time, it is certainly important to realize that regarding metaphysical realism as impossible does not mean that one should adopt Putnam's "internal realism" (this was well demonstrated by Ilkka Niiniluoto[36]).

It should be remarked that when reading the works by J. van Brakel, one cannot be quite sure of whether there exists a third option between metaphysical Realism and internal realism for him; whether he accepts scientific realism which is undoubtedly an account of the world mediated by human practical and theoretical activity (following T. Kuhn, one might call it a paradigm), but this is still not Putnam's internal realism.[37] As a final point, I would like to examine this issue in connection with the relationship between natural kind and objective law or natural law.

First, I will stress once more what I have said before: it is precisely philosophy of science where the natural kinds are clearly definable and, most important, they appear in laws of nature and play an explanatory role.[38] So it is no wonder that most of the problems that have emerged in general philosophy in connection with natural kinds—such as the objective criteria for distinguishing natural kinds, essence, nonlogical necessity, realism, and other issues—have been addressed by philosophers of science considering the issues of natural laws, explanation, theory, models, and so forth. The problem, which starts with definition and classification, takes us, via natural classification, to laws of nature and scientific theory. Here, a relevant classical example from philosophy of chemistry is of course the periodic system of chemical elements based on the periodic law. Chemical elements turn out to be natural kinds precisely for the reason that they are theoretically defined through the periodic system of chemical elements based on the periodic law.[39]

I shall now proceed to the question concerning van Brakel's realism and its connection with objective law. My account will be based on an article by Igor Douven and Jaap van Brakel, entitled "Can the World Help Us in Fixing the Reference of Natural Kind Terms?"[40] The authors answer the question posed in the title of their paper in the negative; however, my main concern here will be the aforementioned, narrower question. Namely, the authors claim that "objective laws" cannot be read in the traditional realist way, that is, as "OBJECTIVE LAWS" (another name for this traditional realism would apparently be metaphysical realism.[41] They attack Putnam's account of natural kind words by pointing out the fact that, as internalist, he cannot allow for "OBJECTIVE LAWS," but nevertheless actually refers to "objective laws," presumably in the sense of laws of nature. So, for example: "What it is rational to include in the extension of "gold" depends on objective laws, governing the behavior of gold."[42] In connection with this, a more general question has been raised: Is it possible to construe "objective laws" in a way that would not be internalist (Putnam does in fact have some difficulties in retaining his internalism) and that, at the same time, would not mean accepting the traditional realist way, that is, "OBJECTIVE LAWS."

Here, the question arises as to what is objectivity. According to these authors, distinguishing laws of nature from objective regularities is no easy task. Attempts have been made to define laws of nature as those laws that are about natural kinds, but such proposals lead to circularity in definition. As to this point, I think, one has to agree with Alexander Bird[43] that an acceptable way of interdefining natural laws and natural kinds might still be possible.[44] Douven and van Brakel, however, whose article I am considering here, undertake to explore other options.

Metaphysical necessity being excluded and Putnam's internal notion of objectivity-for-us being untenable as well, maybe one should interpret it as meaning "objectivity as physical necessity"? Then one should regard the laws of physics as primitives in the definition of physical necessity, the authors find. It has to be noted here that this analysis of physical necessity tends to be unjustifiably derived from laws of physics, instead of being based—as it should be—on laws of nature or scientific, that is, φ-scientific, laws; "physical" should be taken here as synonymous with "natural," and instead of physical necessity it would be better to speak of natural necessity. Douven and van Brakel state: "On realist grounds this is justified because we hope these laws to reflect the OBJECTIVITY of THE WORLD. However, this is not an option for the internalist."[45] That's fine—there is no need for us to be internalists; but what about scientific realism? Our authors find that this presents a problem, since "The question 'What is there to be found OBJECTIVELY, that is, independent of us?' is mistaken, for what we will actually find will necessarily depend upon *our* values."[46]

It seems to me that we are dealing with some kind of misunderstanding here, if what was said is considered from the viewpoint of scientific realism. What is stated above may be true of traditional realism. Critical scientific realism admits that the scientific account of the world is mediated by our practical and theoretical activity, together with our aims and values, which means that our descriptions of the world, our "world-versions" are always relative to us. This does not imply, however, that the world itself (we can call it THE WORLD) is relative to us or that our "world-versions" cannot be versions of THE WORLD. Our scientific "world-versions" are not what we believe about the world, but the way the world is relative to our conceptual framework. Our scientific "world-version," although it represents the world through theories we have constructed, which, in their theoretical models, contain experimentally substantiated idealizations, still do tell us something about THE WORLD, since theoretical models are similar to the real systems in specified respects and to specified degrees. As Niiniluoto writes: "Conceptual frameworks are selected on the basis of our cognitive and practical purposes, and they can always be improved and made descriptively more complete."[47] If we use the cookie cutter metaphor we can say: "A cake [THE WORLD] can be sliced into pieces in a potentially infinite number of ways, and the resulting slices [say, natural kinds and laws of nature identified by us] are human constructions made out of the parts [unidentified (complex, inexhaustible) objects, their properties and relations] of the cake."[48]

CONCLUSIONS

(1) The problem of natural kinds proves to be different in general philosophy and in philosophy of science. This problem, which originally emerged at the borderline of science and philosophy, belongs to the domain of philosophy of science. Philosophy in general cannot hope that scientific knowledge will help to explain philosophical issues.

(2) Chemistry has to be taken very seriously in philosophy of science. Not only do empirical arguments indicate that chemistry should be regarded as a typical science, but it is also relevant for elaborating a theoretical model of science.

(3) Chemistry as the science of substances is well suited for the philosophical analysis of the role of the concept of natural kinds in science.

(4) The viability of the concept of a natural kind in philosophy of chemistry derives from the explanatory role of chemical natural kinds. The problem of explanation in its turn relates natural kinds to the laws of nature and scientific theories, that is, to classical issues in philosophy of science and to discussions on their (supposed) specificity in chemistry.

(5) In philosophy of science, one can demonstrate that there exists a third option between metaphysical realism and internal realism (to use Putnam's terminology). According to this third type of realism our "world versions" (including natural kinds identified by us), but not the world itself, are relative to us.

ACKNOWLEDGMENTS

An earlier version of this paper was read at the 6th Summer Symposium of the International Society for the Philosophy of Chemistry, Georgetown University, Washington D.C., August 4–8, 2002. I would like to thank the participants of that symposium for discussion. I am especially grateful to Rom Harré and Jaap van

Brakel for the comments they kindly sent me after the symposium, and to the anonymous reviewer who helped me to improve this article.

NOTES AND REFERENCES

1. HARRÉ, R. 2001. Review of Jaap van Brakel (2001), Philosophy of Chemistry: Between the Manifest and the Scientific Image. Leuven University Press. Leuven, Belgium. Hyle **7**: 179.
2. PUTNAM, H. 1975. The meaning of "meaning." *In* Mind, Language, and Reality (Philosophical Papers, vol. 1). Cambridge University Press, Cambridge, UK; id. 1981. Reason, Truth, and History. Chapter 2. Cambridge University Press, Cambridge, UK; id. 1983. Realism and Reason. (Philosophical Papers, vol. 3). Cambridge University Press, Cambridge, UK; id. 1990. Realism with a Human Face. Chapters. 4, 21. Harvard University Press. Cambridge, MA/London.
3. KRIPKE, S. 1980. Naming and Necessity. Oxford University Press. Oxford, UK/New York.
4. MELLOR, D.H. 1991. Matters of Metaphysics. Chapters 7, 8. Cambridge University Press. Cambridge, UK.
5. VAN BRAKEL, J. 1986. The chemistry of substances and the philosophy of mass terms. Synthese **69**: 291–324; id. 1997. Chemistry as the science of the transformation of substances. Synthese **111**: 253–282; id. 1999. On the neglect of the philosophy of chemistry. Found. Chem. **1**: 111–174; id. 2000. The nature of chemical substances. *In* Of Minds and Molecules. N. Bhushan & S. Rosenfeld, Eds.: 162–184. Oxford University Press, Oxford, UK; New York; id. 2000. Philosophy of Chemistry: Between the Manifest and the Scientific Image. Leuven University Press. Leuven, Belgium.
6. LAPORTE, J. 1996. Chemical kind term reference and the discovery of essence. Nous **30**: 112–132.
7. HARRÉ, Hyle **7**: 179.
8. HARRÉ, R. 1986. Varietes of Realism: A Rationale for the Natural Sciences. Basil Blackwell. Oxford, UK. pp. 97–99.
9. VAN BRAKEL, Philosophy of Chemistry, pp. 191–198.
10. GIERE, R.N. 1988. Explaining Science: A Cognitive Approach. University of Chicago Press. Chicago.
11. GIERE, R.N. 1999. Science without Laws. Chapter 4. University of Chicago Press. Chicago.
12. GIERE, R.N. 2001. Scientific perspectivism. *In* Ontology Studies. Physis. (International Ontology Congress. Proceedings) San Sebastian. pp. 85–88.
13. See, e.g., VIHALEMM, R. 1999. Can chemistry be handled as its own type of science? *In* Ars mutandi—Issues in Philosophy and History of Chemistry. N. Psarros & K. Gavroglu, Eds.: 83–88. Leipziger Universitätsverlag. Leipzig; id. 2001. Chemistry as an interesting subject for the philosophy of science. *In* Estonian Studies in the History and Philosophy of Science. (Boston Studies in the Philosophy of Science. Vol. 219). R. Vihalemm, Ed.: 185–200. Kluwer Academic Publishers. Dordrecht/Boston/London; id. 2003. Are laws of nature and scientific theories peculiar in chemistry? Scrutinizing Mendeleev's discovery. Found. Chem. **5(1)**: 7–22.
14. VAN BRAKEL, J. 1997. Chemistry as the science of the transformation of substances. Synthese **111**: 253–254; see also, Pechenkin, A.A. 1986. The interaction of physics and chemistry: philosophical and methodological problems. Mysl. Moscow [in Russian], pp. 20–24.
15. SCHUMMER, J. 1997. Challenging standard distinctions between science and technology: the case of preparative chemistry. Hyle **3**: 90–91.
16. SELLARS, W. 1963. Science, Perception, and Reality. Routledge and Kegan Paul. London.
17. HARRÉ, Hyle **7**: 179.
18. VAN BRAKEL, Philosophy of Chemistry, p. 42.
19. Ibid., p. 47.

20. For the crucial difference between physics-like science and natural history, see: TOULMIN, S. 1967. The Philosophy of Science: An Introduction. Hutchinson. London; cf. also: MCALLISTER, J. 1997. Laws of nature, natural history, and the description of the world. Int. Stud. Philos. Sci. **11(3)**: 245–258.
21. VAN BRAKEL, J. 1986. The chemistry of substances and the philosophy of mass terms. Synthese **69**: 291.
22. In his comments to the draft of this paper, J. van Brakel agreed that he did not make a clear distinction between philosophy and philosophy of science, hence misunderstandings may have arisen because of this difference. In his claim that philosophical theory of natural kinds should be independent of any possible scientific explanation of the structure of matter, he took "philosophy" to include "philosophy of science."
23. See the subtitle of: HARRÉ, R. 1986. Varietes of Realism: A Rationale for the Natural Sciences. Basil Blackwell. Oxford, UK.
24. HARRÉ, R. 2001. Hyle **7**: 178.
25. In his comments to the draft of this paper R. Harré indicated, among other things, that philosophers of science pay little attention to classification systems, but they are at least as important as explanatory systems.
26. HARRÉ, Varietes of Realism; id. 1993. Laws of Nature. Duckworth. London.
27. Cf., however: ROSENFELD, S. & N. BHUSHAN. 2000. Chemical Synthesis: Complexity, Similarity, Natural Kinds, and the Evolution of a "Logic" *In* Of Minds and Molecules. N. Bhushan & S. Rosenfeld, Eds.: 202. Oxford University Press. Oxford, UK: "…chemical kinds are not the unproblematic, paradigmatic instances of natural kinds they have been taken to be in the philosophical literature."
28. HARRÉ, Varietes of Realism, p. 121. In his comments to the draft of this paper R. Harré claimed that phenotypes in biology are loosely tied to genotypes; only in chemistry does one find a systematic relation between revisable nominal essences of elements and compounds, and revisable theoretical claims about the constitutive structures that explain the usability of the criteria for making classifications.
29. HARRÉ, Hyle **7**: 178.
30. HARRÉ, Varietes of Realism, p. 99.
31. Ibid.
32. VAN BRAKEL, J. 1986. The chemistry of substances and the philosophy of mass terms. Synthese **69**: 291.
33. Ibid. In his comments to the draft of this paper R. Harré remarked, concerning this quotation, that probably van Brakel was giving here a high redefinition of "true," meaning something like absolute, final, and universal characterizations of a substances. The notion of an essence, in a strong sense, has no place in general philosophy since there are indefinitely many ways individuals can be picked out and classified into kinds. Sciences have revisable taxonomies.
34. See: NIINILUOTO, I. 1999. Critical Scientific Realism. Oxford University Press. Oxford, UK.
35. PUTNAM, Realism with a Human Face, p. 28.
36. NIINILUOTO, Critical Scientific Realism, Chapter 7.
37. In his comments to the draft of this paper J. van Brakel remarked that, if pressed, he would admit there is a third option; but a better way to put the question is to say that anomalous monism transcends the dichotomy between metaphysical realism (universalism) and internal realism (relativism).
38. See for it, e.g.: BIRD, A. 1998. Philosophy of Science. UCL Press. London.
39. It is important to note here that Mendeleev's periodic law is a normal law of nature, i.e., a φ-scientific law. See: VIHALEMM, R. 2003. Are laws of nature and scientific theories peculiar in chemistry? Scrutinizing Mendeleev's discovery. Found. Chem. **5(1)**: 7–22..
40. DOUVEN, I. & VAN BRAKEL, J. 1998. Can the world help us in fixing the reference of natural kind terms? J. Gen. Philos. Sci. **29**: 59–70.
41. Ibid., p. 59.
42. Ibid., p. 62.
43. BIRD, Philosophy of Science, pp. 114–120.

44. Cf. also: HARRÉ, Varietes of Realism; id. Laws of Nature; ROTHBART, D. 1993. Discovering natural kinds through inter-theoretic prototypes. Methodol. Sci. **26:** 171–189.
45. DOUVEN, I. & J. VAN BRAKEL. 1998. Can the world help us in fixing the reference of natural kind terms? J. Gen. Philos. Sci. **29:** 63.
46. Ibid., p. 68.
47. NIINILUOTO, Critical Scientific Realism, p. 216.
48. Ibid., p. 222.

The Primary Properties?

S.H. VOLLMER

Department of Philosophy, The University of Alabama at Birmingham, Birmingham, Alabama 35294, USA

ABSTRACT: Science is concerned not just with objects but also with their various aspects, such as their colors, temperatures, sizes, and shapes. These aspects, or properties, are generally thought to be of at least two kinds: primary properties, such as shape, size, and motion and secondary ones, such as temperature and color. However, there is little agreement on just what the difference is between these two kinds of properties. An argument has recently been put forth that assumes the two kinds of properties differ only in that the secondary ones are dependent on the conditions under which they are observed. This paper suggests, however, that since the primary properties depend on conditions, too, any argument based on this assumption is flawed. There are two main traditional accounts of the distinction. The first account of the distinction—the ideas of primary qualities resemble the properties that cause these ideas, whereas those of the secondary ones do not resemble them—relies on what to many is an implausible assumption, that some of our ideas resemble features of the world and we can speak intelligibly about such resemblances. This paper suggests that we cannot, and hence we have no more reason to believe that the ideas of the primary qualities resemble aspects of the world than we do that those of the secondary qualities do so. The second traditional account—primary properties are those a body has, however small it might be, whereas secondary properties are a consequence just of relations between bodies—fares better than the other two accounts. However, the notion of force posed a serious problem for this second basis for the distinction, one that has, astonishingly, hardly found its way into the philosophical literature.

KEYWORDS: properties; qualities; forces; primary qualities

INTRODUCTION

Science is concerned not just with objects but also with their various aspects, such as their colors, temperatures, sizes, and shapes. These aspects are sometimes separable, that is, they can be observed, and therefore studied, independently. Scientists thus study objects and their effects in terms of these various aspects, that is, their properties. Science, furthermore, is ordinarily concerned not with properties unique to individual objects, but with properties of classes of objects. In this paper, the word property refers specifically to these classes.[1] For example, the property of having a certain temperature refers to a class of objects or parts of them, that is, all those hav-

---

Address for correspondence: Department of Philosophy, The University of Alabama at Birmingham, Birmingham, AL 35294. Voice: 205-934-4805; fax: 205-975-6610.
Vollmer@uab.edu

ing that temperature. Some properties are jointly defined by others, for example, the property of being an electron is defined by such properties as having a certain charge and a certain mass; the property of being a goat by such properties as having a certain sort of phenotype and genotype. However, this paper will be concerned with the simpler properties, such as temperature, color, shape, size, and motion, which can be observed independently of the others. "Properties" as it is used in this paper has the same denotation as "qualities"; "qualities" has the added connotation that what is of concern is primarily, though not exclusively, properties as they are perceived by the human senses.

Properties have been viewed historically and continue to be viewed as being of two kinds, primary properties and secondary ones. Their division into two kinds has, however, been justified on quite different grounds in the various accounts given of them.[2] Nevertheless, the assumption that there are two kinds of properties that are on all accounts divided into more or less the same two groups is rarely questioned.

This paper begins by taking up a recent account of molecular shape which assumes that the difference between the two kinds of properties is that only the secondary properties are dependent on the conditions under which they are observed. After showing that a distinction on this basis is not a good one because the primary properties are also dependent on conditions, this paper takes up the first of the two main traditional accounts of the distinction: (1) the ideas of primary qualities resemble the properties that cause these ideas, whereas those of secondary qualities do not. This paper suggests that this account is wrong because there is no more reason to believe that the ideas of the primary qualities resemble the corresponding properties than that those of the secondary ones do. An account of the second traditional distinction is then given: (2) the primary properties are, as suggested by Locke, those a body has, however small it might be. Finally, this paper argues that, on the second traditional account of the distinction, when the notion of force was introduced, size and shape, surprisingly, lost their status as primary. This change in the status of the properties of material bodies occurred centuries before the development of quantum physics, when their status was to change again, even more profoundly.

Yet another account of the distinction, the primary properties are those that do not possess an intensive magnitude, defended by some as *the* traditional account, is beyond the scope of this paper.[3]

DEPENDENCE ON CONDITIONS

It has recently been argued that, because the shapes of molecules depend on the conditions of observation, shape is not always a primary property.[4] This argument, put forward by Jeffry Ramsey, relies on an assumption that is commonly made, that is, that whereas the secondary properties, such as temperature and color, do not appear the same in every instance but vary from occasion to occasion, the primary properties always appear the same.

The variation in shape of concern is related to the fact that when a molecule is moving fast it sometimes appears as a blur. When this is the case, Ramsey says, it has a different shape than when it is stationery.[5] Its shape, then, depends on conditions—and hence it is a secondary property, not a primary one.[6]

Ramsey argues, then, that molecular shape depends on the amount of blurring, or the extent to which multiple states appear as one. Examples he gives of instrumental methods used to observe molecular shape in which blurring can occur are IR, Raman, and EPR. Another example would be x-ray crystallography. He adds that a low-temperature beam, when used in combination with any of these instruments, decreases the kinetic energy or motion of molecules and therefore can change the amount of blurring.[7]

To understand the implications of Ramsey's account, consider a methyl group, freely rotating at room temperature, which can be slowed down or "frozen out" through the use of a low-temperature beam. When the methyl is observed using an X-ray crystallographic instrument in combination with a low-temperature beam, the rotation of the methyl group is relatively show and the position of each hydrogen as well as the tetrahedral shape of the carbon may be apparent. In contrast, in the absence of the low-temperature beam, the methyl, freely rotating, would appear as a blur. In Ramsey's account, the shape of the methyl is, approximately, the shape its atoms sweep out in the course of an observation. This means that, in his view, the methyl has a different shape when it is in the low-temperature beam than it does when it is at room temperature. Molecular shape, therefore, depends on conditions.

Ramsey goes on to suggest that, because a property that depends on conditions is a secondary one, molecular shape is a secondary property. One difficulty with Ramsey's account is that, in it, the shape of a baseball that is thrown through the air, depending on conditions, is sometimes a long streak and other times a small sphere. The shape of a baseball, however, at least as we use the word, ordinarily is a small sphere and not, depending on conditions, a long streak. Therefore, Ramsey's use is confusing. If there is reason to adopt a different account of shape—when it comes to molecules—than the ordinary one, then we need an explanation of why, as well as an explanation of just how the new account of shape differs from the old one.[8]

There is, however, a second difficulty with Ramsey's account. This is that the common assumption on which his argument is based, that only the secondary properties change with conditions, the primary properties being independent of them, is not as straightforward as it might seem.

To see this, one must realize that if the primary properties were completely independent, that is, independent *in every sense* from conditions, this would mean that, in observing the shape of an object, we would have a God's-eye or perspectiveless view of the object. Only then would the shape of the object always be seen in the same way—regardless of the spatial relation of the viewer to the object. However, things are not seen in this way, that is, in their entirety in a moment. Instead, they appear in different ways depending on the relationship between the viewer and the object.

The notion that a change in perspective causes a difference in a property of an object, however, is counterintuitive. We might, then, attempt to correct this problem by suggesting that, although the primary properties appear different in a way that depends on conditions, that is, on perspective, differences of this sort are of a categorically different kind, different in that they are "immediately and utterly comprehensible." As immediately and utterly comprehensible, then, these dependencies are to be distinguished from ones that are more opaque. For example, the temperature of a reaction mixture is dependent on such things as the ambient temperature, the nature of the chemical reaction occurring, and the amount of heat exchanged with the environment. The reaction mixture's color is dependent on such things as the spectrum and intensity of illuminating light and the energy levels of the relevant electrons. The dependences of temperature and color on conditions are, then, relatively opaque.

On this basis, then, differences in the way the primary properties such as shape appear as a result of perspective are differences that are, significantly, immediately, and utterly comprehensible—whatever exactly that might mean. Properties that vary with conditions in a way that is opaque, in contrast, require special knowledge and interpretation of the details of the conditions and are, on this account, secondary.

However, there is a problem with drawing a distinction between the primary and secondary properties on this basis. That is, there may be no objective fact as to what ought to count as a special knowledge and interpretation. Although differences due to changes of perspective may seem to require no special interpretation, this may be due to the fact that some functions—the primary property ones—are the ones upon which our cognitive system routinely depends. It might, then, be because of this routine dependence that these functions are immediately and utterly comprehensible. Drawing the distinction on the basis of which properties are immediately and utterly comprehensible, then, risks drawing it in a way that is based on a prejudice that is a result of the kind of cognitive creatures we are.

Dependence on perspective, to make matters worse, is not the only kind of dependence the primary properties show. The shapes of rarified stars, for example, are a function of the strength of the gravitational fields exerted by their neighbors and so of the densities and locations of their neighbors at the time of, and so under the conditions of, observation. Therefore, the shapes of stars are dependent on these various conditions. This shows, then, that the idea that the primary properties of an object are independent of the conditions of an observation does not bear out. Admittedly, in a region of the universe like ours, where there are no rarified stars, the sizes and shapes of objects, with few exceptions, appear constant—except for changes due to perspective—to varying conditions. That they do so is a fact about the part of the universe in which we live and is a consequence of a local fact about the relative strengths of the forces on objects from within and without. That the secondary properties depend on the conditions of observation and the primary ones do not cannot, then, be what distinguishes the primary properties from the secondary ones except, perhaps, in a part of the universe like ours. Drawing the distinction on this basis, then, risks drawing it in a way that is based on a prejudice of another sort, this time on one that is a result of the part of the universe in which we happen to live.

Therefore, although it is commonly assumed that a distinction between the primary and secondary properties can be made on the basis of the fact that the primary properties but not the secondary ones are independent of conditions, if this seems to be the case, it is because of the kind of cognitive creatures we are and the part of the universe in which we happen to live. It does not, in any obvious way, reflect a difference between the primary and secondary properties themselves. Interestingly, the traditionalists such as Locke noted the tendency for the secondary qualities and not the primary ones to depend on conditions; however, they did not offer differing dependence on conditions as an analysis of the distinction.[9]

That only the primary properties are independent of conditions cannot, then, be used as the basis of an argument that claims to show that a property, such as "molecular shape" when defined as the shape its atoms sweep out in the course of an observation, is not a primary property. Not, that is, without also giving an account of how, in some nonobvious way, changes in the way properties appear are, indeed, of two different kinds, one associated with the primary properties and another with the secondary ones. Some such accounts have been given, including the two traditional accounts discussed below. In addition, it has been argued recently (as well as in antiquity) that only the primary qualities are common to the many senses.[10,11] It has also been argued recently that the primary ones have far more complex interconnections with our experiences that do the secondary.[12] These accounts, however, are beyond the scope of this paper.

DISTINCTION 1: THE FIRST TRADITIONAL ACCOUNT AND HOW IT FAILS

The notion that the ideas of primary qualities resemble the properties that cause these ideas, whereas those of secondary qualities do not, is perhaps the major traditional account of the primary–secondary quality distinction; it has been suggested that there is no doubt it was accepted by the major early modern philosophers.[13]

However, the account relies on what, to many today, is an implausible assumption—some of our ideas, those of the primary qualities, resemble features of the world, and we can speak intelligibly about such resemblances. (There is some disagreement about just what is meant by "resemblance" and whether Locke's theory was, indeed, a resemblance theory or whether it was of another kind, attributing objectivity uniquely to the ideas of the primary qualites on the basis something like accuracy, likeness, or covariance with the world rather than resemblance.[14])

Without going into the question in detail, this paper suggests that there is a very general way to show that, in fact, we have reason to be highly skeptical that the primary properties, any more than the secondary ones, resemble aspects of the world. This is that, in observing a property, as argued forcefully by Kant, what we observe is the appearances of the property. There is, however, most of us think, a distinction to be made between the way an object or property *is*, its *ding an sich*, and its appearances. We cannot know about the property in and of itself because we know only its appearances. Therefore, we cannot know, nor can we assume, that the property in itself, behind the appearances, is identical with or even that it resembles its appearances. Because we cannot know, when we observe an object or primary property, that what we are observing is anything like the way it, in fact, is, we cannot assume that the idea we have of it is identical with or resembles an aspect of the world. For this reason, when we describe the properties we observe, what we are describing is, straightforwardly, our ideas of these properties and talk about whether or not these ideas resemble the world is more or less unintelligible. For this reason, the first traditional account of the primary–secondary quality distinction fails.

DISTINCTION 2: THE SECOND TRADITIONAL ACCOUNT

There is, however, a second traditional account of the distinction, one that arises directly as a consequence of the commitment of its proponents to a mechanistic science. In this account, as expressed by Locke, primary properties are those utterly inseparable from a body, however small the body might be. This suggests that, were an object to be divided into its smallest parts, or the finest bits of matter possible—corpuscles—an inventory of the properties would show them to be only the primary ones. The secondary ones, since they no longer exist when matter is divided into its smallest parts, are not, therefore, really in the matter. Hence, this account, it is suggested, is essentially a rejection of the forms and real qualities of the Scholastics.[15] Locke's empiricism did not, evidently, prevent him from holding this corpuscularian view, on which the primary properties are not only apparent to the senses in matter that has "bulk enough to be perceived," but they are also in the finer matter, that is, the particles that are too small to be "singly be perceived by our senses."[16] Locke suggested, then, that size, shape, and motion, or as he put it, bulk, figure, texture, and motion of the sensible parts were in the corpuscles themselves, whereas properties such as color were not and, if they were in the corpuscles in any sense, were just in them as "powers."

This account of the primary–secondary property distinction was put forward at a time when it was thought that one, or at most a few, basic kinds of matter would be found. All corpuscles were expected, then, to be composed of these basic kinds of matter with differences between substances being due to differences in the sizes, shapes, motions, and arrangements of corpuscles of the basic matter.

When it was thought that the corpuscles were the atoms, however, it became hard to see how differences between the increasing numbers of newly discovered elements, in the first half of the 19th century, could be explained on the basis just of differences in the shapes, sizes, motions, and arrangements of the corpuscles of the basic matter. It was also judged unlikely, on the principle of Occam's razor, that there were, in fact, a great variety of different kinds of basic matter, each of the elements discovered representing a different one. Subsequent results led to the belief that there were, indeed, corpuscles more basic than the atoms and that these more fundamental corpuscles were, initially, of just two kinds, electrons and protons. Such changes in theory with respect to which objects were, in fact, the fundamental corpuscles caused, then, changes in which objects could have properties such as color. For example, when the corpuscles were identified with the atoms, color, considered a secondary property, could not be a property of the atoms. However, when the corpuscles became identified, instead, with electrons and protons, whereas the atoms could have color or be colored, the electrons and protons could not. If color was in the electrons and protons, then, it was in them as a power, for example, of the electrons to drop toward protons or other systems and so to produce the color that was in the atom.

FORCES

In suggesting that the paradigm primary properties are size, shape, and motion, we have put to the side a critical aspect of the question of what is meant by the primary properties. One way to see this critical aspect is to consider Locke's suggestion that it is impossible for two material objects to occupy the same place at the same time but quite possible for two geometrical figures to do so. Therefore, size, shape, and motion—the properties of the geometrical figure—while sufficient to account for all its properties, are insufficient to account for the properties of the material object. The property of solidity or hardness—which caused a change in motion when two objects would otherwise have come to be at the same place at the same time—was needed to account for what is unique about matter as opposed to just geometrical form. The property of solidity, then, was needed as an additional property if the primary properties were to be sufficient for the basis of a complete science. Locke and others added solidity as an additional primary property. Some suggest that the notion that the primary properties provide an explanation for all the properties is, indeed, what Locke, as well as others, such as Boyle and Reid, meant by the primary qualities.[17,18] Locke did say that, if any of the secondary properties do not depend on the primary ones, they depend on something "yet more remote from our understanding."[19] Locke refers, then, to unknown properties on which the others might depend. However, this does not mean that he and others of the new scientific age did not assume that, if the world were transparent and a complete science possible, the secondary properties would, indeed, depend on those few properties that are inseparable from a body, however small the body might be.[20]

Solidity was at first thought of as what makes the difference between body or material and empty space. In the centuries following Newton, however, solidity came to be understood specifically as an effect due to a force. Scientists lacked a reasonable explanation for how this force—or any other—apparently acting at a distance, could, in spite of appearances, be caused by impact, such as the impact of hypothetical objects. Therefore they attributed solidity and forces more generally—for all intents and purposes—to matter itself as acted on by the laws of nature.[21] The notion of force, then, as a property inherent in matter, was conceived.

Accepting the concept of force among the primary properties, however, involved no slight change. This is because the forces of the most finely divided matter, such as the force of gravity and Coulomb's force, were hypothesized to act outward from a point through empty space, equally in all directions. The most finely divided matter, thus, could not vary in shape. This had the consequence that the shapes of the corpuscles could no longer, in any straightforward way, determine the shapes of perceptible objects. The force fields, furthermore, had no size in the ordinary sense. At a maximum at the center of the field and decreasing radially away from it, such fields never, in principle, fall to zero, and therefore the smallest bit of matter would, in a sense, be infinite in size. Therefore, the most fundamental properties of matter were, as a consequence of the introduction of the notion of force, no longer shape, size, and motion, but force and motion. A fully developed account of this kind was, in fact, put forth by the humanist, philosopher, and scientist Roger Boscovich in the 1760s.[22] Recognition of the effect of Boscovich's account on the primary–secondary quality distinction, however, has, astonishingly, hardly appeared in the philosophical literature.

The radical shift in the notion of the primary properties, of course, could be averted were one simply to reject solidity as a primary property. This tack, indeed, has been suggested and justified on the grounds that observation of an object's solidity is different in kind from observation of its shape. Unlike observation of shape, observation of solidity, the suggestion goes, requires a theoretical or quasi-theoretical inference from the impenetrability of objects, as seen on impact, to their solidity.[23,24] However, as explained above, if the traditionalists had denied solidity as a primary property, they would have given up this second account of the distinction altogether.

SUMMARY

This paper has argued that Ramsey's suggestion that molecular shape is secondary does not hold up for two reasons: first, because it employs a notion of shape different from the one used in ordinary observation and, second, because the common assumption on which the suggestion relies, that the primary properties are uniquely independent of conditions, does not, in fact, distinguish between two kinds of properties. Of the two traditional accounts of the distinction, the first—the ideas of only the primary qualities resemble those properties—fails because, since we can know only how a property appears, we cannot know how it is, and therefore cannot know whether an idea of a quality resembles an aspect of the world. The second traditional account—the primary properties are those a body has, however small it might be, whereas the secondary ones are a consequence of relations between bodies—fares better than the the first. However, when forces, historically, came to be seen as properties of the finest bits of matter extending outward equally in all directions, size and shape lost their status as primary properties. The primary properties became, then, just the strength and kind of force-field emanating from a point at its center and motion. This means that centuries before the introduction of quantum physics, the sizes and shapes of objects—as well as all their other properties except force fields and motion—arose only as a consequence of their force fields and motion and, therefore, were assumed to be less fundamental than they had formerly been thought to be.

REFERENCES

1. LEWIS, DAVID. 1999. New Work for a Theory of Universals. *In* Papers in Metaphysics and Epistemology. David Lewis: 1–55. Cambridge University Press. New York.
2. WILSON, MARGARET D. 1992. History of philosophy. *In* Philosophy Today and the Case of the Sensible Qualities. Philos. Rev. **101:** 191–243. See also MCKITRICK, JENNIFER. 2002. Reid's Foundation for the Primary/Secondary Quality Distinction. Philos. Q. **52:** 478–494.
3. FALKENSTEIN, LORNE. 1998. A double-edged sword? Kant's refutation of Mendelssohn's proof of the immortality of the soul and its implications for his theory of matter. Stud. Hist. Philos. Sci. **29:** 561–588.
4. RAMSEY, JEFFRY L. 2000. Realism, essentialism, and intrinsic properties: the case of molecular shape. *In* Of Minds and Molecules: New Philosophical Perspectives on Chemistry. Nalini Bhushan & Stuart Rosenfeld, Eds.: 118–128. Oxford University Press. New York. p. 122.
5. Ibid., p. 119.
6. Ibid., p. 125.
7. Ibid., p. 119.
8. See also, VOLLMER, S.H. 2003. The philosophy of chemistry reformulating itself: Nalini Bhushan and Stuart Rosenfeld's *Of Minds and Molecules: New Philosophical Perspectives on Chemistry*. Philos. Sci. **70:** In press.
9. See, for example: ALEXANDER, PETER. 1977. Boyle and Locke on primary and secondary qualities. *In* Locke on Human Understanding; Selected Essays. I.C. Tipton, Ed.: 62–76. Oxford University Press. New York. p. 74.
10. MACKIE, JOHN. 1976. Problems from Locke. Clarendon Press. Oxford. p. 28 ff.
11. See also, SMITH, A.D. 1990. Of Primary and Secondary Qualities. Philos. Rev. **XCIX:** 221–254, p. 242.
12. BENNETT, JONATHAN. 1965. Substance, Reality, and Primary Properties. Am. Philos. Q. **ii**.
13. WILSON, "History of philosophy," p. 234.
14. See, for example, CUMMINS, ROBERT. 1996. Meaning and Mental Representation. MIT Press. Cambridge.

15. WARNOCK, G.J. 1983. Berkeley. University of Notre Dame Press. Notre Dame. p. 158.
16. LOCKE, JOHN. 1965/1815. An Essay Concerning Human Understanding. J.W. Yolton, Ed. Everyman Library. London. II. viii, p. 9.
17. SMITH, "Of primary and secondary qualities," p. 234.
18. See also WILSON, "History of philosophy," p. 220.
19. LOCKE, "An essay," IV.iii.II.
20. For an opposing view, see MACKIE, Problems from Locke, p. 27.
21. For early versions of force, see HARMAN, PETER M. 1982. Concepts of inertia, Newton to Kant. *In* Religion, Science, and World View: Essays in Honor of Richard Westfall. Margaret J. Osler & Paul Lawrence Farber, Eds.: 119–133. Cambridge University Press, New York.
22. BOSCOVICH, ROGER J. 1966/1763. A Theory of Natural Philosophy. MIT Press. Cambridge, MA.
23. See also SMITH, "Of primary and secondary qualities, pp. 245, 249.
24. MACKIE, Problems from Locke, p. 25.

Varieties of Properties

An Alternative Distinction among Qualities

JOSEPH E. EARLEY, SR.

Department of Chemistry, Georgetown University, Washington, DC 20057, USA

> ABSTRACT: The traditional distinction between primary (observation independent) and secondary (observation dependent) qualities is not based on a difference that can be sustained in the full light of contemporary scientific understanding. An alternative division of physical and chemical properties is proposed. Like the traditional division of qualities, the alternative system has two main categories. Properties of compound particulars that result from simple combination (e.g., addition) of the properties of their component parts constitute the first class: properties that depend on details of interactions between component parts (e.g. cooperative effects) make up the second type. Application of the alternative dichotomy is considered for the cases of mass (traditionally a primary property) and color (a secondary quality, in the usual division). Both these types of properties can fall in either of the two classes of the alternate division of qualities, depending on the nature of the interaction that occurs between components. Both mass and color show that intermediate cases occur. Application of the alternative categorical scheme is straightforward, but not always simple. The proposed system shows that in chemical combination (and, perforce, in the many more complex systems common in human culture) interactions profoundly influence properties of entities that enter the interaction. This is not adequately treated by philosophical theories of wholes and parts (mereology), which should be extended to apply to such important cases.
>
> KEYWORDS: properties; qualities; mereology; primary qualities; chemical combination; mixis; cooperative interactions; autocatalysis; dissipative structures; mixed valence; IVCT

Philosophers often use the term "property" in a very wide sense—to designate whatever can truthfully be said of something. For instance, Mary Kate McGowen holds that "[P]roperties are just what members of an extension share."[1] In a classic 1969 paper,[2] Hilary Putnam pointed out that such indiscriminate understanding of the concept of property conflates two notions that historically were carefully distinguished. He uses the term "predicate" for one of these two ideas, and (while soliciting suggestions for a more apposite name) employs the designation "physical property" for the other (by analogy with the well-established usage of "physical magnitude").

Address for correspondence: 6450 North 27th Street, Arlington, VA 22213. Voice: 703-532-5238; fax: 703-532-5238.
earleyj@georgetown.edu

In the first (predicate) way of using the word, a property is whatever may be said (predicated) of something, in the other (physical property) mode of discourse, a property is an intrinsic characteristic of an entity. The term "quality" seems close in meaning to Putnam's physical property. Examples of both types of usage can be found in this volume. When Paul Needham[3] discusses "substance properties," he clearly means the condition of being a substance—a predicate rather than a physical property or quality. In contrast, when S. H. Vollmer considers[4] the distinction between "primary and secondary qualities," and G. Krishnan Vemupalli[5] discusses the origin of molecular properties, they are dealing with what Putnam calls physical properties, rather than with mere predicates.

Charles Sanders Peirce (1839–1914) (who worked as a consulting chemical engineer after his retirement from government service)—held that a property is best understood as how a thing behaves, or would behave, in some operation.

> Consider what effects, that might conceivably have practical bearings, we conceive the object of our conceptions to have. Then, our conception of these effects is the whole of our conception of the object....let us ask what we mean by calling a thing *hard*. Evidently, that it will not be scratched by many other substances. The whole conception of this quality, as of every other, lies in its conceived effects.[6]

This characterization clearly does not apply to properties considered as applicable predicates, but does describe every sort of physical property (in Putnam's sense). It is important to note that (for Peirce) every property involves some definite operation—a procedure more or less clearly specified, or at least capable of being specified. The remainder of this paper concerns physical properties or qualities, rather than the larger class of true predicates.

Putnam also accepts[2] a further distinction—that between $physical_2$ properties (those dealt with by physics proper) and $physical_1$ (all other) properties. Properties that necessarily involve mention of two or more chemical species in their specification (e.g., hydrogen gas has the property of flammability, that is to say, dihydrogen burns in dioxygen to produce water) are what chemists call "chemical properties"—a subdivision of what Putnam calls "$physical_1$ properties."

THINGS ARE RECOGNIZED BY PROPERTIES

A maxim (apparently of unknown origin) that is now current in American politics makes a deep philosophical point: "If it walks like a duck, swims like a duck, and quacks like duck—*it's a duck*."

As Alfred North Whitehead (1861–1947) observed, "For physics, the thing itself is what it does."[7] It seems that this also applies in general, so far as physical properties (in Putnam's sense) are concerned. This point seems related to the Principle of the Identity of Indiscernibles of Gottfried Leibniz (1646–1716):

> No two substances are completely similar, or differ solely in number....This implies: if $x = y$, then from $G(x)$ we may infer $G(y)$ for any G, and conversely.[8]

This idea was clearly not original with Leibniz. Consider the Principle of Difference of Ramanuja (India, c. 1070 C.E.)

> Should anyone...assert [that] the theory of a substance free from all difference is immediately established by one's own consciousness, we reply that he is...refuted by the fact...that all consciousness implies difference.[9]

In various ways, each of the four statements quoted above indicates that we recognize things (and get to "know" them, in so far as we do) by their properties.

ARE THINGS *MADE OF* PROPERTIES?

The point made in the previous section is sometimes interpreted in what seems to be a very odd way. Some contemporary philosophers consider that things are *nothing but* aggregations of properties. For instance, one of the central doctrines of a work in which Robert Neville aims to set out a philosophy of nature that would also ground hermeneutics is: "...determinate things are harmonies of essential and conditional features."[10] Also, David Weisman recently set out to frame "a metaphysics"—consistent with what he takes to be current physical science—that will also apply to concerns of the social sciences. One of his basic statements is similar to Neville's presupposition quoted above, but even more clearly (though perhaps less cautiously) phrased: "Everything is constituted, exclusively, of its properties."[11]

Notions of this sort have a long history. George Berkeley (1685–1783) held that all objects are "collections of ideas."[12] Bertrand Russell (1872–1970) held that things are bundles of universals. D. M. Armstrong concludes that the "bundle of universals" doctrine is not likely to be sound, since no one has yet devised a plausible account of "...a bundling principle that holds together the properties 'of' a particular."[13]

Whatever else may be the case, the idea that things are composed of properties is not a feature of contemporary physical science. An example of the cool reception that this notion receives in scientific circles is that a main justification for postulating the Higgs Field in high-energy physics is to avoid having free-floating properties.[14] No matter what sort notion of property one employs, it seems clear that (both in general speech and in scientific discourse) properties are generally *ascribable to entities*. In both popular and scientific usage, every property has reference to at least one thing. In that sense, properties are abstractions from existing (or imagined) things. On that basis, it is difficult to understand *how* a property or a feature could be *prior* (either temporally or ontologically) to *that of which* it is a feature. If properties require some entity in which to inhere, it might seem that Weisman, and possibly Neville as well, have fallen into *the fallacy of misplaced concreteness*: "...the error of mistaking the abstract for the concrete."[15]

TRADITIONAL DISTINCTION BETWEEN PRIMARY AND SECONDARY QUALITIES

There is an ancient tradition of dividing properties into a small number of classes. Aristotle (384–382, B.C.E.) states:

> Quality then seems to have practically two meanings, and one of these is more proper. The primary quality is the differentia of substance.... Secondly, there are the modifications of things in motion *qua* in motion....[16]

Galileo Galilei (1564–1642) distinguished two sorts of properties: primary qualities (e.g., spatial extension)[a] were held to be intrinsic characteristics of the entity—secondary properties (color, taste, etc.) depended on interaction with some perceiving subject.

Now I say that whenever I conceive some material or corporeal substance, I immediately feel the need to think of it as bounded, as having this or that shape, or being large or small in relation to other things, and in some specific place at any given time; as being in motion or at rest; as touching or not touching some other body, as being one in number or few or many. From these conditions, I cannot separate such a substance by any stretch of my imagination. But that it must be white or red, bitter or sweet, noisy or silent, of sweet or foul odor, my mind does not feel compelled to bring in as necessary accompaniments.... Hence I think that tastes, odors, colors, and so on are no more than mere names so far as the object in which we place them is concerned, and that they reside only in the consciousness. Hence if the living creature were removed all these qualities would be wiped away and annihilated.[17]

A somewhat similar division was employed by John Locke (1632–1704):

The ideas *that make up our complex ones of corporeal substances* are of three sorts. *First*, the *ideas* of the primary qualities of things, which are discovered by our senses, and are in them even when we perceive them not; such are the bulk, figure, number, situation, and motion of the parts of bodies which are really in them, whether we take notice of them or no. *Secondly*, the sensible secondary qualities which, depending on these, are nothing but the powers these substances have to produce several *ideas* in us by our senses; which *ideas* are not in the things themselves otherwise than as anything is in its cause. *Thirdly*, the aptness we consider in any substance to give or receive such alteration of primary qualities, as that the substance, so altered should produce in us different ideas from what it did before....[18] [Locke's emphasis]

The traditional primary–secondary distinction clearly refers to what Putnam calls physical properties, rather than to predicates.

Locke's primary/secondary distinction was criticized roundly by Bishop George Berkeley,[19] who maintained (as does S. H. Vollmer, in this volume) that even the so-called primary properties depend in some way on perception, so that the primary–secondary distinction collapses for want of a fundamental basis in fact.

A PROPOSAL FOR AN ALTERNATIVE DISTINCTION AMONG QUALITIES

Just before the passage quoted above, Locke makes clear how (and to what extent) he understood *substance*:

So that if anyone will examine himself concerning his *notion of pure substance in general*, he will find that he has no other idea of it at all but only a supposition of he knows not what support of such qualities as are capable of producing simple *ideas* in us; such qualities are commonly called accidents.[20]

Many aspects of the material universe are now much more clearly understood than they were when Locke wrote about that "supposition of he knows not what." In particular, there no longer seems to be any great mystery about the formation and properties of simple chemical compounds—substances, in the chemical sense. Papers in this volume[21] and elsewhere[22] demonstrate that chemists have good understanding of the sorts of closure of networks of relationships between components that generate and sustain chemical substances.

In a contemporary discussion of the meaning of the concept of substance, Ruth Garrett Millikan holds:

---

[a]I am indebted to Prof. Daniel Rothbart for pointing out that Galileo, in contrast to Newton and later authors, did not consider mass as a primary property.

> Substances...are whatever one can learn from given only one or a few encounters, various skills or information that will apply to other encounters..... Further, this possibility must be grounded in some kind of natural necessity.... The function of a substance concept is to make possible this sort of learning and use of knowledge for a specific substance.[23]

Chemists understand that, rather than being made up of properties, features, or qualities, chemical entities are composed of components—constituent entities that interact in specific ways which go to produce (through "natural necessity") integrations that persist through time and sustain repeated interactions—chemical structures.[24] If we adopt this[b] outlook, what becomes of the primary/secondary property distinction?

If we are persuaded by Berkeley (and by Vollmer) that assessing any and all properties depends on the means used for the assessment, so that the traditional dichotomy is no longer tenable, is there a satisfactory alternate basis on which we could we make a distinction among main classes of properties? I suggest that there is firm basis in contemporary science for an alternative distinction between types of qualities (physical properties). This newer dichotomy is similar in some respects to the traditional division between primary and secondary qualities—but has a quite different fundamental basis. The alternative distinction that I favor is between those properties of an individual particular that are the simple resultant (a sum, say) of aggregation of corresponding properties of the component parts of the entity in question and properties that are not such resultants. On the basis of the current state of scientific understanding, this division of qualities—between those that result from simple combination of properties of components (let's call this "the first class") and those that depend on the components in more complex ways ("the second class")—is as clear as the traditional dichotomy, and may well be more useful.

The question might arise as to what meaning such words as aggregation, combination, or summation might have in this context.[c] One general answer to this question involves a "thought experiment" in which a particular is constructed from the same components as compose an entity of interest, in such a way that the combination gives rise to no change or modification in any of the components so combined. (This is what logicians call a "mereological sum.") For many qualities (weight, for example) the properties of the hypothetical mereological sum often can be computed directly from the properties of the components. To the extent that the value of the property of the mereological sum, however reckoned, is the same as the corresponding property value of the actual particular of interest, that property should be considered a property of the first class in the new division. If the quality of the actual particular of interest differs from that computed for the mereological sum, then that property should be regarded as a property of the second class.[d] In the case of some qualities (such as viscosity), it may not be possible (in view of the best scientific understanding available at the time) to compute (or possibly even to conceive) how

---

[b]Contrary to what was generally thought some years ago (and is still maintained by Weisman) there is no theoretical or experimental evidence in contemporary science for considering that there is some fundamental level of description—some set of simple particles (or fields) from which all higher level entities are built. There are reasons that are at least as good for holding that every concrete entity has parts and that each and all of those parts also are composed of components.

[c]I am grateful to Prof. Daniel Rothbart for raising this point.

qualities of individual components might be aggregated to yield the property of the composite particular. In such cases, the property in question should be assigned (provisionally) to the second class.

APPLYING THE PROPOSED ALTERNATIVE DISTINCTION

Some properties of the compound individuals that chemists deal with do, in fact, result from mere summation of the properties of the components of those complex particulars. Other properties depend on specific *interactions* between components—whether those components are like or unlike—rather than simply on the properties of the components (as individual particulars). Cooperative action of chemical individuals occurs quite commonly, and often with great effect. An adequate survey of cooperative activity of chemical species is not possible here, but a few examples may be mentioned.

Hardness and fragility of salt crystals depend in complex ways on relationships between the positive and negative ions that are components of the crystal—and often on concerted action of myriads of ions (*martinsitic* transitions).[26] The pattern of scattered radiation that emerges from a crystal irradiated with X-rays depends on the microscopic structure of the crystal—the relative arrangement of the component atoms.[27] The details of the NMR spectra of organic molecules depends on interactions of all the components atoms of the molecules.[28] Femtosecond spectroscopy is made possible by cooperative interactions of molecules.[29] The density variations that result in ice floating in water cannot be understood without attention to highly cooperative interactions of water molecules both in liquid water and in ice.[30] Origin of complex spatial and temporal order in far from equilibrium open chemical systems[31] depends on complex networks of cooperative action between various components of those mixtures. Colors (spectra) of dissolved species often depend in subtle ways on cooperative interactions of myriads of solvent molecules.[32] All of these qualities are not simply computable from the properties of the individual particulars (components) involved. To the extent that this is the case, all these are examples of properties of the second class in the proposed new division.

As an indication of how the proposed division of properties might work, let us consider mass (a standard example of a traditional primary quality) and color (a frequently discussed secondary property). Every student of chemistry learns to compute the molecular weight of organic compounds and the formula weight of ionic materials. Knowing the weight of all the components of a chemical species, one can easily compute the total (molecular or formula) weight. Within an accuracy that is quite sufficient for the many manipulations carried out in university chemistry laboratories, the mass property is simply additive. On this basis, mass is property of the first class, at least so far as simple chemistry goes. The alert reader has probably al-

---

[d]I thank Prof. Rom Harré for pointing out that the proposed division is similar to the distinction—used in biology and philosophy of biology—between properties ascribable to individual organisms and characteristics that result from interactions between organisms in groups, and that the proposed division is also, if less directly, related to some psychological distinctions used by Polanyi, for example, that between "focal" and "subsidiary" awareness.[25]

ready thought of the objection that helium atoms (such as those formed in the Sun by the Bethe-Weizsacker mechanism[33]) are substantially lighter than the total mass of the four hydrogen atoms that may be regarded as the components of those helium atoms—the units from which the helium atoms arise in the Sun, say. This weight difference is due to the high energies that are characteristic of internuclear bonding—when the helium atom is synthesized from four protons, mass is converted into intranuclear binding energy. Even though mass is not conserved in nuclear reactions, conservation of mass generally applies to a very high degree of precision in chemical contexts. Energies involved in chemical bonds are many orders of magnitude smaller than the energies involved in intranuclear interactions, so that (within the precision of all but the most refined measurements) the total mass of a chemical molecule may be taken as the sum of the masses of the component atoms. Mass is not a quality of the first class (simply additive) in some absolute sense, but may quite properly be regarded as a quality of the first class for ordinary purposes of chemistry.

Color is perhaps a more interesting case. Is the color of a compound simply the sum of the colors of the components? There are some cases in which color is an additive property and some in which color is not additive. Aqueous solutions of salts of transition metals (copper, chromium, iron, etc.) are customarily used for discussing colors with beginning students. Is the color of an aqueous solution of copper sulfate simply the sum of the colors of the components of that solution? Such solutions follow Beer's law—the absorbance of light (at a fixed wavelength near the red end of the visible spectrum) by a solution of cupric ion is strictly proportional to the concentration of cupric ion in solution. In this case, color is just as additive as is the mass of the components of organic molecules. But then what becomes of the distinction—what might correspond to a quality of the second type?

Colors of many intensely colored materials relate to the colors of their components in ways that are quite different from the simple additivity characteristic of cupric ion solutions. Chemicals that contain two or more metals ions in different states of oxidation (mixed-valence species) can undergo what is known as Intervalence Charge Transfer (IVCT). This is a process in which the molecule changes from a state with one particular distribution of electronic charge to a state with a quite different charge distribution. (In some cases it is appropriate to consider that this shift corresponds to the movement of a single electron from one metal center to another.) For reasons that are quite well understood,[34] but outside the scope of this paper, the IVCT transition can correspond to very high absorption of light—some (but not all) mixed valence species are exceedingly highly colored. Although the intensity of the blue color of dilute aqueous solutions of cupric sulfate depends linearly on the concentration of copper ions—indicating simple additivity, the colors of the dark coloring matter of inks (e.g., Prussian Blue, Turnbull's Blue) depend not on single ions but on *interactions* between two different ions. In such cases doubling the concentration of the metal ions may increase the color by factors of about four times rather than causing simple doubling.

Robin and Day[35] famously distinguished three types of mixed valence materials—substances that contain two or more metallic ions in various states of oxidation. For Class I complexes the UV-visible spectrum (color) of the complex is the sum of the corresponding spectra (colors) of the components. (In this case color is a quality of the first type in the new division.) In compounds of Robin and Day's Class II, the spectra of the components are present, but a new set of spectral features (color) is observed that is characteristic of the intervalence charge transfer (IVCT) of the specific combination of metal ions present. In Class III compounds, the new IVCT bands are intense enough to make the colors of the components unobservable.[35] The colors of Robin and Day Class III mixed-valence species are quite different in origin from the colors of corresponding Class I substances. In the case of Class I substances, the color of the compound is directly predictable from the colors of the components—color is a property of the first class. Conversely, there is no direct relationship between the spectra of Class III complexes and the colors (spectra) of their components—in these cases color is a property of the second class on the alternative division.

Notice that it is not possible to say that property F is (simply) a first-type property, while property G is (simply) a property of the second type. Nevertheless, in any concrete case it should be a straightforward task to decide whether a particular property of a certain compound particular corresponds to the appropriate sum of the corresponding properties of all of the components of that individual, or does not so correspond. On this basis, it should be possible to determine in every case, for chemical species x, y, and z, that $F(x)$ is of the first type, $F(y)$ is of the second type—perhaps $F(z)$ might be intermediate between the two classes (e.g., the colors of Robin and Day class II mixed-valence compounds.) To the extent that this is correct, the proposed alternative distinction is clear and practically applicable.

CUI BONO?

So, what's the point? A major question remains: who would benefit by the introduction of this alternative distinction? Scientists make distinctions of this general sort quite routinely and unselfconsciously. There seems to be little for scientists to learn through this distinction. But it seems fair to say that most philosophers have yet to recognize that, when components enter into chemical combination, those components do not, in general, maintain the same identity that they would have had absent that combination. In spite of the recognition of this problem by Aristotle and the scholastics,[37] by Hegel,[38] and by John Stuart Mill,[39] contemporary philosophers often tend tacitly to assume that all properties are properties of the first (additive) class.[40] When interaction between components is strong (as it often is in chemical combination—such as in Robin and Day class III species), properties of the second type predominate. Interactions of parts within wholes also take place in astronomical, biological, psychological, economic, sociological, linguistic, and conceptual systems. These part-whole relationships are all at least as complex as chemical combination.

Integration of such insights with the philosophical study of wholes and parts (mereology) is only in its initial stages.[41] It would be useful to develop a mereology adequate to deal with chemical systems, in order to facilitate future progress in dealing with other and more complex problems. It seems that one major advantage of use of the proposed distinction would be to facilitate development of the theory and practice of this aspect of philosophical logic in this direction—to specify how that could be achieved is clearly outside the scope of this paper, but is indicated elsewhere.[42]

NOTES AND REFERENCES

1. MCGOWEN, M. 2002. The neglected controversy over metaphysical realism. Philosophy **77:** 5–21.
2. PUTNAM, H. 1969. On properties. *In* Essays in Honor of Carl G. Hempel. N. Rescher, *et al.*, Eds. Reidel. Dordrecht. *Reprinted in* 1999. Metaphysics: An Anthology. J. Kim *et al.*, Eds.: 243–252. Blackwell's. Oxford, UK.
3. NEEDHAM, P. 2003. Chemical substances and intensive properties. Ann. N.Y. Acad. Sci. This volume.
4. VOLLMER, S.H. 2003. The primary properties? Ann. N.Y. Acad. Sci. This volume.
5. VEMULAPALLI, G. 2003. Property reduction in chemistry: some lessons. Ann. N.Y. Acad. Sci. This volume.
6. PEIRCE, C. 1878. How to make our ideas clear. Pop. Sci. Monthly *Reprinted in* 1955. Philosophical Writings of Peirce. J. Buchler, Ed.: 31. Dover. New York.
7. WHITEHEAD, A. (1933) 1967. Adventures of Ideas. Macmillan. New York. p. 157.
8. SUPPES, P. 1957. Introduction to Logic. Van Nostrand. Princeton, NJ. p. 103.
9. RADHAKRISHNAN, S., *et al.*, Eds. 1957. A Source Book in Indian Philosophy. p. 543. *Quoted in* Rom Harré. 2000. One Thousand Years of Philosophy. Blackwell. Oxford, UK. p. 66.
10. NEVILLE, R. 1989. Recovery of the Measure. SUNY Press. Albany, NY. p. 104.
11. WEISMAN, D. 2000. A Social Ontology. Yale University Press. New Haven, CT. p. 26.
12. BERKELEY, G. 1710. Section 1. A Treatise Concerning the Principles of Human Knowledge. *Discussed in* LUCE, A.A. 1968. Chapter III. *In* Berkeley's Immaterialism. Russell and Russell. New York.
13. ARMSTRONG, D. 1989. Universals: An Opinionated Introduction. Westview. Boulder, CO. p. 74.
14. EARMAN, J. 2003. Laws, Symmetry, and Symmetry Breaking. (Presidential Address, Philosophy of Science Association, November 9.) Philosophy of Science. In press.
15. WHITEHEAD, A. 1925/1967. Science and the Modern World. Macmillan. New York. p. 51.
16. ARISTOTLE. Metaphysics V, 14: 1020a 14–18. *In* The Complete Works of Aristotle. J. Barnes, Ed. 1984. Princeton University Press. Princeton, NJ. p. 1611.
17. GALILEO, G. 1623. The Assayer. *Reprinted in* STILLMAN DRAKE. 1957. Discoveries and Opinions of Galileo. Doubleday. New York. p. 274.
18. LOCKE, J. 1690. An essay concerning human understanding; edited with an introduction by John W. Yolton. Book II, Chapter XXXIII. *Reprinted in* JOHN Y. YOLTON, Ed. 1964. 249. Dutton. New York.
19. BERKELEY, G. 1710. Sections 10–15. A Treatise Concerning the Principles of Human Knowledge. *Discussed in* LUCE. 1968. Chapter VII. *In* Berkeley's Immaterialism.
20. LOCKE, J. 1690. An Essay Concerning Human Understanding. Book II, Chapter XXXIII. *Reprinted in* YOLTON, p. 245. See also: PETER ALEXANDER. 1985. Ideas, Qualities and Corpuscles: Locke and Boyle on the External World. Cambridge University Press. Cambridge, UK.
21. HEELAN, P. 2003. Paradoxes of measurement. Ann. N.Y. Acad. Sci. This volume.
22. EARLEY, J. 2003, forthcoming. Constraints on the origin of coherence in far-from-equilibrium chemical systems. Chapter 6. *In* Physics and Whitehead: Quantum, Process and Experience. Timothy Eastman *et al.*, Eds. State University of New York Press. Albany, NY. *Idem.* 2003, forthcoming. On the relevance of repetition, recurrence, and reiteration. *In* Chemistry in the Philosophical Melting Pot. Ewa Zielonaka-Lis *et al.*, Eds. Peter Lang. Dordrecht, NL. *Idem.* 2003, forthcoming. Philosophical implications of chemical symmetry. *In* Philosophy of Chemistry: Synthesis of a New Discipline, Boston Studies in the Philosophy of Science. Davis Baird *et al.*, Eds. Kluwer. Dordrecht, NL. *Idem.* 2000. Varieties of chemical closure. *In* Closure: Emergent Organizations and Their Dynamics. Jerry Chandler *et al.*, Eds. Ann. N.Y. Acad. Sci. **901:** 122–131.
23. MILLIKAN, R. 2000. On Clear and Confused Ideas: An Essay on Substance Concepts. Cambridge University Press. New York.

24. HARRÉ, R. 2003. Structural explanation in chemistry and its evolving forms. Ann. N.Y. Acad. Sci. This volume.
25. POLANYI, M. 1965. The structure of consciousness. Chapter 13. *In* 1969. Knowing and Being. Marjorie Grene, Ed. The University of Chicago Press. Chicago.
26. KHACHATURYAN, A. 1983. Theory of Structural Transformations in Solids. Wiley. New York.
27. HAMMOND, C. 1998. The Basics of Crystallography and Diffraction. Oxford University Press. Oxford, UK.
28. ROBERTS, J. 2003. Negotiated identities of chemical instrumentation: the case of nuclear magnetic resonance spectroscopy, 1956–1969. Ann. N.Y. Acad. Sci. This volume.
29. ZEWAIL, A. Ed. 1994. Femtochemistry: Ultrafast Dynamics of the Chemical Bond World Scientific. Singapore.
30. HORNE, R. 1971.Water and aqueous solutions: structure, thermodynamics, and transport processes. Wiley-Interscience. New York.
31. PRIGOGINE, I. 2003. Chemical kinetics and dynamics. Ann. N.Y. Acad. Sci. This volume.
32. WANG, C. *et al.* 1998. Solvent control of vibronic coupling upon intervalence charge transfer excitation of $(CN)_5FeCNRu(NH_3)_5^-$ as revealed by resonance raman and near infrared absorption spectroscopy. J. Am. Chem. Soc. **120:** 5848–5849.
33. BETHE, H. 1939. Energy production in stars. Phys. Rev. **55:** 434–456.
34. HUSH, N. 1967. Prog. Inorg. Chem. **8:** 391. *Idem.* 1968. Electro-chim. Acta. **13:** 1005.
35. ROBIN, M. & PETER DAY. 1967. Adv. Inorg. Chem. Radiochem. **10:** 247. Although these authors phrase (page 257) their formal definition of the three classes in terms of the value of α, the quantum mechanical parameter that measures electron delocalization between the two metal ions involved, they also note (page 259, ff.) that the more "operational" definition given here is fully equivalent to that more theoretical definition.
36. SMITH, P. *et al.* 1971. The crystal and molecular structure of di-ì-oxo-bis (pentaammineruthium)bis(ethylenediamine)ruthenium hexachloride: the ethylenadiamine analog of "ruthenium red." Inorg. Chem. **10:** 1943–1947.
37. WEISBERG, M. & R. WOOD. 2003. Richard Rufus's theory of mixture: a medieval explanation of chemical combination. Ann. N.Y. Acad. Sci. This volume.
38. NOWAKI, M. & W. VER EECKE. 2003. The superiority of "chemical thinking" for understanding free human society according to Hegel. Ann. N.Y. Acad. Sci. This volume.
39. CARTWRIGHT, N. 1989. Nature's Capacities and Their Measurement. Clarendon Press. Oxford, UK. p. 163.
40. GOODMAN, N. 1977. The Structure of Appearance. 3rd Ed. Reidel. Dordrecht, NL. p. 33ff.
41. EARLEY, J. 2003 (expected for December). How dynamic aggregates may achieve effective integration. *In* Advances in Complex Systems (special issue on emergence in chemical systems). Jerzy Mazelko, Ed.
42. NEEDHAM, P. 2003. Chemical substances and intensive properties. Ann. N.Y. Acad. Sci. This volume. This paper makes a start on adapting received versions of mereology to chemistry, but much remains to be done.

Property Reduction in Chemistry
Some Lessons

G.K. VEMULAPALLI

Department of Chemistry, The University of Arizona, Tucson, Arizona 85721, USA

ABSTRACT: By taking molecules and atoms as models for wholes and parts, I explore the relation between the atomic and molecular properties in this article. Molecular properties do not reduce to atomic properties in the sense the former cannot be obtained by adding or multiplying the latter. However, quantum states of molecules are obtained from the quantum states of atoms by addition and multiplication. Because molecular properties are calculated reliably from the atomic states in quantum theory, it may be claimed that property reduction has been accomplished successfully. I show that such reduction entails underlying holism.

KEYWORDS: wholes and parts; molecules; atoms; molecular particles; quantum states; holism

INTRODUCTION

Relating parts to wholes is an important aspect of methodologies used in scientific investigation and in explanation of the results of those investigations. In this context it is understood that parts can either have separate, stable existence or lose their individuality to become a whole, depending on the environmental conditions. It is also understood that parts may be further divisible so that a part may be considered to be a whole in different contexts of scientific investigation. In general a whole is partitioned more than one way in chemical practice. If we consider the molecule H_2O as a whole, one choice for parts is the three atoms. Another choice may be the ions H^+ and OH^-. Scientific methodology often involves exploring the unknown by the most convenient route for the problem at hand, starting with the known. Hence chemists do not always start with most elementary particles in building theories.

Researchers often try to understand the properties of wholes by assiduously studying the properties of parts. (In this article "properties" refer to physical and chemical properties that scientists measure or derive from theory.) Indeed the word "understood" is often used to imply that a relation between properties of parts and properties of a whole has been established. Yet it is often argued that the properties of a whole cannot be reduced to those of the parts and that a new class of properties "emerge" when "parts" combine to form a whole. For example Redhead states,[1] after

Address for correspondence: Department of Chemistry, The University of Arizona, Tucson, AZ 85721. Voice: 520-621-6350; fax: 520-621-8407.
gkv@u.arizona.edu

discussing the characteristics of the quantum wave function for a two-particle system: "The reductive hierarchy founding the whole of science on the properties of individual elementary particles is thus shown to be mistaken."

The word "reduction" appears in varieties of contexts. Thus we read that thermodynamics reduces to statistical thermodynamics, chemistry to physics, mental states to physical states, just to cite a few examples of usage. In this article I will use the word reduction solely in the context of parts–whole relation. In my use of the word, if a property of the whole is shown to be a necessary consequence of parts having certain properties, then that property of the whole will be said to be reduced to properties of the parts.

If we accept both the scientists' penchant for reductive strategies and the doubts of some philosophers whether such reduction has ever been demonstrated, we have to conclude either (a) scientists are fooling themselves and scientific methodology resembles that of a drunk, searching for a lost object under a streetlight, not because he had lost it there but because he could only "observe" there, or (b) something is missing in our general understanding of the relation between parts and wholes.

I want to explore this puzzle in this article. One of the virtues of philosophy of chemistry and material science is that it allows us immense opportunities for case studies. It may be argued that one can never build a broad theory from the study of individual cases, missing the forest for the trees. It is the case studies, however, that uncover the flaws in a broad theory and keep philosophy—and science—honest. There is a dearth of specific case studies in philosophy of science. Thus in the extensive literature on reduction we see one statement quoted repeatedly: the mantra that temperature is mean kinetic energy, which is only true provisionally. Even among the few case studies available for our perusal, many are concerned with *Gedanken* experiments where the system of interest exists in ethereal world, every real instrument used to probe it banished from discussion. Philosophical analysis of cases of reduction in chemistry and material science can and must remedy the situation.

Another important reason for studying property reduction in chemistry is that this may serve to illuminate the vast literature on reductive and nonreductive physicalism in connection with the mind–body problem. These discussions often assume that a reductive model is valid for chemistry, which is commonly used as a paradigm to contrast with the failure of reduction in biology or in the mind–body problem. Thus Kuppers writes[2]:

> Although the Schrödinger equation cannot be solved for complex molecules, a solid reductionist will have no serious doubts about his working hypothesis that chemistry can be reduced in principle completely to the laws of physics.

Later, while discussing the relation between physics and biology he comments:

> 1. The whole is more than the sum of its parts.
> 2. The whole determines the behavior of the parts.
>
> From the first thesis follows the concept of emergence. From the second thesis follows that of downward causation.

Are there two different principles concerning reduction—one applicable to chemistry and another to biology? Before we can be confident about any generalizations in mind–body problem or in physics–biology reduction shouldn't we thoroughly examine the simpler case of problems encountered in deducing chemical properties from the laws of physics?

Diverse concepts like emergence, supervenience, and holism appear in the discussion of parts–whole relationship. I find the assumed models in which these concepts are applied irritatingly vague when the "mechanism" used to analyze parts–whole relations is not made explicit. In the following section, I will examine the actual mechanism by which properties of wholes (molecules) are estimated and calculated from the properties of parts (atoms).

Much of what is said here would appear to be the standard stock-in-trade for students of physics and chemistry. My aim in reviewing the familiar material is devious. I will show that there are surprises here worth our attention and that, as we are about to corral our ideas of parts–whole relation into confines of one *ism*, they morph into something else.

WARNING AGAINST SOME PRECONCEIVED NOTIONS

Let us first consider the trivial case of a whole with N noninteracting parts. The properties of the whole would simply be the weighted sum of the properties of the parts. This is essentially what is observed in ideal solutions in chemistry. I suspect that this idealized model of weighted sums has been often implicitly assumed as the only acceptable model for parts–whole relation without further examination. For instance, it was assumed at one time that each gene functions by coding for a single protein, so that in heredity the genome could be considered to function as a sum of individual noninteracting parts (genes). From this perspective it was surprising to discover that the number of genes in human genome is only approximately twice that in yeast cells. The surprise, I suggest, was because weighted sum relation was assumed as a model for functioning of whole genome, whose parts being the individual genes.

Before we can intelligently discuss whether a property of whole is reducible to the properties of the parts or not, we must examine the model assumed for the cooperative behavior of the parts. If there is no cooperation, the analysis is trivial, as we have just seen. Let us consider another model in which all parts cooperate to the maximum. Suppose, for illustration, we assume that every gene is involved in the production of a protein, and it is their sequential turning on/off that codes for a protein. Thus genes functioning in the sequence 1–2–3… will lead to production of a different protein than when they function in the sequence 2–1–3…. In that case the number of proteins produced would be N!, where N is the number of genes (parts) in the genome. This is significantly different than N, the weighted sum model predicts. Thus an organism with 30,000 genes could differ quite significantly from another with, say, two fewer genes, and an organism with fewer genes could be a more complex whole. We see from the independent and cooperative models considered here that the property of the whole (genome)—the number and type of proteins whose production it can facilitate—depends on the level of cooperation between its parts. I am not suggesting that such complete cooperation among the genes is the correct model for protein synthesis in cells. The main argument here is that the relation between the properties of the parts and wholes cannot be properly analyzed unless we also examine the model used in relating.

FROM ATOMS TO MOLECULES

My aim is to be very specific before raising general questions. Hence I start with a list of properties that form the basis for my discussion.

- Mass*
- Bond length
- Bond strength
- Geometry (bond angles, isomers)
- Vibrational frequencies
- Rotational constants
- Electronic spectrum*
- Polarizability*
- Dipole moments
- Reactivity*

The properties identified with asterisks (to be referred as class I properties) are common to atoms and molecules. The rest (class II properties) are exhibited only by molecules. Yet it will be inappropriate to call them emergent properties since they are, in some sense, expected in a collection of atoms forming a stable arrangement. When the degrees of freedom of each atom are coupled to those of other atoms, we expect some of the degrees of freedom to manifest as internal degrees of freedom, that is, atoms moving relative to each other around a center of mass. If two nonidentical atoms form a molecule, we expect charge to be nonuniformly distributed, resulting in a dipole moment for the molecule. Class II properties, qualitatively speaking, are relational properties in the sense discussed by Teller.[3] Let us examine how these properties are quantitatively accounted for in chemistry.

H_2^+ MOLECULE

The simplest molecule is the hydrogen molecular ion, H_2^+, which consists of two protons and an electron. In the language of physicists, the molecule presents a three-body problem. Three body problems cannot be solved analytically without approximations. Indeed a two-body problem cannot be solved, unless it is reduced to a one-body problem. And one-body problem cannot be solved unless we ignore the ever-present radiation field. Thus if we insist that property reduction must involve a rigorous mathematical algorithms to connect atoms to molecules, reduction becomes impossible. Yet we know that chemists have been able to successfully calculate molecular properties starting with atoms. What is it they are doing? And how are they doing it?

According to the Schrödinger formulation of quantum theory, all the information we are likely to know about a system in a state is encoded in its wave function for that state. Thus property calculations depend on first finding the wave function for the system of interest in a particular state. This has been successfully accomplished with different sets of approximations, the relative merits of which we need not examine here.

Returning to the hydrogen molecular ion, the Schrödinger equation for an electron moving in the potential of two protons is solved in an elliptical coordinate system. The properties calculated from the wave functions obtained by solving the Schrödinger equation agree quite well with the experimental results. We will not, however, follow this theory, since it is a special case that cannot be extended for other molecules. Besides, the theory is holistic, since it starts with the whole molecule and not with the atoms.

Chemists use two techniques to construct the molecular wave functions from atomic functions. These are the molecular orbital (M.O.) and valence bond (V.B.) approximations. (It is regrettable that each set of approximations is frequently referred to as a "theory" in chemical literature.) According to the M.O. approximation, the wave function for electron in an H_2^+ molecule is a properly weighted sum of atomic orbitals centered on each nucleus. The most general expression for the wave function for electrons in the H_2^+ molecule is:

$$\psi = \sum_i \lambda_i [\varphi_{iA} + \varphi_{iB}] \tag{1}$$

Here φ_{iA} and φ_{iB} are ith identical atomic orbitals centered on protons A and B, respectively. The weighting coefficients are denoted by λ_i. These coefficients depend on the state of H_2^+ molecules. The crux of the above equation is the Fourier theorem according to which any arbitrary function in region R is a sum of a complete set of functions in R.

In actual calculations a truncated set of functions is used instead of a complete set. Other mathematical approximations are also involved. However, based on all the investigations reported so far, we can be confident that the properties of H_2^+ (bond length, bond strength, polarizability, vibrational spectrum, etc.) in its lowest energy electronic state are calculable within the experimental error from ψ shown in the above equation.

What has been said so far should be familiar to any physicist or chemist. If, however, we examine the above procedure for molecular property calculations from a philosophical point of view, we find some ideas that have to be carefully examined. Let me state the results:

- Class II properties are unique to the whole and they are not, even qualitatively speaking, related to any properties of atoms. I do not consider class II properties as emergent for reasons mentioned earlier.
- Among the class I properties, common to whole and its parts, only mass is additive.
- Other class I properties (e.g. polarizability) of the whole cannot be deduced from those of the parts by mathematical operations of addition and multiplication.
- Considering the above three points, we might be tempted to say that properties of whole (except in case of mass) are not reduced to those of the parts.
- Nevertheless, both class I and class II properties of the whole are calculated from the quantum state of the whole. The quantum state for H_2^+ is derived from the quantum states of the parts by simple addition.

We may conclude from these observations that while properties of the whole are not the sums or products of the properties of the parts, the states of the system can be obtained by adding the states of the parts. Because the properties in turn can be derived from the states, it appears that we have shown that properties of wholes are completely determined by parts. But there are two problems here: (1) It is true that the states of the system are composed of the states of the parts, but there are also weighting factors in the composition. These are the constants λ in the linear combination. What factors determine these constants? They are certainly not determined by the states or properties of the parts. (2) Just as in the case of molecular wave function, an atomic wave function may also be represented by a sum of an arbitrary set of functions. Thus one can claim that an atomic function is a linear combination of molecular functions or atomic states (parts) reduced to molecular states (wholes)! In the relation between atomic and molecular states, which are the reduced states? We will return to these questions after examining two other molecules, when we encounter novel features.

H_2 AND H_2O MOLECULES

Consider the hydrogen molecule. Now we have two electrons and two commonly used approximations: the molecular orbital (MO) and valence bond (VB) models. The difference between the two "theories" is not as wide as it appears at first look. For purpose of our discussion, it is sufficient to have a simple picture of these theories. In molecular orbital procedures, atomic orbitals from the two atoms are first added as in the case of H_2^+. Products of the molecular orbitals, one for each electron, are used to represent the state. In further refinements these product functions are added. In valence bond theory, products of atomic orbitals from different atoms are added to give what are often called structures. It is a sufficiently accurate statement to say that we add first and then multiply in MO theory; in VB theory we multiply first and add later. It is easy to see why both procedures give the same state for the molecule if we take a sufficient number of atomic orbitals in the calculation. Hence I will claim that the lowest electronic state of the H_2 molecule is given by a function of the form:

$$\psi = \sum \lambda_i [\varphi_{iA}(1)\varphi_{iB}(2) + \varphi_{iA}(2)\varphi_{iB}(1)][\alpha(1)\beta(2) - \beta(1)\alpha(2)] \tag{2}$$

Here, the numbers in the brackets indicate the electrons; α and β are the spin functions; A and B denote the two protons. The above state function is consistent with the requirement that a multielectron wave function must be anti-symmetric to exchange of electrons (Pauli's principle).

Since the pioneering work of Heitler and London (VB approximation) and Hund and Mulliken (MO approximation), the quantum theory of bonding in the H_2 molecule has been studied extensively. We can confidently expect that a wave function of the above form would allow us to calculate many of the properties of the molecule to the level of accuracy only sophisticated experiments can provide.

We see by examining Eq. 2 that the state of the H_2 molecule is determined by the states of the atoms through a relation that involves both products and sums. In that sense we can say that the state of the whole is reduced to the states of the parts, unless we insist that only additive or multiplicative operations are allowed in reduction. However, we still have to decide on what factors actually determine the weighting coefficients λ in the above equation.

The above equation shows that the two "particles" are entangled. (Strictly speaking, there are no particles in quantum theory, only wave-packets.) As the separation between two H atoms increases to an arbitrarily large distance, the coefficients λ change, but the wave function remains anti-symmetric to exchange and the states of the two particles remain coupled. This is one of the puzzles of quantum theory on which a great deal has been written. We need not concern ourselves with this extensive literature, because our interest is in finding how the properties of parts are related to wholes.

Calculations made with the aid of wave functions of the sort shown above demonstrate that we can indeed accurately predict the properties of the whole (H_2) from its parts (H) through coupling of their states. The procedure for coupling is well established. If by property reduction we mean a procedure by which properties of wholes are deduced from those of the parts, we have achieved that aim, except for the nagging question of the coefficients.

Let us consider one other molecule before drawing conclusions. Water molecules have a triangular shape, with the two lines connecting oxygen to hydrogen atoms making an angle of approximately 104°. In many theoretical calculations, the geometry of the molecule is assumed, and other properties calculated. However, there is no reason to doubt that even the geometry can be calculated by MO or VB approximations, or some combinations of the two, using enormously large numbers of atomic orbitals. The resultant wave function, too complex to write here, will contain combinations of atomic orbitals with weighting coefficients in front of them. Thus we return to one of the original questions: What factors determine these coefficients?

THE WEIGHTING COEFFICIENTS AND HOLISM

Those who are familiar with quantum calculations know the answer: the coefficients are obtained by minimizing the energy of the molecule by variational techniques. There is no mystery—or is there? What law of physics requires that the energy of a molecule must be a minimum? The first law of thermodynamics simply states that energy is conserved. There is no requirement that a molecule be in the lowest energy state. Why, then, do small and medium-sized molecules, under ordinary conditions, exist in the lowest electronic state (with a few quanta of vibrational and rotational excitation)—allowing theorists to practice their art of minimizing energy?

The answer lies with the second law of thermodynamics, according to which the entropy of the system and the surroundings interacting with it tends to a maximum. The molecules (systems) that we study have a lower density of states than their surroundings, say, the walls of the vessel and the material surrounding the vessel. Thus a molecule with energy E above the minimum may have only few electronic states into which that energy can be distributed. When that energy is transferred to the surrounding material, the number of available states increases enormously, leading to an increase in entropy. Except for the interaction with the rest of the universe, which converts excess energy into entropy, our molecules would not be in the lowest energy state, and the variational principle would not help us in deducing the correct wave function or in calculating the properties of molecules from it.

We have noted earlier that the selection of initial states can be arbitrary. Thus one can start with atomic orbitals or Gaussian functions or any other set of functions. None has a greater theoretical claim over the others. They are all equally good candidates for synthesizing molecular wave functions, even though some of them may be better for computational purposes. Thus we are led to conclude that it doesn't matter what the states of the parts are, but it does matter that the surroundings soak up the excess energy of the molecule, increasing entropy, and make the molecule settle down into the lowest energy state. It is that part of the universe coupled to the system, and the varieties of interactions between the system (molecules) and the surroundings, that determines the structure of the molecule. Holism thus appears as the root of the apparent reduction of properties of a molecule to its parts through coupling of states. We are able to follow a reductionist program in calculating molecular properties—but that we are able to do so is a gift of holism.

An example may clarify the situation. Suppose we have two hydrogen atoms in a vessel, and these atoms only interact with the walls very weakly. When the two atoms come together to form a molecule, they will have excess energy, which breaks the molecule apart. The wave function for this pair of atoms remains coupled even though a bond between them has not formed. (The coupling remains unaltered until one of the atoms comes close to the wall. Then the states of the two hydrogen atoms become uncoupled.) A molecule forms only when the wall (or some other substance) is able to absorb the excess energy of two hydrogen atoms. This happens if the energy-absorbing substance has a higher density of energy states than the molecule.

Consider now what happens if the molecule has the same or higher density of energy states than the surroundings. Then the molecule cannot reach the lowest energy state. Because energy flows in and out of the system, there would be no long-time stable configurations. We would not be able to do property calculations, and we are likely to conclude that the properties of the molecule cannot be reduced. I suspect some of the problems with physics–biology reduction arise because the system and the surroundings have comparable density of energy states.

ARE THERE ANY CONCLUSIONS?

Methods for calculating molecular properties from atomic quantum states have been well established. The heart of the procedure is getting molecular quantum states by superposition. If by property reduction, we mean some reliable, logically consistent method of calculating properties of wholes from information on parts, we have here an example of successful property reduction. If, on other hand, we require that whole–part relation must not involve any other factors, such as interactions with the rest of the universe, we have to qualify our concept of reduction and accept a kind of holism. If by emergence, we mean manifestation of any property in the whole that has no qualitative analog in the parts, molecules provide examples of emergence. If, on the other hand, we consider that the different class of properties molecules exhibit are emergent if and only if they are not relational, then there is no case for emergence here.

Because different initial states may be chosen in building molecular states and the initial states need not be atomic states, there is arbitrariness in designating the parts.

ACKNOWLEDGMENT

I am indebted to Michael Weisberg and Professor Henry Byerly for valuable suggestions.

REFERENCES

1. REDHEAD, M. 1995. From Physics to Metaphysics. Cambridge University Press. Cambridge, UK. p. 62.
2. KUPPERS, B-O. 1992. Understanding complexity. *In* Emergence or Reduction. A. Beckermann, H. Flohr & J. Kim, Eds.: 242–243. De Gruyter. Berlin.
3. TELLER, P. 1992. A Contemporary Look at Emergence. *In* Emergence or Reduction, Beckermann *et al.*, Eds.: 139–153.

Chemical Substances and Intensive Properties

PAUL NEEDHAM

Department of Philosophy, University of Stockholm, S-106 91 Stockholm, Sweden

ABSTRACT: Despite the importance molecular structure has acquired in 20th century chemistry, more traditional macroscopic notions in terms of a continuous concept of matter continue to play a role in chemical theorizing. In the light of the extensive and determined criticism of reductionism in recent philosophy of chemistry, it is of interest to see macroscopic ontology treated autonomously. One aspect of this is developed here, namely, the concept of chemical substance. This is characterized by contrast with phases and solutions. The key conception is that of an intensive property, which is defined by appeal to the mereological structure of parts of the entities bearing substance properties.

KEYWORDS: substance; properties; intensive property; mereology; macroscopic ontology

THE MACROSCOPIC PERSPECTIVE

Only relatively recently have atomic theories of matter begun to make a systematic contribution to chemistry. Popular history identifies Dalton's atomic theory as the turning point, at the beginning of the 19th century. I would put the date much later, but this is not the place to argue such details. This paper is concerned with the characterization of what I have called the classical conception of chemical substance.[1] This is a conception based on a continuous view of matter, which was finally captured by classical thermodynamics, and still figures in modern chemistry where thermodynamics and its extensions are used.

In the light of the atomic theory, continuous theories of matter are now thought of as dealing with macroscopic conceptions. But what, exactly, do macroscopic conceptions apply to? Atomic theories deal with entities from the atomic domain—electrons, protons and neutrons, for example—but what kinds of things populate macroscopic ontology?

Explicit reference to macroscopic entities in standard textbooks is rare, but occasional hints are to be found in explanations of the thermodynamic formalism. Callen,[2] for example, introduces an entropy function S as a function of the energy, U, the volume, V, and the amount, N_i, of each component substance, $1 = i = r$, writing:

$$S = S(U, V, N_1, ..., N_r) \qquad (*)$$

Address for correspondence: Department of Philosophy, University of Stockholm, S-106 91 Stockholm, Sweden. Voice: +46-8-16 33 61; fax: +46-8-15 22 26.

paul.needham@philosophy.su.se

But elsewhere he explains the extensive character of the entropy by writing $S = \sum_\alpha S^{(\alpha)}$ (p. 28), where the variable α ranges over parts of the total system ascribed the entropy S, and $S^{(\alpha)}$ is the entropy of the part α. There is thus a systematic ambiguity in the use of S, which in the expression $S^{(\alpha)}$ is a function of one variable ranging over physical objects rather than real numbers and assigning a number to such objects as a measure of their entropy, whereas S in the expression $S(U, V, N_1, \ldots, N_r)$ is a real-valued function of $r + 2$ variables, each ranging over the real numbers. Perhaps the left-hand side of the equality in (*) is to be understood in the first sense, in which case the equality might be rewritten

$$S(\pi) = f_S(U(\pi), V(\pi), N_1(\pi), \ldots, N_r(\pi))$$

distinguishing the two functions, where π ranges over the actual physical entities assigned an entropy, energy, volume, and so forth. As for the expression $S = \sum_\alpha S^{(\alpha)}$, there is clearly a tacit understanding that the parts over which the arithmetic sum ranges do not overlap and that all the parts actually at issue jointly exhaust the whole system. There is, in other words, an implicit algebra governing the structure of the entities concerned, in terms of which we can speak of them as standing in relations such as being part of, overlapping (having a common part), and being separate from (not overlapping) and of operations such as sum, product, and difference. These are not arithmetic operations on numbers and taking numbers as values, but, on the intended interpretation, operations on physical objects and referring to physical objects. The binary product of two such objects is their common part, the binary sum is the totality including each object as a part, and the difference is what remains of one object when the part in common with the other is disregarded. Philosophers refer to this theory of parts as *mereology*. It corresponds to a Boolean algebra without a null element, and can be axiomatised in various ways (one is given in the APPENDIX), but an intuitive understanding of the mereological notions will suffice for much of the present discussion. Mereological concepts are put to work here in elucidating some features of the kinds of entity that are the subject of chemistry.

Chemistry deals, as van Brakel puts it,[3] with substances and their transformations. This involves two broad categories of entity often distinguished, after Johnson,[4] as continuants and occurrents. Transformations are processes belonging to the latter category with a mereological structure that can also be usefully elaborated, but cannot be pursued here.[5] This paper is largely confined to the topic of substances, and in particular distinguishing the category of substances from some closely related concepts.

SUBSTANCES

Chemical substances, I take it, are kinds of matter—for example, water, caustic soda, rubber, and so on. The use of the substantive "substance" doesn't imply that these kinds are things. At all events, since it is macroscopic *ontology* that is at issue here, it is important to see that the claim that chemistry deals with substances doesn't imply that there are substances understood in any other way than that there are things which are substances of one kind or another—that is, have the substance properties of being water, caustic soda, and so forth. Rather than introducing abstract objects like waterhood, named by singular terms, then, substance terms are taken to be properties expressed, in logicians' terminology, by predicates, and what exists is the things these predicates are true of. "Water," for example, is a predicate, and more specifically, a mass predicate, applying to entities that, following usage in the literature on mass terms,[6] I call quantities. Note that the term "quantity" is not being used to specify an answer to a question How much? On present usage, quantities are not amounts of a measurable magnitude, but physical objects whose general character will be better understood as the paper proceeds. Textbook writers often speak of "masses" or "volumes" when referring to bodies or parts of bodies, intending to refer to the bodies themselves rather than the number that is the numerical value of a particular measurable magnitude, although the particular choice of term is governed by the measurable magnitude of interest in the immediate context of the discussion. The term "quantity" is neutral between any particular measurable feature, such as mass or volume, that an object might have.

Viewed macroscopically, substances are said to be continuous. This is a way of saying that what bears a substance property can't be individuated like a particular kind of particle. We speak of *an* atom and say that every *atom* is small or all *atoms* are small. By contrast, we don't say that *a* water is fresh nor that some waters are liquid nor that *every* water is wet, but rather that some *water* is fresh, some *water* is liquid and all *water* is wet. Properties behaving like "water" here, rather than "atom," are said to be mass predicates. It has been suggested that mass predicates can be distinguished from individuating predicates in terms of two claims concerning the mereological structure of the quantities to which they apply. First, writing "$\subseteq$" for the mereological part relation, a mass predicate, φ, satisfies the *distributive* condition:

$$\varphi(\pi) \wedge \rho \subseteq \pi . \supset \varphi(\rho) \qquad (1)$$

(This is read "if $\varphi(\pi)$ and $\rho \subseteq \pi$, then $\varphi(\rho)$," the symbol "$\supset$" denoting implication and "$\wedge$" conjunction.) Second, writing "$\cup$" for the binary mereological sum operation, a mass predicate, φ, satisfies the *cumulative* condition:

$$\varphi(\pi) \wedge \varphi(\rho) . \supset \varphi(\pi \cup \rho) \qquad (2)$$

More generally, for the sum of all those quantities ρ satisfying a condition ψ, written $\Sigma \rho \, \psi(\rho)$, we have

$$\exists \pi \, \psi(\pi) \wedge \forall \pi \, (\psi(\pi) \supset \varphi(\pi)) . \supset \varphi(\Sigma \pi \, \psi(\pi)) \qquad (3)$$

The existential clause, $\exists \pi \, \psi(\pi)$, stating that there is something satisfying the condition ψ, is necessary to ensure the existence of the sum. Given this, if everything satisfying ψ is φ, then so is the sum of whatever satisfies ψ.

Substance properties can be contrasted with predicates of quantities that fail to satisfy at least one of the conditions, such as "occupies such-and-such a region of space at time t," "is more massive than such-and-such a quantity," "has enough energy to melt that piece of ice," "is water and looks blue" (van Brakel), and so on, which are not distributive. Predicates like "weighs less than such-and-such," "is invisible," "is water and looks colorless," and so on, are not cumulative. It will transpire that there are still other properties with characteristics similar to those of substance predicates from which substances are to be distinguished. But it is already clear that the distributive and cumulative conditions won't quite do in the form of

Equations (1) and (2) or (3) as conditions satisfied by a substance predicate φ. For one thing, substances undergo transformations. The ancients thought water was transformed into another substance when it evaporated, and yet another when it froze, just as iron appeared to be transformed into copper when immersed in a solution of blue vitriol. Joseph Black still held onto some of these ideas when he interpreted his discovery of the latent heat required to melt ice in terms of a chemical reaction of ice with caloric generating water. Nowadays, substances are still thought of as participating in chemical reactions, which are processes in which substances are used up in the creation of new substances, although with somewhat different ideas about what goes on in these processes. What is one substance at one time may well be another at another time. There is a time factor to be brought into the account, then, with the recognition that what is, say, water at one time may not be at another, since water is both consumed in some processes and produced in others. Substance predicates should, accordingly, express relations between quantities and times, and are not simple properties (expressed by one-place predicates). When incorporating time into formulations of the distributive and cumulative conditions, times will be treated like quantities of matter in so far as they are entities with a mereological structure of parts.

Second, it might be objected against the distributive condition that sufficiently small parts of what is water are not water. What characterizes a quantity as water is, for example, its state of aggregation (phase) at the quantity's pressure and temperature—water freezes at 0°C at normal pressure. Individual molecules cannot be ascribed a phase, a temperature, or a pressure. Therefore, so the objection runs, it is not true that *every* part of what is water is water.

Irreversible thermodynamics is based on the idea of local equilibrium, according to which thermodynamic features such as temperature and pressure are assigned to points in space at instants of time. Having introduced this idea in their textbook, Kondepudi and Prigogine[7] go on to consider how small a volume it is possible to "meaningfully associate" thermodynamic features. Thermodynamic features undergo fluctuations on the microscale, and the limiting condition is that the fluctuations be small in relation to the magnitude of the feature in question. For an ideal gas at normal temperature and pressure, volumes of one cubic micrometer (10^{-6} m) are safely above this limit, and the limiting size is much smaller for liquids and solids. The formalism, which treats temperature, pressure, and so on, as functions of points of three-dimensional space and instants of time, allowing the application of the operations of mathematical analysis, is therefore inconsistent with the credible interpretation.

Following the interpretation, that is, what the theory is held to say that is true, rather than the formalism, we might seek a way to block quantities ranging over the material occupants of arbitrarily small regions of space with a restriction internal to the present approach. Quantities are whatever can, in principle, bear thermodynamic features, even though, at any given time, they might not actually do so because not at equilibrium. Or, in the spirit of irreversible thermodynamics, even quantities not at equilibrium and consequently not bearing a temperature have parts which do. Having a temperature means having a degree of warmth, that is, being *as warm as* something and *warmer than* something else, which are the fundamental thermodynamic concepts. Writing "π is at least as warm as ρ at *t*" as $W(π, ρ, t)$, the idea is captured with an axiom of the kind

$$\forall \pi \, \exists \rho \, \exists \sigma \, \exists t \, (\rho \subseteq \pi \, \wedge \, W(\rho, \sigma, t)) \quad (4)$$

The universe of discourse—that is, the set of things over which the quantity variables π, ρ, σ, ... range—is thus confined to things satisfying this condition. There is room for discussion about the exact form such a condition should take. Perhaps formula (4) is too strong, and a modal formulation,

$$\forall \pi \, \exists \rho \, \exists \sigma \, \exists t \, (\rho \subseteq \pi \wedge \Diamond \, W(\rho, \sigma, t)) \quad (4')$$

would be safer. ($\Diamond$ is the modal operator read "It is possible that.") I won't take a stand on this issue here, but simply note that the modal alternative would require both times and quantities to range over an outer domain in the standard free-logic account of quantified modal logic. The idea of indestructible quantities might be contended in the light of modern physics, but accords well with the everyday world of chemistry.

Formally, then, by laying down a general condition ensuring that all quantities are macroscopic quantities, it is not necessary to introduce a spatial restriction on the distributivity condition. The question naturally arises of how quantities are related to space. This will be taken up once the distributive and cumulative conditions have been reformulated to take account of the relational character of substance predicates.

REVISING THE DISTRIBUTIVE AND CUMULATIVE CONDITIONS

Substance predicates are now to be recognized as time-dependent. This is not fully accommodated, however, by simply rewriting the distributive condition as

$$\varphi(\pi, t) \wedge \rho \subseteq \pi . \supset \varphi(\rho, t)$$

since times themselves have a mereological structure of parts. It was mentioned above that the formalism of irreversible thermodynamics assumes that thermodynamic properties are assigned to points at *instants*, but just as only claims about finite regions of space are considered true, so too for intervals of time. Accordingly, for a relation between entities with a structure of parts, the distributive condition becomes

$$\varphi(\pi, t) \wedge \rho \subseteq \pi \wedge t' \subseteq t . \supset \varphi(\rho, t') \quad (5)$$

This raises the question of a lower limit on the time during which a quantity can sustain a substance property. On the one hand, it is natural to assume that time is infinitely divisible in the sense that any interval can be divided into two abutting intervals (i.e., any interval is the sum of two abutting intervals).[8] But for very short intervals, thermodynamic properties cease to be applicable. An internal principle analogous to formula (4), restricting the domain to intervals during which a temperature can be sustained by something, would therefore seem appropriate. Or rather, bearing in mind that some intervals may be so long that nothing maintains a warmth relation to anything that long, any interval must contain a subinterval of which this is true:

$$\forall t \, \exists t' \subseteq t \, \exists \pi \, \exists \rho \, W(\pi, \rho, t) \quad (6)$$

Once more, a modal formulation might be thought more appropriate:

$$\forall t \; \exists t' \subseteq t \; \exists \pi \; \exists \rho \; \Diamond \; W(\pi, \rho, t) \tag{6'}$$

Imposing restrictions of this kind is not in the spirit of a simple and elegant formal theory of substances, however, and we may well be tempted to take a leaf from the irreversible thermodynamicist's book and abandon such limiting assumptions for the purpose of developing the theory.

An adequate formulation of the distributive condition is now at hand. A corresponding modification of the cumulative condition is not so straightforward, however. To see the problem, let us begin with the natural observation that the mereological theory of intervals shouldn't allow the summation of arbitrary intervals, but only pairs (and by induction, finite numbers) of intervals that are *connected* (either abut or overlap).[9] On this view, an interval of time contains no gaps in the sense that it contains as a part every interval between two nonoverlapping intervals abutting the given interval. With this in mind, a first shot at generalizing the cumulative condition (2) might read

$$\varphi(\pi, t) \wedge \varphi(\rho, t') \wedge Con(t, t') . \supset \varphi(\pi \cup \rho, t \cup t') \tag{7}$$

where "*Con*" abbreviates the relation of being connected. But this might well not be true. The time-dependence of substance predicates was introduced in recognition of the fact that what is water at one time might not be at another. Suppose, then, that t' is later than t, when π is subjected to electrolysis and gradually dissociates into hydrogen and oxygen. Then it wouldn't be true that the sum, $\pi \cup \rho$, is water during $t \cup t'$, even if π and ρ are each water during t and t', respectively. For if φ satisfies the distributive condition (5), then $\varphi(\pi \cup \rho, t \cup t')$ implies $\varphi(\pi, t')$, which, as just envisaged, may well not be true.

To find a satisfactory generalization of the cumulative condition, note that the original condition in its more general formulation, (3), is (cf. Roeper 1983, p. 259) equivalent with

$$\forall \pi' \subseteq \pi \; \exists \pi'' \subseteq \pi' \; \varphi(\pi'') \supset \varphi(\pi) \tag{8}$$

A proof in terms of the mereological system adopted here is outlined in the APPENDIX.[10] Although the reason for this particular construction of quantifiers is not obvious and can only really be understood by working through the proof, intuitively it might be accepted that "if φ is true of some part of any part of π, then φ is true of the whole of π" expresses the cumulativity idea. Now Eq. (8) can be generalized into a suitable cumulative condition for relational predicates that is not subject to the counterexamples which tell against Eq. (7):

$$\forall \pi' \subseteq \pi \; \forall t' \subseteq t \; \exists \pi'' \subseteq \pi' \; \exists t'' \subseteq t' \; \varphi(\pi'', t'') \supset \varphi(\pi, t) \tag{9}$$

The distributive and cumulative conditions (8) and (9) for dyadic predicates can now be straightforwardly generalized to predicates with arity greater than 2. The three-place predicate "is at least as warm as," for example, is cumulative in the sense that:

$$\forall \pi' \subseteq \pi \; \forall \rho' \subseteq \rho \; \forall t' \subseteq t \; \exists \pi'' \subseteq \pi' \; \exists \rho'' \subseteq \rho' \; \exists t'' \subseteq t' \; W(\pi'', \rho'', t'') \supset W(\pi, \rho, t)$$

and similarly for the distributive condition.

Predicates satisfying the more general distributive and cumulative conditions (5) and (9), or their generalizations for three or more placed predicates, are said to be *intensive*.[11] It seems that being intensive is a necessary condition for being a substance property. But it is not a sufficient condition since warmth relations, as we have just seen, are also intensive, and these are certainly not substance predicates. Other examples of properties that are intensive but not substance properties are the phase properties—being solid, liquid, or gas—and pressure relations such as having at least as high pressure as. How might substance properties be distinguished from these?

A superficial distinction in some of the predicates that have appeared so far is that substance predicates are dyadic, involving a relation between a quantity and a time, whereas warmth relations are triadic, involving a relation between two quantities and a time. But, on the one hand, simple phase properties such as being liquid are dyadic, involving a relation between a quantity and a time, and these, as we have just seen, are to be distinguished from substance predicates. On the other hand, the relation of being the same kind of substance is triadic, involving a relation between two quantities and a time, and is clearly a substance predicate. It might even be suggested that some warmth predicates are not three-place, but in fact four-place predicates, having two places for times. For example, it seems quite coherent to be able to say that this piece of metal is as warm now as that piece was yesterday. This simple-minded syntactic distinction won't work, and a more general ground for drawing the distinction is required. For this purpose, I return to the question left unanswered earlier in the paper of the relation of the part relation between quantities and the notion of spatial parts.

HOW ARE SUBSTANCES RELATED TO SPACE?

The device of internally restricting the domain of quantities rather than explicitly discussing their minimum size made it possible to avoid speaking of spatial parts earlier. But in so far as the question of how the parts of substances are related to parts of space bears on the understanding of parts of quantities, it has been with us from the outset and is even more pressing now that other intensive properties have come into the picture. Let us begin by returning to Callen's equation (*) and noting that in the standard thermodynamic treatment, a quantity may comprise several different kinds of substance in amounts N_i, $1 \leq i \leq r$. A quantity of matter may, in other words, be a mixture of several substances, and the theory has something to say about how changes in some of its features are related to changes in the amounts of the various substances. But although there may be several substances in the quantity, only one volume is specified—the volume of the whole. And when determining the concentrations of the various substances, which are the variables used to follow the course of a reaction, it is the mass of the substance concerned in the volume of the whole that is at issue. There is no justification in the formalism for interpreting the various substances mixed in the original quantity as each having some volume smaller than that of the entire original quantity. Accordingly, a mixture may be partitioned into a collection of nonoverlapping parts, each comprising all and only one of the substance kinds in the mixture. Nonoverlapping quantities have no part in common. But the separate (i.e. nonoverlapping) quantities in such a partition all occupy the same place. They have their location in common and do not constitute proper spatial parts of the whole mixture.

In order to talk about spatial parts, we need a notion of a quantity occupying a region. Spatial regions, $p, q, r, \ldots$, are considered to have a mereological structure of parts.[12] Because what occupies one place at one time might occupy another at another time, occupying is a time-dependent notion expressed by a triadic relation, for which purpose $Occ(\pi, p, t)$ is introduced for "π occupies region p for t." This said, two principles governing "occupies" can be stated. First, every quantity is always somewhere:

$$\exists p\, Occ\, (\pi, p, t) \qquad (10)$$

This expression serves as a kind of temporal existence predicate applicable to quantities—to be at t is to be located somewhere at t. Second, what has just been said about occupying the same region occupied by another quantity presupposed that "occupies" is specifically interpreted to mean "occupies exactly the region in question," and so satisfies the principle

$$Occ\, (\pi, p, t) \wedge q \neq p \,.\!\supset\, \sim\!Occ(\pi, q, t) \qquad (11)$$

Quantities may become dispersed when their parts occupy nonabutting spatial regions. Accordingly, Eq. (10) requires, in contrast to what has been assumed for times, that arbitrary sums of regions be considered as regions. They satisfy the usual mereological criterion of identity, so that $q \neq p$ in Eq. (11) means that q and p don't have all their parts in common. One may be a proper part of the other, or more generally they may *overlap* (have some part in common) and each have a part not in common with the other, or perhaps they don't even overlap (and are entirely *separate*). In view of the existence and uniqueness postulates Eqs. (10) and (11), the occupies relation $Occ\, (\pi, p, t)$ can be expressed as a function, writing $Occ\, (\pi, t) = p$, where $Occ\, (\pi, t)$ is *the* region occupied by π at t.

Although it is natural, perhaps, to regard space like time as infinitely divisible—that is, any region is the sum of two abutting regions—what occupies sufficiently small regions is not a substance. An internal restriction analogous to that imposed on times is appropriate:

$$\forall p\, \exists q\, \exists \pi\, \exists t\, (q \subseteq p \wedge Occ\, (\pi, q, t)) \qquad (12)$$

or the corresponding modal formulation.

It is convenient at this point to distinguish the general mereological relation of *proper part*, written "$\subset$," from the part relation "$\subseteq$" by explicitly excluding the possibility of identity:

$$\text{Def.}\quad \rho \subset \pi \equiv.\, \rho \subseteq \pi \wedge \rho \neq \pi$$

("$\equiv$" stands for biimplication, "if and only if.") One spatial region being a proper part of another is defined analogously. We are now in a position to address the question of how the parts of quantities are related to spatial parts.

Given what was said above about co-occupancy, a quantity ρ might be a *proper part* of a quantity π, and yet occupy the same place. The salt[13] in a brine solution occupies the same place as the brine. Again, some of the parts of the region occupied by the salt are parts of the region occupied by the water in the brine; but the salt and water are separate, and so the parts of the salt in question are not parts of the water. There is in general no implication, then, either from $\rho \subset \pi$ to $Occ\,(\rho, t) \subset Occ(\pi, t)$ or conversely. All we can say is

$$\rho \subset \pi \supset Occ\,(\rho, t) \subseteq Occ(\pi, t)$$

since proper parts certainly do not extend beyond the spatial extent of the whole. Note that this is equivalent with $\rho \subset \pi \supset \forall t\,(Occ(\rho, t) \subseteq Occ(\pi, t))$—proper parts of a quantity never extend beyond the spatial extent of the whole. Because, moreover, identical quantities clearly always occupy identical regions, we have

$$\rho \subseteq \pi \supset Occ\,(\rho, t) \subseteq Occ(\pi, t) \tag{13}$$

In view of this, it seems reasonable to interpret the notion of a *spatial part* of a quantity as a part of a special kind, namely one that occupies a part of the region occupied by the quantity. Because of the time-dependency of the "occupies" relation, this is a time-dependent notion:

Def. $\rho \sqsubset_t \pi \equiv .\, \rho \subset \pi \wedge Occ\,(\rho, t) \subset Occ\,(\pi, t)$

This is just as it should be. No part of the salt dissolved in a quantity of water at t was a spatial part of the water before t (though it was, in accordance with Eq. (13), a spatial part of its sum with the quantity of water in which it is dissolved at t). So it is not in general true that

$$Occ\,(\rho, t) \subset Occ\,(\pi, t) \supset \forall t\,(Occ\,(\rho, t) \subset Occ\,(\pi, t))$$

DISTINGUISHING SUBSTANCE PROPERTIES

A formulation of what it is that distinguishes substance properties from others can now be put forward. Whereas different substance properties may be borne by quantities occupying the same place at the same time, this is not so for phase properties. At all events, any quantities of uniform phase and occupying the same place at the same time have the same phase property. Much the same holds for warmth and pressure relations. What matters here are the mereological relations between the regions occupied by the quantities in question. Accordingly, we might consider versions of the distributive and cumulative conditions formulated in terms of regions occupied:

$$\varphi(\pi, t) \wedge t' \subseteq t \wedge Occ\,(\rho, t') \subseteq Occ(\pi, t') . \supset \varphi(\rho, t') \tag{14}$$

$$\forall \pi'\, \forall t' \subseteq t\, \exists \pi''\, \exists\, t'' \subseteq t'\,(Occ\,(\pi'', \rho'') \subseteq \\ Occ\,(\pi', t') \subseteq Occ\,(\pi, t) \supset \varphi(\pi'', t'')) \supset \varphi(\pi, t) \tag{15}$$

A predicate φ satisfying formulas (14) and (15) is said to be *(spatially) homogeneous*, or to satisfy the *(spatial) homogeneity condition*. This is not the same as being intensive, since in view of formula (13), parthood of regions occupied is weaker than parthood of quantities, so *spatial distributivity* (14) is stronger than the ordinary distributive condition (5). And by the same count, *spatial cumulativity* (15) is weaker than the general cumulative condition, since the antecedent is stronger (its antecedent being weaker). Consequently, the collection of predicates that are intensive and the collection of predicates that are homogeneous cut across one another.

Three possible categories of predicates can be distinguished, then: those that are both intensive and homogeneous, those that are intensive but not homogeneous, and those that are homogeneous but not intensive. From what has been said, it is clear that substance properties are distinguished from phase, warmth, and pressure properties by being *intensive but not homogeneous*. As illustrated above, they are not spatially homogeneous, as are phase properties. Take π to be a liquid salt solution at t, where ρ is the salt and σ is the water in the solution. ρ and σ are separate, as are σ and any proper part ρ' of ρ occupying a proper part of the region occupied by σ, which is not the same kind of substance as σ. But since σ is liquid, and $Occ(\rho', t) \subseteq Occ(\sigma, t)$, then ρ' is liquid by the spatial distributive condition for the phase property of being liquid. Again, a case of a quantity being a substance kind without being a phase kind is provided by a quantity of water comprising several phases in equilibrium—part water and part vapor, for example, or solid, liquid and vapor at the triple point. In this case, any part of the water would be as warm as any other, whatever their phase properties. But a quantity of water might be separated into two adiabatically insulated parts not as warm as one another, so that the quantity as a whole, although falling under a substance property, does not enter any warmth relations.

This settles the classification of phase properties, along with warmth and pressure properties. They are homogeneous *and* intensive, the stronger cumulativity condition holding for them too, as was pointed out at the end of the third section (Revising the Distributive and Cumulative Conditions). Where does this leave the third grouping? Are there any predicates falling into the category of predicates homogeneous but not intensive? Solution predicates fall into this category. Brine is not a substance because any brine has parts that are not brine, such as its water content, and so is not distributive in the sense of formula (5). But it is spatially distributive in the sense of formula (14). And if all the spatial parts of a region are occupied by quantities that are brine, then the whole region is occupied by brine. It is spatially cumulative in the sense of formula (15) too.

ELEMENTS AND COMPOUNDS

Where does this leave the distinction between elements and compounds? As the examples and terminology have been developed so far, it must fall within the category of substances. Water has been taken as a standard example of a substance and contrasted with solutions containing parts of different substance kinds that make them nonintensive, and water is a compound of hydrogen and oxygen. Another way of proceeding, however, is equally compatible with the tripartite distinction of the last section in terms of intensive and homogeneous predicates. This would be to accept that water is not intensive, but has parts that are oxygen and others that are hydrogen. Water would then fall in the category of homogeneous, nonintensive properties, which in deference to normal chemical usage might better be labeled "homogeneous mixtures" rather than "solutions" in that case. Call this approach the Stoic view of compounds. The former course, classifying compound substances as intensive so that none of their parts are elemental—no part of water is hydrogen and none oxygen—implicitly involves a distinctly Aristotelian aspect.[14] Particular elementary substances are characterized by the properties they display in isolation—hydrogen and oxygen are both gaseous at normal temperature and pressure, for example, which water patently is not. Water is potentially hydrogen and oxygen in the sense that a decomposition process could transform what is water into hydrogen and oxygen. To be composed of hydrogen and oxygen is not, in this understanding of compounds, to have parts that are hydrogen and oxygen, but to be the result of a transformation of hydrogen and oxygen into water. Call this approach the Aristotelian view of compounds. Each view can be developed further.

Consider first the Aristotelian view and suppose that a quantity π of water has been formed from a quantity ρ that was hydrogen and a quantity σ that was oxygen. It is natural to follow Lavoisier and assume that "an equal quantity of matter exists both before and after the experiment."[15] Interpreting this to mean that matter is neither created nor destroyed in the transformation of substances, then $\pi = \rho \cup \sigma$. In that case, by the distributive condition, all the parts of what is water at t are water, and so in particular, each of ρ and σ are water at t. If the water is subsequently decomposed into hydrogen and oxygen, there is no reason to suppose that ρ will be all and only the hydrogen and σ the oxygen. Other parts of π, that is $\rho \cup \sigma$, in the appropriate proportion may be hydrogen after the decomposition, and the remainder will be oxygen. A more radical alternative would be to interpret Lavoisier's principle to mean merely that the property of having the same mass as some reference object is preserved, allowing that ρ and σ are each destroyed and a new quantity π created in their place. At first glance, Aristotle's original terminology of generation and destruction suggests as much. On either version of the Aristotelian view, the category of elements is not so very different from that of compounds: element and compound properties are both intensive, and so not distinguishable by the way they apply to their parts. But following Duhem,[16] an approach more suited to modern chemistry than Aristotle's is possible if substitution in chemical reactions is understood along the same lines as decomposition, that is, not the literal replacement of a part, but the transformation of one substance into another whose chemical formulas involve the replacement of one unit by another. Elements are then the ultimate units of substitution in the sense that they can substitute, and are substitutable by, other elements or groups, but contain no constituents that are substitutable.

The category of elements is more straightforwardly characterized on the Stoic view. Quantities of elemental substances are intensive, having no parts that are not of the same elemental kind, whereas compounds are not. Compounds must be distinguished from solutions by an additional relation of combination in which the elemental parts stand. We can speak of the quantity π of water formed from the quantity ρ of hydrogen and the quantity σ of oxygen being *constituted* of ρ and *constituted* of σ during the time π is water[17] and define combination as the time-dependent relation between quantities of which some quantity is constituted at the time in question. Clearly, in this case $\pi = \rho \cup \sigma$, and the stronger interpretation of Lavoisier's principle holds. In contradiction with the less radical version of the Aristotelian view, ρ here is hydrogen and σ is oxygen when π is water. They obviously have some properties markedly different from those of the corresponding isolated elements under the same temperature and pressure, and the problem therefore arises for the Stoic view of what determines the characteristic features of particular elemental properties in virtue of which ρ, for example, is the same substance, hydrogen, both when isolated and combined with oxygen. Modern theory doesn't seem to have improved upon the ancient Stoics in this respect and provided a clear suggestion in nonmodal terms about how this might be done.[18] This leaves us with ρ being hydrogen in virtue of its having the possibility of being isolated and freezing at 14°K and being gaseous at normal temperature and pressure, just as a quantity of isolated hydrogen has the property of being possibly combined with a quantity of oxygen in water. The difference between this and the less radical version of the Aristotelian view is not as great as first appears. Were there no transmutation of elements, however, then the property of being an element of some kind would not be a time-dependent property in the Stoic view. But there would still be relational properties like "being the same element as" that call for the appropriate generalization of the cumulative condition. (If π and ρ are the same element—say hydrogen—and π' and ρ' are the same element—say oxygen—then it doesn't follow that $\pi \cup \pi'$ is the same element as $\rho \cup \rho'$. That would be inconsistent with the distributive condition on being the same element.)

SUMMARY

Macroscopic ontology introduces entities of various kinds. These kinds are specified in terms of groups of properties. One such group of interest in chemistry delimits the classical concept of chemical substance. It includes such predicates as "is water," "is hydrogen," "is the same kind of substance as," "is an element of group I of the periodic table," "is a hydrophobic substance," and so on. We might try to specify necessary and sufficient conditions for the applicability of each of the particular predicates in this group. The question which has been broached here is rather, What general features do these have in common? An answer to this question would provide some insight into the general concept of a substance.

The notion of an intensive property is central to this account. Intensive properties were characterized by appeal to the mereological structure of parts of the quantities bearing them, in terms of a distributive and a cumulative condition, suitably extended to cover relational predicates. But the class of substance properties couldn't be straightforwardly identified with the class of intensive properties, first because closely related properties also studied in thermodynamics, to do with phase, degree of warmth, and degree of pressure are also intensive, and second, because the distinction between elements and compounds must be accommodated. The first point led to a consideration of how quantities are related to space, and in particular what possessing the properties at issue here implies about relations between regions occupied by quantities. It proved possible to define a notion of spatial homogeneity, analogous to that of being an intensive property, but with a stronger analogue of the distributive condition and a weaker analogue of the cumulative condition. Consequently, a tripartite distinction arises between properties that are intensive and (spatially) homogeneous, properties that are intensive but not homogeneous, and properties that are homogeneous but not intensive. Phase, warmth, and pressure properties fall within the first class of intensive and homogeneous properties, whereas substance properties fall within one of the others. Which one depends on how the distinction between elements and compounds is understood. Two principal views were outlined, both with ingredients suggestive of ideas found in the ancients. One alludes to an idea of Aristotle on the potential, nonactual character of elements in a compound, and the other alludes to the Stoic tradition after Aristotle, according to which elements are in compounds.

This by no means exhausts the kinds of entity populating the macroscopic ontology. If we are guided by van Brakel's dictum that chemistry deals with the transformation of substances, then the immediately pressing issue is the category of processes within which transformations fall. But that will have to wait for another occasion.

NOTES AND REFERENCES

1. NEEDHAM, P. 2002. The discovery that water is H_2O. Int. Stud. Philos. Sci. **16**: 201–222.
2. CALLEN, H. 1985. Thermodynamics and an Introduction to Thermostatistics, John Wiley. New York.
3. VAN BRAKEL, JAAP. 1997. Chemistry as the science of the transformation of substances. Synthese **111**: 253–282.
4. JOHNSON, W.E. 1921. Logic, Vol. I. Cambridge University Press. Cambridge, UK. pp. 199–202.
5. A start is made in NEEDHAM, PAUL. 1999. Macroscopic processes. Philos. Sci. **66**: 310–331, and NEEDHAM, PAUL. 2000. Hot stuff. In Facts, Things, Events. Jan Faye, Uwe Scheffler & Max Urchs, Eds.: 421–446. Rodopi. Amsterdam.
6. CARTWRIGHT, HELEN MORIS. 1970. Quantities. Philos. Rev. **79**: 25–42; Idem. 1975. Amounts and measures of amounts. Noûs **9**: 143–164; ROEPER, PETER. 1983. Semantics for mass terms with quantifiers. Noûs **17**: 251–265.
7. KONDEPUDI, D. & ILYA PRIGOGINE. 1998. Modern Thermodynamics: From Heat Engines to Dissipative Structures. John Wiley. London. p. 335.
8. This is the case in the complete first-order theory of intervals of NEEDHAM, PAUL. 1981. Temporal intervals and temporal order. Logiq. Anal. **93**: 49–64.
9. This, at any rate, is a restriction imposed in the theory referred to in the preceding note.
10. Note that the proof of the equivalence relies on the distributive condition, so that any restriction on that condition would obstruct the proof. The policy of restricting the range of quantification over quantities to quantities that can sustain warmth relations allows the distributive condition to be retained.
11. Disjunctions of intensive properties are also intensive. What is water or salt, for example, is either water or salt or a sum of a quantity of water and a quantity of salt. See ROEPER, "Semantics for mass terms," Note 6 for a more systematic discussion of logical compounds of mass predicates.
12. Shortly after Hilbert published his rigorous axiomatization of Euclidean geometry based on the traditional geometrical primitives of point and line, a system with the notions of part and sphere as primitive was published by HUNTINGTON, EDWARD V. 1913. A set of postulates for abstract geometry, expressed in terms of the simple relation of inclusion. Math. Annal. **73**: 522–559.
13. Or whatever it is apart from water that is present in brine.
14. This corresponds in many ways to Duhem's view of substances—see NEEDHAM, PAUL. 2002. Duhem's theory of mixture in the light of the Stoic challenge to the Aristotelian conception. Stud. Hist. Philos. Sci. **16**: 685–708. For an exposition of Aristotle's account of chemical substance, see NEEDHAM, PAUL. 2003. Aristotle's theory of chemical reaction and chemical substances. In Philosophy of Chemistry: Synthesis of a New Discipline. Davis Baird, Eric Scerri & Lee McIntyre, Eds. Boston Studies in the Philosophy of Science. Kluwer. Dordrecht. In press.

15. LAVOISIER, ANTOINE-LAURENT. 1789. Elements of chemistry in a new systematic order, containing all the modern discoveries. Trans. by Robert Kerr of Traité élémentaire de Chimie. 1965. Reprinted by Dover. New York. p. 130.
16. DUHEM, PIERRE. 1892. Notation atomique et hypothèses atomistiques. Rev. Quest. Sci. **31**: 391–457. Trans. by Paul Needham. 2000. Atomic notation and atomistic hypotheses. Found. Chem. **2**: 127–180. DUHEM, PIERRE. 1902. Le mixte et la combinaison chimique: essai sur l'évolution d'une idée. C. Naud. Paris. Pt. II, Chapt. 4–6. *Translated in* Mixture and Chemical Combination, and Related Essays. Trans. and ed. by Paul Needham. 2002. Kluwer. Dordrecht.
17. Axioms delimiting this notion of "constituted of" are given in NEEDHAM, PAUL. 1996. Macroscopic objects: an exercise in Duhemian ontology. Philos. Sci. **63**: 204–223.
18. See NEEDHAM, 2002, Duhem's theory of mixture (note 11), section 5.
19. LEONARD, H. & NELSON GOODMAN. 1940. The calculus of individuals and its uses. J. Symbol. Logic **5**: 45–55.

Appendix

I assume a standard theory of mereology for quantities. Following Leonard and Goodman's classic formulation,[19] the dyadic relation of separation, "|," is taken as primitive, in terms of which the part relation $\rho \subseteq \pi$ is defined by $\forall \sigma \, (\sigma \mid \pi \supset \sigma \mid \rho)$, from which transitivity of "⊆" follows immediately, and the overlap relation $\rho \mathbf{o} \pi$ is defined by $\exists \sigma \, (\sigma \subseteq \rho \wedge \sigma \subseteq \pi)$. The two axioms

M1 $\rho \subseteq \pi \wedge \pi \subseteq \rho \, . \supset \rho = \pi$

M2 $\rho \mathbf{o} \pi \equiv \sim\!\rho \mid \pi$

are supplemented by an axiom providing for the existence of sums:

M3 $\exists \pi \, \varphi(\pi) \supset \exists \rho \, \forall \sigma \, (\sigma \mid \rho \equiv \forall \pi \, (\varphi(\pi) \supset \sigma \mid \pi))$

Uniqueness of sums follows, permitting the definition of the sum of φ-ers, $\Sigma \rho \, \varphi(\rho)$, by

Def.Σ $\exists \pi \, \varphi(\pi) \supset . \, \sigma = \Sigma \pi \, \varphi(\pi) \equiv \forall \rho \, (\rho \mid \sigma \equiv \forall \pi \, (\varphi(\pi) \supset \rho \mid \pi))$

with a binary sum, $\pi \cup \rho$, defined by the condition "is identical with π or ρ." An operation of mereological difference is defined by

Def.− $\rho \mathbf{o} \pi \supset \pi - \rho = \Sigma \sigma \, (\sigma \subseteq \pi \wedge \sigma \mid \rho)$

The condition on the existence of the difference is necessary because of the lack of a null element.

Some straightforward consequences of these axioms, stated here without proof, are:

MT1. $\pi \subseteq \rho \wedge \rho \mid \sigma \, . \supset \pi \mid \sigma$

MT2. $\sim\!\pi \subseteq \rho \equiv \exists \sigma \, (\sigma \subseteq \pi \wedge \sigma \mid \rho)$

MT3. $\exists \pi \, \varphi(\pi) \supset . \, \rho \subseteq \Sigma \pi \, \varphi(\pi) \supset \exists \pi \, (\varphi(\pi) \wedge \rho \mathbf{o} \pi)$

MT4. $\Sigma \pi \, \varphi(\pi) \subseteq \rho \supset \forall \pi \, (\varphi(\pi) \supset \pi \subseteq \rho)$

This puts us in a position to show that, given the distributive condition (1), the cumulative condition,

$$\exists \pi\, \psi(\pi) \wedge \forall \pi\, (\psi(\pi) \supset \varphi(\pi))\, .\supset \varphi(\Sigma \pi\, \psi(\pi)) \qquad (3)$$

is equivalent with

$$\forall \pi' \subseteq \pi\, \exists \pi'' \subseteq \pi'\, \varphi(\pi'') \supset \varphi(\pi) \qquad (8)$$

Proof. To establish Eq. (8), assume Eq. (3) and for an arbitrary π,

$$\forall \pi' \subseteq \pi\, \exists \pi'' \subseteq \pi'\, \varphi(\pi'') \qquad (i)$$

Consider the sum of those things π' with the property $\pi' \subseteq \pi \wedge \varphi(\pi')$. It can be shown that $\pi = \Sigma \pi'\, (\pi' \subseteq \pi \wedge \varphi(\pi'))$. Suppose not. Then by M1, either $\pi \not\subseteq \Sigma \pi'\, (\pi' \subseteq \pi \wedge \varphi(\pi'))$ or $\Sigma \pi'\, (\pi' \subseteq \pi \wedge \varphi(\pi')) \not\subseteq \pi$. Take the first case. By MT2, there is a σ such that

$$\sigma \subseteq \pi \wedge \sigma \mid \Sigma \pi'\, (\pi' \subseteq \pi \wedge \varphi(\pi')) \qquad (ii)$$

Now the first part of this, together with (i), implies that there is a π'' such that

$$\pi'' \subseteq \sigma \wedge \varphi(\pi'') \qquad (iii)$$

By MT1, the first conjunct of (iii) and the second conjunct of (ii) imply $\pi'' \mid \Sigma \pi'\, (\pi' \subseteq \pi \wedge \varphi(\pi'))$. But by transitivity of $\subseteq$, the first conjuncts of (ii) and (iii) together imply $\pi'' \subseteq \pi$, giving us

$$\pi'' \subseteq \pi \wedge \varphi(\pi'')$$

which in turn implies that $\pi'' \subseteq \Sigma \pi'\, (\pi' \subseteq \pi \wedge \varphi(\pi'))$ by the reflexivity of $\subseteq$ and MT4, contradicting $\pi'' \mid \Sigma \pi'\, (\pi' \subseteq \pi \wedge \varphi(\pi'))$

Now consider the second case, $\Sigma \pi'\, (\pi' \subseteq \pi \wedge \varphi(\pi')) \not\subseteq \pi$. By MT2, this implies that for some π'',

$$\pi'' \subseteq \Sigma \pi'\, (\pi' \subseteq \pi \wedge \varphi(\pi')) \wedge \pi'' \mid \pi.$$

Clearly, the first conjunct implies $\pi'' \subseteq \pi$, contradicting the second conjunct. Since neither alternative is sustainable, then, $\pi = \Sigma \pi'\, (\pi' \subseteq \pi \wedge \varphi(\pi'))$. Now (i) implies $\exists \pi'\, (\pi' \subseteq \pi \wedge \varphi(\pi'))$, and since $\forall \pi'\, ((\pi' \subseteq \pi \wedge \varphi(\pi')) \supset (\pi' \subseteq \pi \wedge \varphi(\pi')))$ is essentially a tautology, (3) implies $(\pi \subseteq \pi \wedge \varphi(\pi))$, that is, $\varphi(\pi)$.

For the converse implication, to establish (3), assume the distributive condition (1), (8), and $\exists \pi\, \psi(\pi)$ as well as

$$\forall \pi\, (\psi(\pi) \supset \varphi(\pi)) \qquad (iv)$$

Put $\pi = \Sigma \pi'\, \psi(\pi')$, which exists because there are ψ-ers. By MT3, for every $\pi' \subseteq \pi$, there is a σ such that

$$\psi(\sigma) \wedge \pi' \mathbf{o}\, \sigma$$

This last conjunct implies that there is a part, π'', common to π' and σ, and by the distributive condition (1), the first conjunct then implies $\psi(\pi'')$. Accordingly, $\forall \pi' \subseteq \pi\, \exists \pi'' \subseteq \pi'\, \psi(\pi'')$, and so by (iv), $\forall \pi' \subseteq \pi\, \exists \pi'' \subseteq \pi'\varphi(\pi'')$. By (8), this implies $\varphi(\pi)$, that is, $\varphi(\Sigma \pi'\, \psi\, (\pi'))$. Q.E.D.

Paradoxes of Measurement

PATRICK A. HEELAN

Department of Philosophy, Georgetown University, Washington, DC 20057, USA

ABSTRACT: Applying Husserl's Eidetic Phenomenology to the transcendental structure (or "phenomenological reduction") of scientific data "given" in measurement to a first-person individual observer (the experimenter, S_1) and a related third-person individual observer (the observer of the measurement process, S_3), and comparing the outcomes, two paradoxical theses, "paradoxes of measurement," are derived. Thesis I: classical science necessarily entails "complementarity," "uncertainty relations," and the "entanglement" of observers and the data they observe. This situation is analogous in structure to that of quantum physics. Thesis II: a quantum object is a physical object with footprints in the perceptual world, but lacking a space–time "body"; it exists ontologically before the constitution of the perceptual world of the laboratory.

KEYWORDS: Husserl; phenomenology; measurement; complementarity; entanglement; uncertainty principles; classical data; quantum data; laboratory space-time; intuition; Primas; Atmanspacher; group theory; Wigner

INTRODUCTION

My scientific field is theoretical physics.[a] But in the philosophy of chemistry I have been influenced by the work of Michael Polanyi,[14] Eugene Wigner,[15] Jaap van Brakel,[13] Hans Primas,[11] and Harald Atmanspacher.[12] This paper addresses some of the core problems addressed by Primas and Atmanspacher. My philosophical orientation, however, is phenomenology, especially hermeneutical phenomenology. This is a tradition of thinking that for historical and cultural reasons has scarcely been applied to the philosophy of natural science.[b] There is an urgent need today to broaden the philosophical understanding of natural science.

Address for correspondence: Department of Philosophy, Georgetown University, Washington, DC 20057. Voice: 202-687-8021; fax: 202-687-8039.
heelanp@georgetown.edu

[a]My scientific field is theoretical physics. During my post-doc at Princeton, I came under the influence of Eugene Wigner who always referred to himself as a chemical engineer. He was the founder of the group theoretic formulation of the quantum theory. For the group-theoretic structures of observational data, see Refs. 1 and 2; for the praxis-ladenness rather then the theory-ladenness of scientific data see Refs. 3 through 10. For the philosophy of chemistry, I have been influenced by the works of Hans Primas[11] and Harald Atmanspacher[12], who refuse to exclude the subjective dimension from their analysis of natural science. This paper attempts to look at roughly the same core range of problems as do Primas and Atmanspacher, but from a simpler point of view inspired by Husserl. I want especially to thank Jaap van Brakel, whose book first raised my awareness of the differences between physics and chemistry.[13]

UNDERSTANDING MEASUREMENT

The specific topic of this paper is to find the right balance among theory, experiment, and what Husserl called "*die Sache selbst*" or the "givenness" of the scientific objects as experienced, recognized, and measured in the activity of measurement.

A factor in this "givenness" is what Primas calls "intuition." He calls it "the immediate grasping of evidence." This "grasping of the evidence" is not just the grasping of the evidence of a formal deduction. The formal structure is accompanied by an ideal inner "picture" in terms of which the order of the external reality is "mirrored" in the order of the internal reality. But he does not specify how such a "picture" is composed and how it functions to "mirror" the external reality. The "internal reality" has to explain the ideal unity of multiplicity and intentionality that constitutes the "picture."[c] It is Husserl who points the way as to how this is done.

About the terminology used in this paper: by *scientific objects*, I mean events that occur, usually in the laboratory, taken to manifest the local presence and measure of a named element of a scientific explanatory account. Such scientific objects can be called *data*, *phenomena*, or *protocol events*. They are particular, local, observed, and recorded events, described in scientific terms.

The starting point for my reflection on natural science is an application of Husserl's eidetic phenomenology of perception to measurement.[d] From this I go on to the analysis of data "constitution,"[e] and thence, to two theses. The first establishes "uncertainty relations," "entanglement," and "complementarity" in classical physics within the perceptual world of the laboratory. These turn out to be analogous in formal structure to those in quantum physics. The second explores the givenness of quantum objects also in the perceptual world of the laboratory. It establishes that, though quantum systems are physical objects, they are not "embodied" in the perceptual world of the laboratory; their existence is attested to only by their footprints in that world according to when and where measurement-like operations are performed; this implies that the existence of quantum systems is ontologically prior to the constitution of the space–time of the experimenter.

I will show in this analysis that not one, but two theories are involved in the constitution of data. One governs the constitution of the ideal object; the other governs

---

[b]Phenomenology is the tradition beginning with Hume and Kant that eventually wended its way through Hegel, Schleiermacher, Dilthey, to Husserl, the so-called modern founder of "phenomenology," and later to Heidegger, the founder of "hermeneutical phenomenology" (for the history of phenomenology, see, for example, Spiegelberg[16]). In my understanding of perception, I owe much to my good friend Karl Pribram,[17] with whom much of this paper was discussed.

[c]Primas[11] (Chapter 2, especially, p. 32). See also Atmanspacher,[12] particularly the articles by Atmanspacher, Primas, and Rössler. I find the prefixes "endo-" and "exo-" confusing insofar as they tend to nominalize moments in the constitution of scientific knowledge that are better understood when treated phenomenologically—which is what I do in my paper.

[d]I have listed in the references some relevant titles from my published papers. A complete list will be found on the web site, www.georgetown/heelan. Heelan[4] is an early attempt to deal with these problems on the broadest scale; see also Heelan.[5]

[e]The term *constitution* is a technical term with Husserl. An object "constituted in perception" means that it is structured by the perceiving subject "intentionally," that is, for the purpose of presenting to the perceiver a named (or nameable) object of perception different from and over against the perceiving subject. Note: scientific and functional orderings can be incompatible with one another; consequently there is only a contingent connection between the scientific and the functional orderings.

the production of data. Both can be expressed group-theoretically in the spirit of Einstein and Wigner.[15] My study will show that, despite the traditional consensus on the "theory-ladenness" of data, neither of these theories play a definitive role in the constitution of data, but that data are instead contingent and praxis-laden. Such a conclusion would also seem to be in general agreement with Primas.

How is "group-theoretical" to be understood here? This term presupposes that the measurable quantities of the physical system being investigated are modeled by a system of equations in space–time that relate them to one another in a way that defines the essential or ideal structure of the system. In classical physics, the values of such quantities and their mutual relationships are preserved if the system is relocated relative to its original position in space/time, or if it were moved in accordance, say, with Newton's first law. It is possible moreover that other changes imposed on the system would also leave particular quantities and relationships undisturbed, but only space–time transformations will concern us here. Such motions and changes are represented by sets of transformations of the equations that preserve their mathematical form. Each such set of transformations must have the properties of a group, and the inner or essential structure of the system under investigation is given by the invariants and covariants, also called, the *symmetries* of the groups. Expressed as such, the transformation laws are "group-theoretical."

Of the two theories referred to above, the one governing the ideal scientific object is what is usually referred to as the "theory" of the system. The other is a theory governing the scientific observer's essential contribution to the constitution of data. This is Husserl's perceptual "*eidos*" or "essence" and arguably, Primas's "internal picture." The so-called "theory-ladenness" of scientific data involves only the former theory.

Husserlean phenomenology is a philosophy of what is "given" in perception. It is also a "science" and a "scientific philosophy," by Husserl's reading of the standards of the Göttingen science that transformed physics in the first half of the twentieth century. What is perceived in science are data. Phenomenology then can claim to be a scientific philosophy of scientific data. I will be concerned with the phenomenology of measured data. While this may seem to narrow the notion of experimentation and to fall short of giving recognition to other traditional modes, say, of chemical experimentation based more on the observation of quality-changes than of quantity-changes, the argument to follow holds for quality-changes too. In fact it would be easier to make the argument for quality-changes alone. I am aware, however, that modern chemistry uses qualitative data as well as measured data, but taking this into account does not change the argument or its outcome.

Husserl was trained in mathematics as well as in philosophy. He taught philosophy at the University of Göttingen from 1901 to 1916 at a time when German experimental physics was being converted into international theoretical physics, under the influences of the Göttingen mathematicians and physicists, or "natural philosophers," as they were then called, within his own Faculty of Philosophy—to name some, David Hilbert, Felix Klein, Richard Courant, Emmy Noether.[1,2] Also powerfully influenced by the Göttingen orientation were, among others, Einstein, Heisenberg, and the physical chemist, Eugene Wigner. This new notion of science privileged mathematical models based on group theoretic invariance and covariance. Husserl, influenced by the intellectual and scientific climate of Göttingen, set about trying to cure the positivistic crisis in the philosophy and psychology of his time by re-doing philosophy on the model of Göttingen science. Following his earlier works, *Logical Investigations* (orig. pub. 1900 and revised in 1913),[18] *Ideas I* (orig. pub. 1913)[19] and *Ideas II* (orig. draft. 1913),[20] *Cartesian Meditations* (orig. draft 1929),[21] and *The Crisis of European Sciences* (orig. draft 1929)[22]—he claimed that phenomenology was to be a "scientific philosophy." One can read his work then as attempting to marry the new definition of science as necessarily theoretical and group-theoretic with the notion that science had to be essentially about phenomena, that is, data as perceptual objects. His clue would have been to find or construct a theory or model of the perceptual object as a covariant spatial structure of the inhomogeneous Galilean spatial group. This is the group of transformations that represents relocations of an object and/or movements given to it in accordance with Newton's first law. At the center of all space–time representations is the origin of coordinates that can be taken to represent the point from which the system is viewed by an observer. All coordinate transformations can be interpreted in two ways, either actively as representing a relocation or movement of the object, or passively as a relocation or movement of the observer. Since this transformation group, in both its active and passive interpretations, maintains the form invariance of the equations under the transformation group, one of the invariants or covariants of the group is the stable space–time relationships of the system as ideally modeled by theory.

The theoretical model, of course, purports to represent the real perceptual object, which, however, is an empirical, not an abstract, object; it is an intuited sensible object, an observed datum seen in and against the background of that part of the real, lived world that is the laboratory. The perceptual object or datum and its theoretical model are then two different objects: one is something presented in the perceptual world of the laboratory, the other is a mathematical model that purports to reflect accurately the bare objective but ideal structure of the former. We are entitled then to ask: With what justification or within what limits do we equate the scientific model with the scientific phenomenon? The question may sound odd to scientific ears because scientific training orients scientific thinking and reasoning almost exclusively towards the model as, towards that which is universal and impersonal in the object, forgetful of the fact that it is only through data that the object is given as a stable object of inquiry. Scientists also tend almost as a matter of principle to use the same scientific term for the phenomenon and for its theoretical model. This last practice is disturbing because, to use possibly an extreme example, it risks confusing, say, Number 10 Downing Street with the number, here 10, within the numbered system used to number the houses on Downing Street, which may tell you nothing about what is most important to know, namely, that it is the British Prime Minister's official residence. These two numberings belong to different categories. Clearly, some kind of explanation is necessary of how we observers use measurement to link mathematical models with given practical and presumably stable scientific objects, such as data. Going to Husserl, we find an answer for perceptual objects in his earlier works just referred to and a parallel one for scientific objects in his posthumous work, *The Crisis of European Sciences and Transcendental Philosophy*.[22] Let me briefly summarize what is essential.*f*

---

*f*See Petitot[23] or Heelan[3,4] for a more technical account.

Husserl asks: what is involved, essentially, in perception? We see, hear, feel, smell, (etc.) something; let's give it a name, "O." O can be recognized as a stable something, even though it shows itself in time only by and through a multiplicity of different perspectives and ever-changing appearances. He calls these appearances "*Abschattungen*."[g] We usually translate the German word either as "appearances (of something)" or "profiles (of something)." Literally, "*Abschattungen*" means "a shadowing-forth (of something)." Remember the "shadows" and "images" of reality in Plato's *Myth of the Cave*! We never see a perceptual object as a simple unity, since it presents itself to our senses only as a unity distributed over an indefinite multiplicity of ways of appearing in typical situations. Not just the ancient Greek philosophers, but later empiricists and rationalists alike, were puzzled by the unity in multiplicity of a stable, sensible object. Consider its space–time appearances, these change either because the object moves or is moved in relation to the observer (an active transformation), or because the observer moves in relation to the object (a passive transformation). Despite all this dynamic change, the perceptual object remains one, stable, and essentially unchanged. Husserl was the first to note that the appearances are not just connected by mere association in the empiricist sense. He concluded that they are produced dynamically by a set of active and passive transformations among themselves. Under these transformations, the object remains unchanged.[h] The mathematical model of such a system is a group-theoretic one, where the transformations constitute a group, where the active and the passive groups are identical, and where the object is represented by the invariant of the transformation group that organizes the appearances. The objectivity of the object, then, is not constituted by the mere association of appearances with a stable object, but by the symmetry or invariance of its dynamic transformation properties under some relevant group of space–time movements.

Husserl's essential point was that the same identical transformation group governs the possible movements and acts of the subject and possible movements and acts of the object within their common space and time. Whether the subject moves in accordance with the group or the object moves or is moved independently in accordance with the group, one and the same identical object is constituted and pre-

---

[g]In this paper the term *appearance* will have the sense of Husserl's term "*Abschattung*," which implies a "shadowing forth" of some to-be-found perceptual object. This process may or may not be successful; when successful it will be an "*Erscheinung*." There are terms in English to express this difference in connotation. Other terms more or less synonymous with *appearance* are *perspecitve* and *profile*.

[h]Physics recognizes three different physical space–time transformation groups: the Galilean group characteristic of Euclidean geometry (assumed by classical mechanics), the Lorentz group (assumed by Maxwell's Equations and Special Relativity), and the Continuous Group (assumed by Gravitation and General Relativity). While it is generally assumed that Euclidean scientific space–time is the unique idealization of perceptual space–time, Heelan[4] has criticized this assumption claiming that perception and pictorial space, unassisted by physical techniques of measurement, is better described by the family of 3-D constant negative curvature Riemannian geometries. These are incommensurable and incompatible spatial orderings. Although Euclidean geometry is taken to be the normative model for perceptual/scientific objects such as crystals, plants, colors, and so forth, it is often (mistakenly) taken to be essentially (or eidetically) normative for such objects. Two dubious assumptions tend to lead to this conclusion: (1) that classical (measurement-dependent) objects are simple idealizations of perceptual objects and (2) that Euclidean space–time (based on rigid rods and standard clock measurement) is a unique idealization of perceptual space–time. See Husserl.[22]

sented for observation. Husserl was thus able to claim that the objectivity of perception is linked to an essential dynamic symmetry linking the subject and the object of perception. The phenomenological constitution of a stable perceptual object involves, then, a normative group-theoretic structure or model that is linked objectively and subjectively by active and passive space–time transformations to a suite of contingent empirical data by the appropriate practices of measurement. Such a conjunction of necessity and contingency is what in fact characterizes scientific explanation; it is neither the regular association of merely contingent data, nor is it the mere abstract presence of a mathematical norm, it is the symmetry of the data when and to the extent that they fall practically under the interpretation of the norm. For that reason, scientific explanation is rightly called *hermeneutical*.

So far, so good! What the explanation so far given lacks is the capacity to explain the fact of "objectification" as a reality in the world. "Intentionality" is the name Husserl used for the human capacity to objectify as real, where the object of perception is taken to exist in the world separately from and independently of the subject. This means that in the world of our experience, the subject exists, the object exists, and the subject is not the object. For Husserl, this capacity of human observers to objectify is a primitive given ontological capacity of humans and a universal condition of possibility of all human inquiry. An object so objectified is said to be present to the perceiver as a datum structured by its perceptual "*eidos*" or "essence." Such e*idoi* are retained by the subject and used for the habitual uses of recognition, description, and categorization. Their existence supposes the possession, construction, and retention of what seem in Kant's philosophy to play the role of "schemata" that play a mediating role between concepts and sensible intuition. Husserl called the subject's constitutive activity in perception "*noesis*"; it is one of probing and seeking objects in the environment by giving meaning to group-theoretic invariants of sets of possible appearances according to an implicit dynamic plan or "schema" structured by the space and time transformation group of the perceived world. Husserl called the object's self-manifestation in the world according to its eidetic form, the "*noema*," or the "object normed by its proper law among its appearances." To what extent the activities of *noesis* and the discovery and constitution of *noemata* are historical, developmental, and potentially creative is a question to which Husserl and Heidegger gave different answers: Husserl chose fixed transcendental norms for both *noesis* and *noemata*; Heidegger chose to take the historical and developmental view.[23]

DATA:
ARE PERCEPTUAL OBJECTS THEORY-LADEN OR PRAXIS-LADEN?
BY ABSTRACTION OR INTERPRETATION?

You are all familiar with the apparent truism that "all scientific data are theory-laden."[i] This phrase was originally coined in the 1950s by N.R. Hanson[24] to refute the then widespread view that data, qua perceptually given, are facts, free of interpretation before they are networked by theory. He showed that scientific data make no sense antecedent to theoretical relationships that mutually define their theoretical scientific essence. As far as the logical analysis goes, so far, so good! But what about a phenomenological analysis of data? How do practical data fulfill this logical analysis? How are thought and perception put together?

One answer is that the logical analysis is an "abstraction" from what is already there in the practical data and is achieved by choosing what is relevant and eliminating the merely irrelevant in the data. Or are data original creations, artifacts of human invention, perhaps, of whimsy? I take them to be none of these. The data of most scientific explanations are not "given" first and then analyzed; they are produced with the aid of sophisticated theories and elaborate technologies in a special laboratory environment. Data are what are "given" at the end of a piece of basic research; not at the beginning; they are understood as phenomena only when the research is completed. Hence, philosophical reflection begins at the end only when the phenomenon can finally be presented to the philosophical inquirer for his/her reflection. Its aim is to understand the phenomenon in terms of how it is "constituted" as an object of human scientific knowledge. This is the goal of a phenomenology of a scientific phenomenon. The conclusion here defended is that a scientific phenomenon is an object in the world long concealed as to its possibility of becoming an object of human knowledge and eventually revealed by theory-based scientific practices as an object for human contemplation and cultural use.

How does an observer come to recognize in the given outcome of a measurement the presentation of a stable object in the experimental laboratory? The practical answer seems to be that the scientist *learns* to do this. The fact is that a well-skilled experimenter accepts a scientific datum unquestioningly, often on the occasion of just one measurement, that is, of one glimpse of what he/she then unhesitatingly pronounces to be there in the world. Such an observer experiences a "given," or what Husserl calls, "*die Sache selbst*," by what Primas calls an "intuition" of a scientific object. We have this experience ourselves every day with things familiar to us. For example, we glimpse a familiar face ahead of us and immediately get ready, say, to greet the person in question, but we are mistaken because a moment later the familiar face turns out to be just an immobile and singularly flat life-size cardboard snapshot. One single measurement is no more than a single snapshot of something that could turn out to be something quite other than what at first sight it appears to be. Let me compare this situation with the duck/rabbit illusion. In this we are presented with an image that by design is at the intersection of two possible sets of associated images, a set of duck images and a set of rabbit images. The image presented is at the intersection of these possible but imagined sets and is so drawn as to suggest that it could belong to the intersection of the two sets of images. In relation to one set it can be seen as a duck while in relation to the other set it can be seen as a rabbit. The imagination provides the alternate sets. The eye now sees a duck, a moment later it sees a rabbit, but it never sees both together. This illusion illustrates the view that there exists a specific underlying *noetic* structure operating in ordinary perception for perceptual images such as (in this case) those of a duck and those of a rabbit. A skilled experimenter has likewise developed a similar *noetic* structure for those laboratory measurements with which he has acquired familiarity and skill on account of which

---

[i]In the practice of science, the term "theory" usually implies a model insofar as it is related to the world. One needs to be reminded that within the model the relationships are mathematical, whereas within the world, and between the model and the world, the relationships are just factual. There is much confusion in scientific and philosophical literature about this, particularly where science is viewed from a Pragmatist perspective (see Refs. 4, 7 and especially 7, 9, and 10) on the praxis-ladenness of scientific phenomena).

he/she can effortlessly recognize even in a single measurement event either a blind instrumental response or a scientific datum.

Husserl wanted to tie down philosophically and scientifically the theory of the intuitive givenness of a perceived object. While Husserl asked the question of everyday perception, I am asking it of measurement. Both experience the "givenness" of perceptual objects, everyday in one case, scientific in the other. Both are given as stable objects revealed as present in intuitive sensibility only through infinite manifolds of possible appearances structured in such a way as to reveal the presence of a single stable object in the space and time of sensible intuition. It is through this multiplicity of quantitative and qualitative appearances that we come to recognize (what Husserl calls) the "core meaning" of the kind of being that is "shadowed forth" in perception.

Revisiting for a moment the seeming truism that scientific data are "theory-laden," and given now that there are two theories in question, one for the observer as such and the other for the observed datum, I can ask: to which theory does this truism refer? My answer is, to neither, for the recognition of a measured object always occurs in the lifeworld of the observer as a contingent empirical act dependent on experimental skill, the discernment of "all things being equal" in environmental circumstances, and the assessment of the purpose of inquiry. Data then are primarily praxis-laden, based on measurement and on their circumstantial "givenness" or "intuitiveness," that under doubtful circumstances is checkable with reference to the two theories just mentioned. I wish to point out that these conclusions belong to the genre of philosophy; not sociology, history, psychology, anthropology, or even theoretical science. Similar conclusions views have been expressed by Hacking,[25] Latour,[26] and some others, but argued more on the grounds of social science or common sense than of philosophy.

PARADOXES OF MEASUREMENT, THESIS I

Using the above analysis of scientific phenomena and data, I will briefly summarize two fascinating, but paradoxical, philosophical principles about the natural sciences to which they lead. These are the "Paradoxes of Measurement" of the title of this paper. The first thesis is about the existence of some basic similarities between classical phenomena in natural science and quantum phenomena, such as "uncertainty principles," "entanglements," and "complementarity."

Thesis I. *Classical Uncertainty Principles: There are basic Uncertainty Principles for classical phenomena in classical physics and natural science that are analogous to those of quantum physics. These are due to the "entanglement" of the observer and the observed in the constitution of empirical data, and possible "complementarities" in the dynamic play of noesis and noema in relation to the use of measuring instruments.*

To explain what this thesis means: consider two observers, S_1 (a first-person observer) and S_3 (a third-person observer), looking at the same measurement process but from different points of view. S_1 is the individual skilled experimenter who is observing the datum of O_1; he takes a certain response M of the measuring instrument to manifest the presence of O_1 under the aspect of a certain measured quantity. About this S_1 makes a first-person report. S_3 is someone who is observing S_1's engagement with the physical process of measurement but is blind to S_1's interpretive act; he just

$$\boxed{S_1 S_3 M O_1}$$

FIGURE 1. Two subjects, S_1 and S_3, and two objects, M and O_1.

$$\boxed{\begin{array}{c}(S_1 + M) \text{ observes } O_1 \\ (\text{but M is not an object for } S_1)\end{array}}$$

FIGURE 2. S_1 observes O_1.

$$\boxed{\begin{array}{c}S_3 \text{ observes } M \\ (\text{but } O_1 \text{ is not an object for } S_3)\end{array}}$$

FIGURE 3. S_3 observes M.

sees the instrumental response M but not the scientific datum O_1. S_3 could be the same person as S_1 but in the role of a technician or engineer. Alternatively, S_3 could be a social scientist, a cognitive scientist, or even a reductionist philosopher. S_3 makes a third-person report about the response of the instrument.

FIGURE 1 lists the two subjects, S_1 and S_3, and the two possible perceptual objects, M and O_1—each given through one of a set of its appearances without "entanglements" deriving from perceptual relationships. In FIGURE 2, S_1 observes O_1, the measured datum; the instrumental response M is in this case a functional part of the operating subject S_1, since it brings into play the measured datum in which O_1 makes its appearance to an observer in the laboratory. In FIGURE 3, S_3 observes M in one of its appearances, but O_1 is not present to S_3 because the one appearance of M that is evidence of the measured datum cannot be, at the same time, both an appearance of M. and an appearance of O_1 to the same observer. The reason for this is the same as that given for the duck/rabbit illusion; the ambiguous image cannot be seen at the same time as an image of a duck and as an image of a rabbit because an object is perceived only if the entire range of its connected appearances is virtually present through the dynamic *noetic–noematic* schema in which objective information is virtually exchanged between the observer and the world. This relationship is a kind of dynamic hermeneutical "entanglement" between the observer as a *noetic* agency and the observed as a noematic responder.

The analogy with quantum physics can be pursued further. Let the dynamic world of multiperspectival classical human perception be modeled by a Hilbert space Ψ where the states of the dynamic world of perception are represented by vectors in this space. Let S_1 and S_3 generate projection operators, $P(S_1) = P_1$ and $P(S_3) = P_3$ on the space Ψ. P_1 generates $P_1\Psi$, the subspace of Ψ that represents the dynamic world of S_1's perception, and P_3 generates $P_3\Psi$, the subspace of Ψ that represents the dynamic world of S_3's perception. The subspaces, $P_1\Psi$ and $P_3\Psi$, are theoretical representations of the empirical *noetic–noematic* perceptual horizons of S_1 and S_3, respectively, in the Hilbert space representation of the dynamic world of multiperspectival human perception. In the subspace $P_1\Psi$, O_1 is represented as an object but not M; in the subspace $P_3\Psi$, M is represented as an object but not O_1. Thus, the subspace $P_1P_3\Psi$ contains O_1 but not M, while the subspace $P_3P_1\Psi$ contains neither M nor O_1. Forming the commutation operator $[P_1P_3 - P_3 P_1]$, we find that the commutation operator

$$[P_1P_3 - P_3 P_1]\Psi \neq 0$$

that is, the commutation operator is not zero, but contains at least a representation of O_1.

This establishes that there is a formal analogy between the classical phenomena and quantum phenomena. This becomes apparent only when it is understood that data recognition assumes an unconscious structure in the viewing subject that can be group theoretically modeled identically in the datum and in the observer. This common structure describes the "entanglement" of S_1 with O_1 and precludes their separation while observing. The reason is that S_1 embodies a *noetic* orientation towards O_1 that shapes and is shaped by O_1's *noematic* structure as the object known. Neither can exist apart from the virtual flow of information that establishes S_1 and O_1 as a functioning unity of perceptual knower and perceptual known. Each needs the other to establish its respective existence. The same can be said for the cognitive "entanglement" between S_3 and M. On this Husserlean account of perception, the basis of the analogy between perception and quantum physics can be expressed in the following way: S_1 is dynamically *entangled* with O_1, and S_3 with M, in such a way that subject and object are dynamically inseparable during observation; moreover, the respective horizons of S_1 and S_3 are *incommensurable* within the world of human perception in a way analogous to complementary observables in quantum physics, that is, they are constrained factually and hermeneutically by *Uncertainty Principles.*[j]

Returning to the real world, we ask: Who in real life should be concerned with the results just obtained? Who are S_1 and S_3 and what roles do they play? Clearly S_1 is a scientific researcher in his/her native habitat, the "enclosed garden" of the laboratory. S_3, however, could have several roles; for example, (1) a scientist reflecting critically on the foundations of scientific thinking, or (2) an interdisciplinary scholar concerned to know how to evaluate cross-disciplinary factual data, or (3) a philosopher reflecting on the hermeneutic paradoxes of scientific thinking, or (4) a cognitive scientist puzzling how to link the theory/practice methods of modern science to human consciousness. There are lessons that each can draw.

Regarding the foundations of scientific thinking, it seems that, contrary to the traditional expectations of scientists, the thesis that a universal objective space–time exists onto which all factual data can simultaneously be mapped from a single universal point of view that is theoretical and at the same time human and practical proves not to be the case. It may, of course, turn out to be a useful fiction or postu-

---

[j]This is another way of saying that quantum logic is a nondistributive logic of contexts (see Heelan[4] on this topic).

late—Plato's "likely story"—for certain purposes, for example, for the solution of classes of problems for which the models and practices of, say, classical mechanics are found to be *de facto* successful. However, the thesis stated above is true for any science that is based on the theoretical modeling of factual data, and what science is not so structured? The root of the classical uncertainty is in measurement, where instrumental data are converted into scientific data, not by a textual hermeneutics (or reading) as in the Cartesian view, nor by deriving the higher from the lower by "abstraction," but by a human embodied object-constituting interpretative process that Husserl called "a *noetic-noematic* intentionality structure." This is not just a scientific thesis but, according to Husserl, a transcendental philosophic thesis.

PARADOXES OF MEASUREMENT, THESIS II

Quantum physics provides an especially interesting case for phenomenological analysis. This will be the subject matter of my second thesis.

Thesis II. *Quantum Systems as Disembodied Physical Objects: An individual quantum object is intuitively given to an observer only by the actual isolated footprints it leaves in the perceptual world of the laboratory, the record of individual measurements. It is not just a conceptual object, nor is it an "embodied" object in its own right. It is, however, physical and material because of the footprints it makes in the world. It seems then to exist and function ontologically prior to and in some way independently of the phenomenological constitution of classically scientific laboratory space-time.*

There is an entrenched notion among many scientists and philosophers that subjectivity and objectivity in science can be defined in terms of S/O cuts that separate subject (S) and object (O) from one another spatially and temporally. This account is incompatible with Husserl's account of perception as structured by a *noetic-noematic* intentionality structure, the reason being that subject and object are "entangled" in the cognitive process of measurement. The fact of entanglement was already mentioned in the analysis of the cognitive process of measurement in classical physics. S's entanglement with O precludes a spatio-temporal separation, because S embodies a dynamic noetic orientation towards O that shapes and is shaped by O's dynamic noematic structure as the object known. Each in some sense is working "inside" the other. *Neither subject nor object can exist apart from the flow of information that constitutes S as a perceptual knower and O as a perceptual known.* Each needs the other to establish its "objective" existence vis-à-vis the other; in other words, both subject and object exist, and they are distinct existentially. The core meaning of "objectivity" is: S exists, O exists, and S is not O. But this does not necessarily involve spatio-temporal separateness; or rather it raises the question of the relationship between spatio-termporality and objectivity in the experimenter's world. The core distinction of S/O cuts is existential and ontological rather than spatial.

In contemporary terminology, the S and O are energetic systems that interface with each other in a physical and dynamic zone that is called the "Heisenberg cut,"[k] This cut can be imagined as a space-time zone of interaction, but this imagery is

---

[k]See Atmanspacher and Primas,[12] pp. 21, 181, 187, 318.

merely a metaphor for the dynamic interplay of two distinct things that are shaping each other mutually.

'To be embodied" means "to have a stable extension in some perceptual space–time." Scientific objects disclosed by measurement would normally be displayed with the anticipation of a classical body. Quantum systems turn out then not to be "embodied"—they are "disembodied," they are not "bodies." Although they have properties, such as mass, charge, spin, and so forth, which are usually defined relative to the classical space–time of the laboratory—they lack an embodied presence in this space–time. What relationships they have to this space–time, they acquire only by measurement, and these relationships are episodic because they are quantized and isolated seemingly from the dynamic ordering of laboratory space–time. There is then only the residue of a bodily presence in the potential set of isolated appearances that are, as it were, its footprints in the laboratory world when the quantum system is actually measured. These isolated appearances do not constitute a body that fulfils the Husserlean protocol for a body, that is, where the body is an invariant or covariant of a representation of the relevant perceptual space–time group. They are nevertheless more than just signs of an abstract or nonphysical presence, they show a momentary local presence in the "*place*" and at the "*time*" where they appear. Perhaps it would be useful to adopt Aristotle's notion of an object's "place," since it is subject-related rather than object-related. Aristotle defines *place* as the smallest closed surface (in the observer's perceptual world) that contains the (observed) object. A quantum system can then be said to have occupied a "place" in the neighborhood of any footprint, whenever and wherever a measurement occurs. But except during measurement, quantum systems seem to exist and function in a manner that is in some way independent of and antecedent to the constitution of an experimenter's laboratory space–time. Even though quantum systems are physical and material objects in the world of the scientist's laboratory, even though, despite their disembodied state, they can be present in a "place" in an observer's perceptual space, even though on this account they possess a title to being called "real," they do not show themselves as classical bodies.

Consider some examples: (1) The ionized trajectory of a particle in a cloud chamber is not the track of an embodied charged particle, it just "speaks" of the presence in laboratory space of a succession of appearances of a disembodied charged particle. (2) The context of preparation "tells" us that an electron is prepared, say, with a vertical spin, but the context of measurement "tells" us that when detected it can appear in laboratory space as disembodied but with a horizontal spin. (3) The silver particle in a photo emulsion "tells" us that a charged particle or photon has been absorbed; that particle existed in this "place" (in laboratory space) for a moment and then disappeared. In none of these cases is the quantum system described as an embodied object; its presence in the laboratory world, however, is manifest in the context of measurement.

To get a better intuitive sense of what this implies, we look for things in the dynamic field of human perception that could throw light even by analogy on the character of a quantum system as a real but disembodied perceptual object? Music seems to be something of this kind; it displays itself in a circumscribed "place" within perceptual space–time, filling every part of this "place" with identical sequences of sounds. The form of the music, however, is dynamic and non-reversible; consequently it is not preserved when the temporal order is reversed. The musical object then is not an embodied object under the conditions set by Husserl. Nor are speech, taste, smell, in general, embodied phenomena. All of these display themselves in perceptual space but fail to satisfy the Husserlian space–time protocols for the phenomenological reduction of a body to an *eidetic* form. Each, however, has its underlying embodied space–time physical field that is the classical explanatory ground of the phenomenon in question, but it follows from Thesis I that such fields, insofar as they are given only by measurement, are also affected by uncertainty principles, entanglement, and complementarity.

REFERENCES

1. HEELAN, P. 1988. Husserl, Hilbert and the critique of Galilean science. *In* Edmund Husserl and the Phenomenological Tradition. R. Sokolowski, Ed.: 157–173. The Catholic University of America Press. Washington, DC.
2. HEELAN, P. 1989. Husserl's philosophy of science. *In* Husserl's Phenomenology: A Textbook. J. Mohanty & W. McKenna, Eds.: 387–428. CARP & University Press of America. Pittsburgh, PA; Washington, DC.
3. HEELAN, P. 1989. After experiment: research and reality. Am. Philos. Q. **26:** 297–308.
4. HEELAN, P. 1983 (1987 pbk.). Space Perception and the Philosophy of Science. University of California Press. Berkeley, CA.
5. HEELAN, P. 2002. Afterword. *In* Hermeneutic Philosophy of Science, Van Gogh's Eyes, and God: Essays in Honor of Patrick A. Heelan. S.J.B.E. Babich, Ed.: 445–459. Kluwer. Dordrecht, the Netherlands; Boston, MA.
6. HEELAN, P. 2002. Phenomenology and the philosophy of the natural sciences. *In* Phenomenology World-Wide. A-T. Tymieniecka, Ed.: 631–641. Kluwer. Dordrecht, the Netherlands; Boston, MA.
7. HEELAN, P. (with JAY SCHULKIN). 2002. Hermeneutical philosophy and pragmatism: a philosophy of science. *In* Philosophy of Technology, The Technological Condition: An Anthology. Robert Scharff & Val Dusek, Eds. Blackwell's 2002. Oxford, UK.
8. HEELAN, P. 2002. Lifeworld and scientific interpretation. *In* Handbook of Phenomenology and Medicine, Series in Philosophy and Medicine, vol. 68. S. Kay Toombs, Ed.: 47–66. Kluwer. Dordrecht, the Netherlands; Boston, MA.
9. HEELAN, P. 1998. Scope of hermeneutics in the philosophy of natural science. Stud. Hist. Philos. Sci. **29:** 273–298.
10. HEELAN, P. 1997. Why a hermeneutical philosophy of natural science? Man World **30:** 271–298.
11. PRIMAS, H. 1983. Chemistry, Quantum Mechanics, and Reductionism: Perspectives in Theoretical Chemistry. Springer. Berlin, Germany.
12. ATMANSPACHER, H. & H. PRIMAS. 1997. Representation of facts in physical theory. *In* Time, Temporality, and Now. H. Atmanspacher & E. Ruhnau, Eds.: 241–263. Springer. Berlin, Germany.
13. VAN BRAKEL, J. 2000. Phiilosophy of Chemistry. Leuven University Press. Leuven, Belgium.
14. See BRANDT, WERNER. 2003. Chemistry beyond positivism. Ann. N.Y. Acad. Sci. This volume.
15. WIGNER, E. 1967. Symmetries and Reflections. Indiana University Press. Bloomington, IN.
16. SPIEGELBERG, H. 1982. The Phenomenological Movement: A Historical Introduction, Vols. I & II. Nijhoff. The Hague, the Netherlands; Boston, MA.
17. PRIBRAM, K. 1991. Brain and Perception. Erlbaum. Hillsdale, NJ.
18. HUSSERL, E. 2001. The Shorter Logical Investigations. J.N. Findlay, trans. & D. Moran, Ed. Routledge. New York, NY.
19. HUSSERL, E. 1983. Ideas: General Introduction to Pure Phenomenology [Ideas I]. F. Kerstan, trans. Kluwer. Dordrecht, the Netherlands; Boston, MA.
20. HUSSERL, E. 1989. Ideas Pertaining to a Pure Phenomenology: Second Book [Ideas II]. Kluwer. Dordrecht, the Netherlands; Boston, MA.

21. HUSSERL, E. 1960. Cartesian Meditations. D. Cairns, trans. Nijhoff. The Hague, the Netherlands.
22. HUSSERL, E. 1976. Crisis of European Sciences and Transcendental Philosophy. Northwestern University Press. Evanston, IL
23. PETITOT, J. 1999. Morphological eidetics for a phenomenology of perception. In Naturalizing Phenomenology: Issues in Contemporary Phenomenology and Cognitive Science. J. Petitot, F.J. Varela, B. Pachoud & J.-M. Roy, Eds.: 330–371. Stanford University Press. Stanford, CA.
24. HANSON, R.N. 1958. Patterns of Discovery. Cambridge University Press. Cambridge, England.
25. HACKING, I. 1983. Representing and Intervening. Cambridge University Press. Cambridge, England.
26. LATOUR, B. 1987. Science in Action. Open University Press. Philadelphia, PA.
27. FRIEDMAN, M. 2000. Parting of the Ways: Carnap, Cassirer & Heidegger. Open Court Press. Chicago, IL.

Chemical Kinetics and Dynamics

ILYA PRIGOGINE

Solvay Institutes for Physics and Chemistry, Brussels, Belgium

The Ilya Prigogine Center for Studies in Statistical Mechanics and Complex Systems, The University of Texas, Austin, Texas 78712, USA

ABSTRACT: Chemical reactions correspond to irreversible processes creating entropy. Chemistry belongs to the class of *nonintegrable* Poincare systems. In general, chemistry is associated with resonances—transitions of quantum states. We have studied some very simple examples of such processes, like decay of an unstable state, in detail. (In such cases, there are always multiple time scales.) We obtain a nonunitary ("star unitary"), invertible, nondistributive operator Λ (which reduces to the unitary transformation operator U for integrable systems). The explicit form of Λ depends on the interaction of each species with all other types of molecules in the system including the solvent. The basic property that results from Λ is that the fundamental description of nonintegrable systems is no longer in terms of Hamiltonian equations, but in terms of kinetic equations with broken time symmetry. Once we have the kinetic equations, it is easy to show that we have irreversible processes and entropy production.

KEYWORDS: irreversible processes; entropy production; kinetic equations; broken time symmetry chemical kinetics; chemical dynamics; nonintegrable Poincare systems; star unitary operator; Zeno time; dressed excited state

PART I

In his famous book on quantum mechanics, Dirac stated that chemistry can be reduced to problems in quantum mechanics. It is true that many aspects of chemistry depend on quantum mechanical formulations. Nevertheless, there is a basic difference. Quantum mechanics, in its orthodox form, corresponds to a deterministic time-reversible description. This is not so for chemistry. Chemical reactions correspond to irreversible processes creating entropy. That is, of course, a very basic aspect of chemistry, which shows that it is not reducible to classical dynamics or quantum mechanics. Chemical reactions belong to the same category as transport processes, viscosity, and thermal conductivity, which are all related to irreversible processes. Chemical processes as well as transport processes are described by kinetic equations and not by canonical equations as would be the case, for example, for a harmonic oscillator. This difference has been noticed for a long time, and as far back as in 1870 Maxwell considered the kinetic equations in chemistry as well as the kinetic equa-

---

Address for correspondence: Prof. I. Prigogine, International Solvay Institutes for Physics and Chemistry, Campus Plaine, ULB/CP.231, Boulevard du Triomphe, B-1050 Brussels, Belgium. Voice: +32-2 650 50 47; fax: +32-2 650 50 28.

njockman@ulb.ac.be

tions in the kinetic theory of gas as *incomplete* dynamics. From his point of view, kinetic equations for chemistry would be the result of adding "ignorance" to the physical description. This is indeed still the opinion of the majority of physicists today. However, it would be paradoxical if chemistry, which plays a fundamental role in the description of nature, were the result of our own mistakes.

This is why we have investigated the role of irreversible processes for many years, and I am glad to accept Prof. Earley's invitation and write a short note on this problem, although it is a large subject and this note will admittedly be somewhat sketchy and incomplete. My hope is that it will present the basic idea behind our work. In our approach, we start with Poincare's famous classification of dynamic systems into integrable and nonintegrable systems. We accept the Hamiltonian description as a basis and, as is well known, this description contains both potential energy and kinetic energy. The Hamiltonian description considered in chemistry is, of course, of rather special character. For example, we have the Heitler London theory, which shows the origin of chemical bounding. For integrable systems we can, by a canonical transformation, eliminate the potential energy. It is true that this can be done completely only for very simple systems, otherwise we can approach this transformation by power expansion in terms of the perturbation. Once we have the canonical transformation eliminating the potential energy and diagonalizing the Hamiltonian description, the integration becomes trivial—that is the reason Poincare called these systems integrable systems.

But integrable systems are the exception. Most dynamic systems are nonintegrable and present different behavioral characteristics. Our point of view is that chemistry belongs to the class of *nonintegrable* Poincare systems. That is what I wish to explain here. Of course I cannot provide any demonstrations—and for the proofs, the reader has to refer to the original papers.[1-13]

PART II

Let us summarize briefly the dynamics of integrable systems. Ordinary, we have two kinds of variables, the coordinates q_i and the momenta p_i, at the initial time $p_i = (0)$, $q_i = (0)$. The value of these variables is supposed to be given. However, the point is that there are many choices for these variables. We can indeed introduce the equations:

$$p' = U^{-1} p \qquad q' = U^{-1} q$$

Canonical variables can take various expressions according to the operator U, which is called the unitary transformation operator. Let us briefly observe some basic properties of U. As I mentioned, U is unitary; that means $U^{-1} = U^+$. It is invertible; that means $UU^{-1} = 1$ and distributive; that means $U^{-1}(AB) = (U^{-1}A) \cdot (U^{-1}B)$. Those are remarkable properties, which correspond to what is called "lie algebra." The operator U is not only of theoretical but also of practical importance. We may start with arbitrary coordinate momenta and choose U so the Hamiltonian equation is diagonalized. We can then go back to the initial variables by the inverse transformation U^{-1}.

Let us now go to the nonintegrable systems. The class of nonintegrable systems that we shall consider is that system when resonances prevent the construction of U

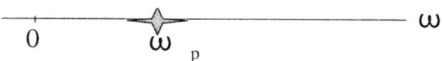

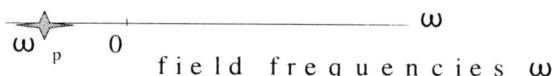

FIGURE 1

by introducing divergences. Consider a simple example: a particle, such as a harmonic oscillator, is submitted to a field. Let suppose that the frequency of the oscillator is ω_p whereas the field of the oscillator forms a continuous set starting with $\omega \geq 0$ (see FIG. 1).

There are obviously two different situations. Either ω_p is below 0 or ω_p is larger than 0. In the second case, the oscillator is dissolved in the continuum of the field. This is the well-known Einstein/Bohr mechanism for the description of spectral lines. The fundamental result by Poincare was to show that these resonances lead to difficulties through the appearance of divergent denominators in the perturbation expansion. Here is an example:

$$\frac{1}{\omega_k - \omega_l} \quad (\omega_k = \text{particle} - \omega_l = \text{field})$$

How do we remove this difficulty? We have shown that we have to perform an analytic continuation. That means that we transform the expression into a distribution or generalized function. For example, we write:

$$\frac{1}{\omega_k - \omega_l \pm i\varepsilon} \quad \varepsilon > 0$$

Of course, there are specific rules and methods that we have to apply to perform the analytic continuation. But that has to be seen in the original papers. Chemistry is in general associated with resonances, transitions of quantum states. Therefore, it is not astonishing that we are in the same situation. We have studied in detail some very simple examples, like the decay of an unstable state, and I may mention that there are always different time scales. First, a time scale corresponding to a dressing of an electron by photons, and then the emission of photons, the first part is nonexponential—this is often discussed in terms of what is now called the "Zeno time." The emission of the photons is, on the contrary, an exponential process. Our theory permits us to define another object, a dressed excited state, which decays exponentially in time. It would be interesting to consider chemical reactions from this point of view.

PART III

After the elimination of the Poincare divergences, we obtain an operator Λ, which for integrable systems reduces to U. As we have shown, Λ has number of remarkable properties:

(1) It is a nonunitary ("star unitary") operator; that means $\Lambda^{-1}(i\varepsilon) = \Lambda^{+}(-i\varepsilon)$
(2) It is invertible: $\Lambda\Lambda^{-1} = \Lambda^{-1}\Lambda$
(3) In contrast with U, it is a nondistributive operator: $\Lambda(AB) \neq \Lambda A \cdot \Lambda B$. This nondistributive operator leads new fluctuations and uncertainty relations even in dissipative classical systems.

The basic property that results from Λ is that the fundamental description of nonintegrable systems is no longer precisely in terms of Hamiltonian equations, but in terms of kinetic equations with broken time symmetry. More precisely, we may start with the well-known Liouville equation: $i\,\delta\rho/\delta t = L\rho$. For integrable systems, we can "reduce" this equation to wave functions or classical trajectories, but for nonintegrable systems, this is not possible. We may, however, write a different equation multiplied by Λ:

$$i\frac{\delta\Lambda\rho}{\delta t} = \Lambda L\rho = (\Lambda L \Lambda^{-1})\Lambda\rho$$

But the operator has a basic physical meaning. It is the operator that appears in the kinetic theory as the collision operator. Once we have Λ, we may use it for nonintegrable systems in a way very similar to the use of U in integrable systems. Let us suppose that we start with arbitrary variables in a chemical system. By performing the Λ transformation, we obtain the basic kinetic equations. If we can solve them, we can go back if necessary to the initial variables by applying the inverse transformation Λ^{-1}.

In this short note, I have remained at a very general and abstract level, but I hope that this new tool (the construction of the Λ transform) will lead to many applications in chemistry. The mathematical side is rather similar to the theory of interacting fields, which has been partly published. The explicit form of Λ depends on the interaction of each species with all other types of molecules in the system including the solvent. It gives the kinetics of dressed particles exactly as has been seen in the decay of unstable particles. Therefore, the problem of the influence of the medium on chemical reaction is in the range of the construction of the operator Λ for nonintegrable systems. Once we have the kinetic equation, it is easy to show that we have irreversible processes and entropy production. It seems to me therefore very natural to consider that chemistry is indeed a very important example of nonintegrable Poincare systems, where the nonintegrability is due to resonances. I hope that this new aspect will continue to be explored by future generations of physicists and chemists.

ACKNOWLEDGMENTS

I would like to thank Prof. Tomio Petrosky and Dr. Gonzalo Ordonez for their important contributions to this subject.

REFERENCES

1. PRIGOGINE, I. 1962. Non-Equilibrium Statistical Mechanics. Wiley. New York.
2. PRIGOGINE, I. 1980. From Being to Becoming. Freeman. New York.
3. PETROSKY, T., I. PRIGOGINE & S. TASAKI. 1991. Quantum theory of nonintegrable systems. Physica A **173:** 175–242.
4. ANTONIOU, I. & S. TASAKI. 1993. Generalized spectral decomposition of mixing dynamical systems. Int. J. Quantum Chem. **46:** 425–474.
5. ANTONIOU, I. & I. PRIGOGINE. 1993. Intrinsic irreversibility and integrability of dynamics. Physica A **192:** 443–464.
6. PETROSKY, T. & I. PRIGOGINE. 1996. Poincare resonances and the extension of classical dynamics. Chaos Solitons Fractals **7:** 441–497.
7. PETROSKY, T. & I. PRIGOGINE. 1997. The Liouville space extension of quantum mechanics. Adv. Chem. Phys. **99:** 1–120.
8. ORDONEZ, G., T. PETROSKY & I. PRIGOGINE. 2001. Quantum transitions and dressed unstable states. Phys. Rev. A **63:** 052106.
9. PETROSKY, T., G. ORDONEZ & I. PRIGOGINE. 2001. Space–time formulation of quantum transitions. Phys. Rev. A **64:** 062101.
10. KARPOV, E., G. ORDONEZ, T. PETROSKY & I. PRIGOGINE. 2002. Quantum transitions in interacting fields. Phys. Rev. A **66:** 012109.
11. PRIGOGINE, I., S. KIM, G. ORDONEZ & T. PETROSKY. 2001. Stochasticity and time symmetry breaking in Hamiltonian dynamics. Proceedings of the XXII Solvay conference in Delphi, Greece. Submitted (see Ref. 12).
12. ORDONEZ, G., T. PETROSKY & I. PRIGOGINE. 2001. Microscopic entropy flow and entropy production in resonance scattering. Proceedings of the XXII Solvay conference on Physics: "The Physics of Communication," held in Delphi, Greece in November 2001. Journal of Quantum Computers and Computation (2003) and World Scientific (2003): In press
13. KARPOV, E., G. ORDONEZ, T. PETROSKY & I. PRIGOGINE. 2002. Microscopic entropy and nonlocality. Proceedings of the Workshop on Quantum Physics and Communication held in Dubna, Russia, on May 16–17, 2002 (QPC 2002). Dubna, Russia. In preparation.

Reaction Mechanisms and Chemical Explanation

GIUSEPPE DEL RE

University of Naples Campus "Federico II", Monte S. Angelo, I-80126 Napoli, Italy

ABSTRACT: The methodological and ontological issues arising in the organic chemists' explanation of transformations of matter by means of reaction mechanisms are presented based on the general view that chemistry is at the same time an autonomous discipline with its own paradigm and an integral part of the unitary body of science. Among the issues relevant to research is the nature and actual observability of activated complexes, which have by definition a vanishing lifetime. That issue illustrates the interplay of molecular chemistry and quantum physics at the submicroscopic level of sensible reality. A moral concerning the planning and strategy of research is briefly outlined.[1]

KEYWORDS: reaction mechanisms; molecular structure; molecular lifetimes; ontological problem; specificity of chemistry; unity of science

INTRODUCTION

There is today, in spite of increasing specialization,[2] a wide consensus that science is a unitary body of knowledge concerned with producing a coherent description of the structure and becoming of the physical world.[3] Its ultimate criteria of truth are experimental verification and internal self-consistency.

Beyond these very general features, however, it is also admitted that the various disciplines differ by their fields of inquiry (the kind of objects and properties they focus their attention on), their universes of concepts (the specific concepts they use to express their findings), and their programs (the kind of cause–effect chains their explanations of facts involve). These specific features of individual disciplines are not in contrast with the idea that ultimately all phenomena can be described in terms of particle motions, and the special place physics has in science is not jeopardized by the existence of different approaches to the understanding of reality.

THE SPECIFICITY OF CHEMISTRY

The case of chemistry, which is our concern here, illustrates the scope and significance of the above general remarks.[4] The field of inquiry of chemistry is delimited by a particular class of transformations of matter, namely those that do not take place

---

Address for correspondence: University of Naples Campus "Frederico II", Monte S. Angelo, I-80126 Napoli, Italy. Voice: +39.081.674208; fax: +39.06.3973.0607.
Giuseppe.Delre@unina.it

Ann. N.Y. Acad. Sci. 988: 133–140 (2003). © 2003 New York Academy of Sciences.

reversibly upon changes in the external conditions, but can be induced directly or indirectly by temperatures of the order of that of fire. The concepts of chemical substance, chemical reaction, and related ones are specific to chemistry, inasmuch as they are required in any attempt to put order in experimental data concerning chemical transformations; other concepts, primarily that of *molecule*, were generated in succession as the discipline developed. The program of chemistry consists in discovering and classifying all possible (pure) chemical substances so as to account for their properties and behavior in terms of suitable laws. If the one-to-one correspondence between pure substances and molecules (including ions) is accepted, then chemistry can be characterized as the science of molecules.

Explanations in terms of molecules have recourse to general rules and laws holding within the sensible world, and therefore they are perfectly legitimate scientific explanations; but they differ from other kinds of scientific explanations. Because it is concerned with molecules becoming other molecules and not with looking for the ultimate constituents of matter, chemistry treats physical explanations (i.e., explanations in terms of forces and motions of particles) as belonging to a different level of complexity or layer of being—indeed, the deepest one[5]—much in the same way as a neurologist would normally ignore the molecular level in discussing the transmission of signals in the nervous system.

To grasp the difference between chemical and other explanations of facts, consider the question: Why are most glass bottles green? A technologist would respond by saying that their production cost is lower; a physicist would answer that their glass absorbs light selectively, so that the transmitted light has a different spectrum from that of sunlight; but a chemist would answer that it is because ordinary glass contains ferrous ions. This instance of a chemical explanation will do as a concrete reference for most of the following considerations, provided it be kept in mind that ferrous ions are to be seen as simple molecules, in agreement with the usage of chemistry, for which the name "molecule" also applies to atoms or ions participating in chemical reactions on their own.

What characterizes explanations of the sort under consideration here? In the frame of a complexity approach, they belong to the class of "horizontal" explanations, meaning by this attribute that they are the outcome of a special type of analysis. One starts by considering the atomic level of reality, that is, material bodies seen as nothing but collections of electrons, atoms, and molecules. One then tries to approach the transformations of matter as if they were processes involving the formation, transformation, and disruption of entities belonging to the molecular level of reality, all interactions requiring a higher degree of complexity being treated as interactions between molecules. In this way, the becoming of material bodies is reduced to events and processes that take place within a single level of reality: atoms and electrons are seen as "elementary objects," that is, only as building blocks of molecules.[6]

The perspective under which chemistry looks at the becoming of the world is thus characterized by the fact that it treats as causes entities of a special class, the molecules, which correspond to chemical substances at the macroscopic level, that is, the level of reality of our direct experience. In other words, chemistry first reduces the transformation under study—a chemical reaction—to a transformation of molecules, and then makes use of its specific conceptual tool, molecular structure.[7]

THE CONCEPT OF MECHANISM

In the frame of the program of chemistry, the specifically chemical understanding of processes in matter conforms to C. K. Ingold's view that theoretical chemistry is essentially the science of reaction mechanisms.[8] Of course, also chemical kinetics and thermodynamics have theoretical aspects that are indispensable for chemical research, but those aspects may be seen as applications of physical concepts, particularly equilibrium, even at the molecular level, where statistical mechanics applies. Reaction mechanisms may require a quantum mechanical approach, but this is the case only when certain quantitative estimates (particularly lifetimes) become essential. One such problem, on which we shall have to pause below, is posed by activated complexes.[9]

What is a reaction mechanism? In its ideal form, it is a sequence of particular geometric configurations of an atomic system from a given initial configuration to a final one. For a more detailed answer, let us consider the question: Why does the reaction between the reactants A and B give certain products C and D and not others? A typical answer is: because A and B give A' and B', A' and B' give A" and B", and so on until C and D are formed. This means that the answer is given in terms of entities that form a chain of causes of a very special type, namely, causes in the sense that each entity is a necessary and sufficient condition for a step in the transformation under study. A classical example is the nitration of benzene, where the formation of a short-lived electron transfer complex is a widely discussed step.[10] In general, the intermediate structures will be branching points of a probability tree, and the probabilities will depend on the conditions of reaction; but this is a complication which, as far as we can see, is not especially significant for the present general considerations.

In general, a reaction mechanism has been established when those successive configurations of the reacting molecules have been found that characterize the steps of the reaction and show how the general laws of molecular structure and reactivity determine the final product-molecule configurations. This definition applies to organic chemistry proper as well as to borderline fields such as adsorption and desorption processes in surface science, phase transitions, and receptor substrate interactions in biochemistry.

The explanation of chemical reactions in terms of mechanisms conforms to the standard scheme of scientific explanation: modeling the given system (e.g., by considering only one or a few molecules in a standard environment) and showing how certain general rules (say, the laws of valence) work. With respect to physical explanations, there are, however, many differences. Those that interest us here are two: the real existence and nature of the intermediates and the absence of any correlation with motion. In connection with them several epistemological considerations that have important consequences on research issues are necessary.

In the literature, the spatial configurations that can be used for specifying a mechanism are of two different types: the reaction intermediates proper, that is, structures that may be short-lived, but obey the rules of valence and may therefore be considered akin to genuine molecules, or "transition states," which are, so to speak, virtual molecules. The meaning of this distinction becomes clear if one just thinks of the reaction between an alkyl halogenide and a strong base:

$$R-X + OH^- \rightarrow R-OH + X^-$$

As is well known, a reaction can be discussed in terms of a one-dimensional path on a potential-energy hypersurface (PES) defined in accordance with the Born-Oppenheimer theorem.[11] This will give a potential-energy profile if a suitable reaction path is defined. The minima of the profile usually correspond to minima of the PES, in which case the geometries associated with them are those of stable or quasi-stable molecular species.

In the frame of the representation of the Walden inversion in terms of a potential energy profile versus the appropriate reaction coordinate the following considerations apply:

- The SN1 mechanism, which is expected to account for the first-order kinetics, essentially consists in the assumption that a carbocation will be formed before linkage to the OH⁻ ion. The carbocation can be considered a real intermediate that obeys the rules of valency and corresponds to a relative minimum of the potential energy profile; in principle, it is susceptible of isolation or at least observation under appropriate conditions.

- The SN2 mechanism, which accounts for a second-order kinetics, is based on the assumption that the reaction takes place in a concerted mode. For a certain configuration, more precisely a certain value of the reaction coordinate, the potential energy profile (or more generally the potential energy hypersurface) presents a maximum. The latter defines what is called a transition state or transition complex T*: in the example given above, the well-known bipyramidal complex of Walden inversion T* is [Cl–R–OH]⁻. It is a virtual intermediate geometry that has a zero lifetime inasmuch as it decays immediately into the reactants or into the products.[12] Nevertheless, the transition complex is a key structure in the sequence that is the complete reaction mechanism under consideration, because the subsequent evolution of the system depends on the details of its structure.

ONTOLOGICAL PROBLEMS POSED BY REACTION MECHANISMS

An important ontological problem underlies the above considerations.[13] Are the structures postulated in a reaction mechanism representations of real molecules, or are they mere entities of reason, that is, fictitious entities to be used as aids in reasoning? This question is by no means an idle one, because chemistry is interested in knowledge as much as the other natural sciences, and the knowledge it aims at is knowledge of reality.

Existence in a physical sense calls into play the modern notion of observability, because it is now generally accepted that in science *real* is what is observable or participates in events (e.g. a collision) that are in principle observable. Galileo's famous statement that science is only involved with making sense of *sensate esperienze* (sensible experiences, i.e., observable facts) led in due course to Bohr's fundamental (and much debated) claim that for science the criterion of existence is observability. Unfortunately, observability is by no means a simple property. All those who have a background in physics know that the only direct measurements are those that involve length, mass, and time and do not rely on any theory. Therefore, direct observations are only possible in very simple cases in the world of our ordinary experience, and otherwise a theoretical argument stands between the output signal of the measuring apparatus and the phenomenon which that signal should reveal.

Let us grant that there are fully reliable theories, in which case one may speak of observations as distinct from conjectures also for submicroscopic layers of reality. Even with this premise, observability implies a nonvanishing lifetime of the entities under consideration, for an observation or, more generally, the interaction of a given system with another takes time. Now, in reaction mechanisms, there is normally no serious problem for intermediates, because by definition they behave as ordinary molecules; they may be short-lived, either because of high reactivity in the given environment or because of the tendency to dissociate, but unless their lifetime lies below the picosecond threshold they are molecules to all effects, because stretching vibration periods are in the range of 0.01 picoseconds.[14] Activated complexes, on the other hand, are unstable by definition and will break down as soon as they are formed. Does it make sense to consider them real entities? If it does not, how can we treat them as, so to speak, the ancestors of real intermediates or final products, possibly participating in thermodynamical equilibrium with other species? An entity whose existence lasts for an infinitely short time cannot participate in any physical process, let alone the repeated collisions needed to ensure an equilibrium distribution.

Yet, the activated complex theory works. The reason might be that it is a simplified form of a quantum statistical approach in which the transition complex is treated as an instantaneous quantum state of the given reaction system. Then also the intermediates should be treated as quasi-stationary quantum states of the reaction system under consideration. There would be nothing wrong in accepting such a shift in perspective, but the chemical perspective would be lost altogether. To treat a molecule as just a particular stationary state of an atom cluster would imply at least unnecessary complications in analyses based on reaction mechanisms, especially in connection with the traditional statistical view of chemical reaction theory. The latter rests on the ontological view of molecules as entities belonging to different species and having an existence per se, as is the case with planets, animals, and so forth; to reformulate it in terms of equilibria among states of a single system is certainly an interesting challenge, but it is for mathematical physicists to work at it until it becomes clear and easy to handle.

To ensure coherence with existing theories, maybe the best choice is to extend the notion of observable molecular system to spatial configurations whose rate of change is very small during the time intervals required by fast observations (or the most sensitive interactions). This will be the case if the energy maximum to which an activated complex corresponds is rather flat. If the frequency of collisions between dissolved molecules is sufficiently high, there will be no difficulties with treating an activated complex as a very short-lived molecule. As to observability, it is certainly ensured by recent advances in experimental techniques; in fact, observation of certain reaction intermediates became possible in virtue of ultrafast techniques several decades ago: suffice it to mention the observation of the nitronium ion by Raman spectroscopy.[15] The conclusion thus reached is also supported by the consideration that the chemists' procedure for establishing a reaction mechanism is essentially a combination of isolation of intermediates and comparison with known instances of a similar type. The search for reaction paths on an energy hypersurface, on the other hand, is practically unfeasible because of the size of the numerical computations involved. Even if it were feasible, it would be but a preliminary step for two reasons: (a), it is only a preliminary processing of information and not the answer to a "why" question; (b), even the best computations may not be sufficiently accurate to describe a system with weak bonds,[16] especially considering that what counts is not internal energies, but free energies in the thermodynamical sense.[17]

CONCLUDING REMARKS

Regardless of how the data are obtained and processed, the understanding of reactions at the molecular level begins with two steps: hierarchical ordering from general to particular in terms of simplified models, and ad hoc explanatory schemes.[18] If taken in Ingold's sense, a reaction mechanism is precisely one such explanatory scheme, inasmuch as it transfers to the submicroscopic plane the temporal relation of reactants to products, and gives the chain of structures that determines the path from the initial molecular species to the final ones.

From the point of view of philosophy of science, all this stands in favor of the claim, made all through this paper, that in general the explanation of submicroscopic dynamical processes is given by chemistry in terms of entities, not of forces and fields. This is fully in line with the idea that molecular structure is the key concept of molecular (organic) chemistry, whose fundamental tasks consist in trying to find the relations among structures in two dimensions:

- a synchronic (time-independent) dimension: find the rules that decide which atomic structures correspond to molecules that are stable (in the chemical sense) and apply them to the classification of known molecules and to the synthesis of new ones;
- a diachronic (time-dependent) dimension: find the sequence of structures that constitute, as it were, the decisive branching points in the evolution of certain reactants into certain products.

As to the place of chemistry in the edifice of science, it appears that the sort of simplifications a chemist makes, the kind of results he looks for, the rationalization at which he aims are chosen on the basis of specific epistemological criteria, in agreement with the view that each discipline has its own explanation schemes, which are an integral part of its paradigm.

All this, however, does not imply that an approach to chemistry according to the paradigm of physics is not useful. Quite to the contrary, it may provide essential information, that is, estimates of the lifetimes of intermediates of a reaction mechanism and conjectures about possible transition states. It is the formulation of the problem and the final description of processes that is different. Similarly, nuclear physics adopts a chemical point of view in discussing nuclear reactions.

If we look at the above conclusions as a case study in contemporary science, then they appear to touch upon a very critical aspect of contemporary research, namely the decline in creativity due to loss of interest in combining specialization with flexibility in methods and viewpoints. In outline, the general moral consists in the suggestion that research planning and strategies should fulfill two conditions: on the one hand, they should be tuned to the needs of each discipline rather than borrowed from other disciplines; on the other hand, they should be such that the explanations of facts they provide be consistent with (and make proper use of) the explanations other disciplines give of the same facts. The example given above of different answers to the simple question: "Why is the glass of most bottles green?" is very instructive in this connection, and might be worth a much deeper analysis.

NOTES AND REFERENCES

1. A preliminary, abridged version of this paper was published in Italian, in collaboration with P. Severino, in *Atti del primo convegno di storia della chimica*—Gruppo Nazionale di Fondamenti e Storia della Chimica. Turin, Italy, 1985.
2. SEITZ, F. 2000. Decline of the generalist. Nature **403:** 483–484.
3. DEL RE, G. 2000. The Cosmic Dance. Templeton Press. Philadelphia, PA.
4. One of the first philosophers of science to discuss chemistry as an independent discipline was Mario Bunge: BUNGE, M. 1982. Is chemistry a branch of physics? Z. Allg. Wiss. **13/2:** 209–223; cf. also DEL RE, G. 1987. The historical perspective and the specificity of chemistry. Epistemologia (Genoa) **10:** 231–240. An excellent review of the whole subject has been given by J. VAN BRAKEL. 2000. Philosophy of Chemistry. Leuven University Press. Leuven, Belgium.
5. THEOBALD, D. 1976. Some considerations on philosophy of chemistry. Chem. Soc. Revs. **5:** 203–213.
6. LIEGENER, C. & G. DEL RE. 1987. Chemistry vs. physics, the reduction myth and the unity of science. Z. allg. Wiss. **18:** 165–174; LIEGENER, C.M. & G. DEL RE. 1988. The relation of chemistry to other fields of science: atomism, reductionism, and inversion of reduction, Epistemologia (Genoa) **10:** 269–284.
7. DEL RE, G. 1998. The ontological status of molecular structure. Hyle **4:** 1–23.
8. INGOLD, C. 1969. Structure and Mechanism in Organic Chemistry. Cornell University Press. Ithaca, NY.
9. GLASSTONE, S., K.J. LAIDLER & H. EYRING. 1941. The theory of rate processes. McGraw-Hill. New York. (Particularly p. 185.)
10. PELUSO, A. & G. DEL RE. 1996. On the occurrence of electron transfer in aromatic nitration. J. Phys. Chem. **100:** 5303–5309, and Refs. therein; *cf.* INGOLD, C.K., D.J. MILLEN & H.G. POOLE. 1946. Nature **158:** 480.
11. BORN, M. & R. OPPENHEIMER. 1927. Zur Quantentheorie der Molekeln. Ann. Physik **84:** 458–488; NIKITIN, E.E. 1970. Teorija Elementarnykh Atomno-Molekuljarnykh Protsessov v Gazakh [Theory of Elementary Atomic–Molecular Processes in Gases]. Izd. Khimija. Moscow. pp. 9–27. We write "theorem" instead of "approximation," because the latter name has misled some researchers into believing that the Born-Oppenheimer study has no physical content: actually, it is the proof that quantum mechanics is compatible with the separation of nuclear motions from electronic motions as revealed by observed molecular spectra; and novelties are only found when two hypersurfaces cross.
12. In spite of this, the theory of rate processes assumes that statistical equilibrium is established between the transition state and the products, thus implying for T* a lifetime longer than the mean collision time.
13. We call "ontological" a discussion that tries primarily to establish what distinguishes things that exist from things that do not. The major lines of thought in this connection go back to Aristotle; in the English speaking world ontology has long been obscured by pragmatism, but the logicians have become again interested in it. See, e.g., QUINE, W.O. 1953, 1961. From a Logical Point of View. Harvard University Press. Cambridge, MA.
14. The importance of the lifetime of an intermediate molecule was already pointed out in 1937 by E.D. HUGHES, C.K. INGOLD & K.D. SCOTT. 1937. J. Chem. Soc. 1201.
15. *Cf.* ATKINSON, G.M., Ed. Transform Raman Spectroscopy. 1983. Academic Press. New York.

16. See, e.g., RASSAT, A., A. PELUSO & G. DEL RE. 1999. Neutral mixed-valence organic monoradicals. Chem. Phys. Lett. **313:** 582–586.
17. Computation of the partition function is open to reservations because of the very fact that the variational principle guarantees only the quality of the ground-state energy, not the slope of the energy well.
18. *Cf.* DEL RE, G. 2000. Models and analogies in science. Hyle (Karlsruhe) **6:** 1–15.

Explanation in Organic Chemistry

WILLIAM GOODWIN

Group in Logic and the Methodology of Science, University of California, Berkeley, California 94720-2390, USA

> ABSTRACT: In this paper, a model of a subclass of the explanations given in organic chemistry is developed. This model is supported by three concrete examples. The model suggests that in this discipline, laws, theories, and causal reasoning are interrelated in interesting and heretofore unexplored ways. The model also reserves a prominent place for idealizations and capacities ascribed on the basis of the structural features of molecules. The author hopes to have established that philosophical reflection on the methodology of organic chemistry can yield interesting and valuable new insights into classical issues in the philosophy of science.
>
> KEYWORDS: explanation; capacities; mechanism; organic chemistry; theory; laws; potential energy surface; philosophy of science

There are a variety of ways in which one might approach the issue of scientific explanation. One might begin with a particular image of the nature of scientific inquiry and then infer what a scientific explanation must be in order to fit one's comprehensive vision of science. This approach, though it promises a unified account of explanation, risks providing a philosophical analysis of a concept that is rarely, if ever, applicable to the scientific practice. On the other hand, one might begin by investigating, discipline by discipline, the explanations that scientists provide, both to their peers and to those being trained to enter the discipline. An investigation of explanation at the local level can be expected to provide a more accurate reflection of actual scientific practice; however, it risks finding that there is nothing interesting to say about a global notion of scientific explanation. This paper is intended to be a contribution to the latter approach. I will sketch a model of a class of the explanations that are provided in organic chemistry and suggest some of the implications that this model might have for more global accounts of scientific explanation.

FOCUSING THE PROJECT

In order to approach the project of characterizing explanation in organic chemistry in a manageable way, I want to limit the sort of explanations that I will consider

---

Address for correspondence: William Goodwin, Group in Logic and the Methodology of Science, University of California, Berkeley, Berkeley, CA 94720-2390. Voice: 510-642-2722; fax: 510-642-4164.
  wmgoodwin4@aol.com

to an important subset of those explanations offered in the discipline. One way to characterize a class of explanations is by describing the sort of 'Why' questions in response to which those explanations would be offered.[1] A *why*-question can be analyzed into three components: a topic, a contrast class, and a relevance relation. The topic of the question is the proposition to be explained, while the contrast class is the set of alternatives from which the topic is to be distinguished. Lastly, the relevance relation characterizes what sort of information is explanatorily relevant to the question at hand. The class of explanations that I want to consider is those explanations that would be given in response to the following sorts of questions:

(1) Why do [reactions of type A occur faster than reactions of type B] (rather than the other way around)?
(2) Why do [reactions of type A have the products B, C, etc.] (rather than products E, F, etc.)?
(3) Why do [reactions of type A have product distribution D] (rather than product distributions E, F, or G, etc.)?
(4) Why does [reaction of type A proceed by pathway M] (rather than pathways P, Q, or R, etc.)?

The topics of the above questions are given in square brackets, while the contrast classes are specified in the round brackets. The classification of a reaction as "type A" might involve not only consideration of the particular compounds being reacted, but also the conditions under which that reaction is run. These conditions might include, though they need not be limited to, the temperature at which the reactions is run, the solvent in which the reaction is run, and the presence of any catalysts, etc. I will argue in the next section that all of these questions, when construed in a standard way, share the same relevance relation; that is, the same sort of information will be brought to bear in order to answer the question.

A SKETCH OF EXPLANATION IN ORGANIC CHEMISTRY

A natural way to begin to model explanations in organic chemistry is to consider what sort of propositions would constitute answers to questions of the above sorts. All of the questions posed above are about transformations. The first sort of question is about the speed of transformations, the second and third are about the products of transformations, and the fourth is about the pathways of transformations. It should not be surprising that an important set of questions that are asked and answered in organic chemistry are questions about transformations. Organic chemistry is, after all, supposed to be the study of organic molecules and the reactions or transformations that they undergo. As a start then, the answer to a question of one of the above sorts will be an assertion about a transformation, or a class of transformations.[2]

The next natural step is to articulate how organic chemists think about transformations. This can be done, I believe, by characterizing what sort of description of a transformation would constitute an ideal account of the transformation from the point of view of an organic chemist. I will refer to this ideal account of a transformation as an Ideal Explanatory Account, or IEA. If an IEA were available, then the chemistry would possess a complete characterization of a chemical reaction and so could answer any relevant questions about that transformation in a rigorous way. The notion that comes closest to that of an IEA in organic chemistry is that of a potential energy surface for a transformation. If one had a quantitative description of the energy of both products and reactants and all the possible pathways connecting the two, then one could employ quantitative theories such as thermodynamics and/or transition state theory to answer any of the questions of the above sorts for a particular reaction. Because the potential energy surface for a transformation would, in general, be multi-dimensional, it is usually easier to think of transformation in terms of their potential energy diagrams. In a potential energy diagram, the potential energy of a transformation is plotted against the reaction coordinate, or the measure of progress through a transformation. The potential energy surfaces, or diagrams, bring the chemical transformation into quantitative theories. This is just to say that if one is able to give such descriptions of a reaction, then the rigorous (or quasi-rigorous) machinery of thermodynamics and/or a quantitative theory of reaction rates can be brought to bear on questions about the transformation.

Of course, chemists generally do not have access to IEAs; nonetheless when answering questions about transformations, these IEAs form the background theoretical model against which the answer is formulated. Even if one cannot measure or otherwise quantify the activation energies of a set of reactions, the differences in the rates of these reaction are still explained by making assertions about differences in their energy barriers. More generally, an answer to one of the questions of the above sorts is a claim about the IEAs of the reactions mentioned in the topic of the question. The answers to these questions will generally not state quantitative facts about the IEA because these sorts of facts are not readily available. Instead, one would answer a question of one of the above sorts by making a qualitative assertion about the class of IEAs of the reaction type in question. In terms of the anatomy of questions, then, the sort of information that is explanatorily relevant to the questions being considered is a fact about a class of IEAs.

An assertion about the class of IEAs corresponding to a type of reaction offered in response to a question of one of the above sorts should, if it is to form part of a good explanation, allow one to conclude that the topic of the question is true, while the alternatives mentioned in the contrast class are not true. Thus if one wishes to explain why a particular substance, X, is the product of a reaction rather than the members of some set of alternative products, Y, Z, etc., one would have to assert enough about the IEA of the reaction in order to infer that X would in fact be the product, while Y, Z would not. So, one might offer an explanation of the following sort: the activation energy along the pathway leading to product X is significantly lower than the activation energy leading to either Y or Z, etc. This claim, if true of the IEA of the reaction in question (under the appropriate conditions and so forth), would entail that X is the product of the reaction and that neither Y nor Z would be produced. Thus the assertion about the IEA that constitutes the direct answer to the question is perfectly adequate to distinguish the topic of the question from the alternative mentioned in the contrast class. Even though the assertion about the IEA offered as a direct response to the question is adequate, in a certain sense, for answering the question, it would not, in itself, constitute a good or interesting explanation of the phenomena being inquired about.

In order to provide a good or interesting answer to a question of the sort described above, it is not sufficient to assert something about the IEA of the reaction(s) in question. One must also provide what I will call a "structural account" of that direct answer to the question. Loosely speaking, a structural account begins with the information provided by the structural formulas of the types of molecules involved in the reaction type in question and argues for the direct answer to the question on the basis of that structural information. For example, if the direct answer to a question is an assertion about the relative energies of some products, say in response to a question about the product distribution of a reaction under "thermodynamic control," the structural account of this direct answer must provide for the energy differences of the products in terms of the structures of those products.

A structural account generally proceeds by applying robustly applicable concepts that allow one to convert recognizable structural features of the molecules or classes of molecules involved in the question into qualitative energy differences between those entities. These robustly applicable concepts form a bridge between structure and energy. They allow one to infer qualitative relative energy differences between structures by appealing to easily memorized trends or readily accessible data. Some of these concepts can be applied on the basis of recognizing individual atomic symbols, or patterns of such symbols—such as functional groups. For example, individual atoms or more complex substituents can be categorized as "electron-withdrawing groups" on the basis of either their atomic symbol (in the case of chlorine for instance) or the functional group to which they belong. Once a group has been recognized as an electron-withdrawing group, this fact can in certain circumstances be used to infer facts about the relative stability of the species to which it belongs (consider the stability of carbanions, for example). There are also some concepts that llow one to infer relative energy differences on the basis of more general structural features. For example, one can infer that there will be "resonance stabilization" in cyclic alkenes with particular numbers of alternating double bonds. Collectively, these concepts serve the purpose of converting the structural information given in a question into the sort of information about the IEAs that can provide the basis for a direct answer to the question.

Sometimes the relative energy differences recognizable in virtue of these structural concepts will be sufficient to account for the direct answer to the question. So, if one happened to know that a reaction was under thermodynamic control, then the relative stability of the products would be sufficient to account for their distribution. In general, however, the relative energy differences between species involved in the transformations being explained are not sufficient to account for the assertion about the IEA that constitutes a direct answer to the question. In many cases, a question is framed in terms of the structures of the reactants and the products of a transformation; however, the answer to the question depends upon information about the pathway of that transformation. In cases such as these, information about the relative energies of the reactants or products—which can be discerned on the basis of structural features of these species—must be converted into information about the relative energies at various points along the pathways of the transitions. For example, in order to explain the relative rates of reactions, one generally needs to compare the energies of the transition states of the reactions. When information about the pathways of transformations is required in order to answer the question, the structural account of the answer will, in general, project energy differences between the reactants or the products into energy differences along the pathway of the transformation. This projection will, in most cases, require an understanding of the mechanism of the reaction. Furthermore, even when one understands the mechanism of the transformation, an account of how structural changes in the reactants or products of the reaction (or other aspects of the reaction conditions) translate into energy differences in the transition state will involve additional theoretical assumptions. For example, one might invoke the Hammond Postulate, or some similar theory, in order to explain why a shift in the energy of the products would have an important effect on the transition states, and thus on the rates, of some endothermic reaction.[3] These theories about how the transition states of reactions respond to changes in the reaction conditions become especially important when there are competing influences on the relative energy of the transition state and when there is a competition between several mechanisms by which a transformation might take place.

To recapitulate: Many of the interesting questions asked and answered in organic chemistry are about transformations. There is a basic theoretical background structure in which such questions are addressed. Transformations are understood as trajectories along a potential energy surface and questions about such transformations are answered by making assertions about that trajectory and its potential energy. The direct answer to a question about a transformation is a claim about the relevant potential energy surface (or diagram). Direct answers are ideally supplemented by a structural account. A structural account supports the direct answer by arriving at it from structural information provided in the question. This process involves the application of concepts that convert structural features of species involved in the transformation into relative energy differences. In most cases, the relative energy differences indicated by these concepts are differences in the energy of the products or the reactants. Energy differences at the endpoints of a transformation can be converted to energy differences along the trajectory by appealing to the mechanism and/or additional abstract theories.

EXAMPLE 1: RELATIVE RATES OF NUCLEOPHILIC SUBSTITUTION REACTIONS ON HALOALKANES

In this type of reaction, a nucleophile, or chemical species with an unshared electron pair, displaces a halogen from a substrate. The rates of these reactions will vary with a variety of factors including the nature of the nucleophile, the structure of the alkyl group, and the polarity of the solvent in which the reaction is run. In this example, I will focus on one such factor—the particular halogen being displaced. It has been found that all else being equal, these reactions proceed more quickly and to a greater extent when certain halogens are displaced rather than others. The relative ease with which a substituent can be displaced from a substrate in a nucleophilic substitution reaction is known as the "leaving-group ability" of that substituent; thus the greater the leaving group ability (all else being equal), the greater the rate of the substitution reaction. An ordering of the leaving group ability of the halogens will therefore amount to a qualitative comparison of the relative rates of nucleophilic substitution reactions on haloalkanes. The *why* question that I wish to consider in this section is, "Why is the order of the leaving group abilities of the halogens I > Br > Cl > F rather than any of the other possible orderings?" This is a shorthand way of asking for an explanation of the relative rates of a class of reactions. The direct answer to this question will be a general claim about the IEAs of this class of reactions. Because this question is about the rates of chemical reactions, its answer will be an assertion about the relative activation energies of these reactions (or more generally about the paths of these transformations). The answer would be something like the following: For a given alkyl group, a given nucleophile, and a given set of reaction conditions, the activation energy for the nucleophilic substitution reaction on the iodoalkane is less than that for the bromoalkane, which is less than the chloroalkane, which is less than the fluoroalkane.

While this answer is perfectly adequate in the sense that it would allows one to employ the theoretical apparatus of the IEA to distinguish the phenomena described in the topic from the members of the contrast class, it has not yet been developed from the structural features of the molecules in the reactions in question. Because the only features that vary between the reactions being compared are the halogens being displaced, a structural account of the differences in the activation energies of the substitution reactions must be developed in terms of the nature of the halogens alone. The structural account of the differences in activation energy asserted by the answer to the question depends on an identification of leaving group ability with the group's "capacity to accommodate a negative charge."[4,5] The greater this capacity, the more stable the products of the nucleophilic substitution reaction will be. Because this same capacity is responsible for the varying acidity of the hydrogen halides (HF, HCl, HBr, HI), it is inferred that the ordering of leaving group ability will parallel the ordering of the corresponding acid strengths. In the case of the halogens, this capacity, and the consequent acid strength, is itself sometimes analyzed in terms of more fundamental notions, for example: electronegativity and atomic size, which are both taken to correlate with the relative stability of the anion. These two considerations are at odds with each other in the case of the halogens (F, Cl, Br, I) because as you proceed down this list, the anions get larger and less electronegative. It turns out that the size effects dominate so that the capacity to accommodate a negative charge increases as you proceed down the periodic table.

To this stage, the structural account of the answer to the question has, whether one stops at the analysis in terms of acidity or proceeds through to electronegativity and size, taken readily available data obtained by considering a different reaction(s) and assumed that the reaction in question depends on the very same capacity that was responsible for the measured differences in energy in the prior case. This allows one to compare the capacity to accommodate a negative charge between the halogens on the basis of the previously obtained results. Since the capacity to accommodate a negative charge is supposed to correlate with an increased stability of the products in nucleophilic substitution reactions, one expects the halogens that demonstrate this capacity to a larger extent to have lower, relatively speaking, product energy than those halogens with a lesser capacity. By evaluating this capacity in terms of data that can be located on the basis of the structural information contained in the question, one would only seem to get information that allows for a comparison of the stability of the products of the reactions in question. Because the answer to the question requires kinetic information, the information about the end results of this transformation must be converted into information about the paths of the transformations in question.

The second step in accounting for the answer to the question is to convert the information about the products that was gleaned from the structures given by the question and readily available data into information about the activation energies of the relevant reactions.

Such conclusions are reached on the basis of what is primarily information about the stability of the products of the reaction by employing certain assumptions about the relationship between thermodynamic stability and activation energy. In this case, the assumption is that energy differences between the products of the reactions will translate into energy differences in the transitions states of the reactions. Thus, because iodine has a greater capacity to accommodate negative charge, the energy of the transition state for the substitution reaction wherein iodine is displaced will be less. This inference from the energy differences between the products of the reaction and the transition states of the reaction is supported by an appeal to the mechanism of the reaction. In this case, it is because the leaving group will have a partial negative charge during the transition state, and thus resemble the products to a certain extent, that one expects the activation energy to decrease with increased capacity to accommodate negative charge.

This example demonstrates the general structure that I outlined earlier in this paper. The explanation begins with an assertion about a class of IEAs. This assertion is itself supported by a structural account. The structural account employs a robust concept that allows conversion of readily accessible experimental data (obtained for different, simpler reactions) into qualitative facts about energy differences in the IEAs. These energy differences are then converted into differences in activation energies by invoking assumptions about the shapes of potential energy curves that are supported by information about the proposed mechanism of the reaction.

EXAMPLE 2: REGIOSELECTIVITY AND STEREOSELECTIVITY IN THE FORMATION OF ALKENES

The type of reaction that I wish to consider in this example is the formation of alkenes from haloalkanes by bimolecular elimination. In these reactions, a carbon–carbon double bond is formed by "removing" a hydrogen halide (such as HCl, HBr, or HI) from a halogen-containing alkane. It is often the case that there are several different hydrogens that might be removed in a manner compatible with the proposed mechanism. The question then becomes, "Why is one particular hydrogen preferentially removed rather than other hydrogens that are compatible with the mechanism?" The preferential removal of one particular hydrogen rather than others can lead to the regioselectivity and/or stereoselectivity of the reaction. If the preferential removal of a particular hydrogen leads to a product distribution favoring one location for the double bond over others, then the reaction displays regioselectivity. If the removal of particular hydrogen favors the formation of a particular stereoisomer (a *cis* double bond vs. a *trans* double bond), then the reaction is stereoselective.[6]

Both regioselectivity and stereoselectivity are explained by relative differences among the activation energies of the pathways leading to the various possible products. The question is not a direct comparison of the rates of different reactions; nonetheless, the product distribution of essentially irreversible reactions is to be explained by the relative rates at which the reactants proceed down the various possible reaction coordinates. Reactions in which product distribution is determined by the differences of rate of formation are referred to as being under "kinetic control." In such circumstances, questions about product distributions can be converted into questions about the activation energies of various reaction pathways. Thus the direct answer to the question goes as follows: For any given haloalkane, the hydrogen that is preferentially removed will be that hydrogen whose removal has the lowest activation energy. As it stands, this answer is too general to be of much interest because the hydrogen whose removal has the lowest activation energy depends on a variety of structural factors. In the cases where the reaction is run with a "non-bulky" base, the selectivity can be accounted for in terms of the relative thermodynamic stability of the distinct products that result from the removal of particular hydrogens. Because these products differ only in the location and stereochemistry of the double bond within the molecule, their relative stabilities can be compared by comparing how much heat is given off when this double bond is hydrogenated to yield an alkane (it will be the same alkane in each case). This is known as the "heat of hydrogenation" and it correlates to the stability of the alkene (the larger the heat of hydrogenation, or the less negative, the more stable the alkene). The results of these sorts of experiments can be generalized: "The relative stability of alkenes increases with increasing substitution, and *trans* isomers are usually more stable than their *cis* counterparts."[7] This rule of thumb is itself explained (or partially so) as follows, "the first trend ... is due in part to hyperconjugation. ... The second finding is more easily understood by looking at molecular models. In *cis*-disubstituted alkenes, the substituent groups frequently crowd each other."[7] Hyperconjugation is the name of a phenomenon that is generally described in terms of molecular orbital theory. Essentially, bonding pairs of electrons from neighboring atoms interact with the orbitals that make up the carbon-carbon double bond in such a way that there is a net stabilizing effect.[8] The crowding described in the quote above is an appeal to a phenomenon known as "steric hindrance." Vollhardt says, "To a first approximation, this effect can be attributed to bulk: Two molecular fragments cannot occupy the same region in space."[9] When two bulky groups are positioned near each other in a molecule, there is a loss of stability because the electron clouds repel each other. Appeals to steric hindrance are extremely common in analysis of relative stabilities.

Both the hyperconjugation and the steric hindrance can be recognized in a structural description of the molecules and they facilitate inferences about the energy of the structures in which they are recognized. The relative extent to which the hyperconjugation stabilizes the alkene can be evaluated by comparing the number of alkyl groups that are attached to the carbons in the double bond. The steric hindrance can also be visually identified in the structural formula of the alkene in that the larger substituents appear on the same side of the double bond in the figure. Both of these are robustly applicable concepts that allow one to convert structural features of the possible products of the elimination reaction into qualitative information about the relative stability of the molecules. In this case, as opposed to the last example, these robustly applicable concepts depend on geometrical features of molecules represented in the structural formula, not just in the individual atomic symbols. The phenomenon of hyperconjugation and steric hindrance are also distinct from basicity or the capacity to accommodate a negative charge in that they are both accounted for by appeal to a theoretical model of chemical bonding. The results of many elimination reactions can be explained by appeal to the generalization that the thermodynamically preferred product (the most stable) will dominate and the side products will be produced in ratios that reflect their relative stability. Ultimately, however, the answer to this question must be in terms of relative differences in activation energies because the reactions are essentially irreversible. Again, the kinetic conclusion is justified by extrapolating from the energy difference of the products to the energy differences of the transition states. For example, Vollhardt explains, "The more stable product is formed faster because *the structure of the transition state of the reaction resembles that of the products to some extent.*"[10]

These elimination reactions can be run in such a way that they violate the general rule of favoring the most stable product by changing the reaction conditions to employ a "bulky" base rather than a "non-bulky" one. In these cases, the less thermodynamically stable alkene may be the dominant product. Since the mechanism is the same, the transitions states of these reactions will still resemble the product of the reaction, but this tendency for the thermodynamically more stable product to have the lower activation energy must be overwhelmed by other factors. The transition state of reactions of this type involves a base removing a hydrogen while the halogen is departing. When the base is "bulky," steric hindrance destabilizes the transition state of the mechanism leading to the thermodynamically more stable product relative to the transition state leading to the thermodynamically less stable product. Removing the hydrogen that results in the more stable product requires the base to attack at a position that is more "internal" in the haloalkane. The "bulky" base will encounter more steric hindrance when attacking this internal hydrogen than it would attacking the terminal hydrogen. As a result, this reaction will produce the less thermodynamically stable alkene at a faster rate than the more stable alkene. This is taken to explain the regioselectivity in such cases.

The structural account of regioselectivity and/or stereoselectivity in alkene formation again makes essential use of the mechanism of the reaction in order to extrapolate from energy differences that can be discerned on the basis of structural features of the molecules in the reaction to differences in the energy of the transition states of the reactions. The mechanism provides the link between the facts about the IEA that provide a direct answer to the question and the structural features of the molecules involved. In the case of a "bulky" base, it is by directly considering the geometry of the transition state postulated by the mechanism that one can recognize the structural feature (steric hindrance) that is responsible for the difference in activation energy. Another interesting feature of this example is that there are competing influences on the energy of the transition state (and there are others not addressed in this example); thus the outcome of any particular reaction will depend on the balance between (at least) two features of the transition state: the steric hindrance encounter by the base in abstracting the hydrogen and the stability of the partially formed double bond. In general, without significantly more information about the potential energy surface of the reaction in question, one could not predict on the basis of structural features which of these two effects would dominate. However, given the product distribution of a particular reaction, that outcome could nonetheless be explained by asserting that the relevant feature of the transition state was responsible for the outcome.[8] The process of comparing the activations energies of reactions is not generally a straightforward projection from the energies of the products to the energies of the transition states. Instead it involves a detailed consideration of how the energetics of the transition state varies with changing reaction conditions. This process become even more complicated when there are competing mechanisms by which the same transformation might occur. Much of the work in theoretical organic chemistry, such as the Hammond Postulate and related approaches, is devoted to devising ways to infer the effects of changes in reaction conditions on the location and energy of transition states. Generally these inferences begin with structurally identifiable energy differences in the species involved in the reaction.

EXAMPLE 3: STEREOSPECIFICITY IN THE FORMATION OF ALKENES

Stereospecificity in the formation of alkenes by elimination occurs when the haloalkane from which the hydrogen halide is to be removed has adjacent stereocenters, one containing the halogen and the other the preferred hydrogen. In such cases, under appropriate circumstances, one of the stereoisomers (either the E or Z, which is a fancy version of the *cis* vs. *trans* distinction) of the alkene will be formed exclusively. The stereoisomer that is formed need not be the stereoisomer that is thermodynamically most stable. The question that I wish to consider is, "Why is one particular stereoisomer produced exclusively rather than the other, or a mixture of the two?" Because the phenomenon of stereospecificity can occur even when the less thermodynamically stable alkene is formed, this question must also receive an answer that explains why the rate of the reaction that occurs is much faster than the rate of the formation of the stereoisomer that is not produced. Thus the direct answer to the question will be something like: For any haloalkane of the appropriate sort, under the appropriate conditions, the activation energy of the pathway leading to the stereoisomer that is produced is much less than the activation energy leading to its alternative. Again, the interesting part of this explanation will be an account of how one can recognize on the basis of the structure of the haloalkane that one elimination pathway will have a much larger activation energy than its alternative.

The structural account of this answer depends in a more direct way than our other examples on knowing the mechanism of the reaction. Essentially, by following the bond breakage and formation that occur during the mechanism of the elimination one can "see" why one particular geometric isomer is created to the exclusion of the other. The mechanism that is appealed to in the structural account of this answer is known as the E2 mechanism. This is a concerted mechanism, which means that several processes occur simultaneously in the transition state: (1) the base removes the hydrogen; (2) the leaving group (the halide) departs; and (3) the carbons begin to form a double bond. The E2 mechanism requires that in the transition state the hydrogen to be removed and the leaving-group are arranged in an *anti* configuration. This essentially means these two atoms are 180 degrees away from each other on adjacent carbons. If you build a model of this configuration, or draw it accurately, you can see which of the possible products (E or Z) will be formed from a transition state in this configuration. Thus, given that the elimination occurs by the E2 mechanism, one can account for why one particular stereoisomer is formed to the exclusion of the other—basically, only one of the stereoisomers is compatible with the mechanism.

The direct answer to the question considered in this example is a statement about activation energies, thus the appeal to the *anti* configuration of the transition state of the E2 mechanism must be supplemented by some structural account of why the E2 mechanism has a much lower activation energy than other available elimination pathways. Both the kinetics of the reaction and the fact that it is stereospecific indicate that the mechanism of the reaction must be concerted. This would rule out other elimination pathways that either have first-order kinetics or proceed through a symmetrical intermediate. The remaining alternative would be a concerted elimination in which the hydrogen and the halogen are lined up on the same side of the partially formed double bond. An elimination reaction that proceeded through a transition state of this type would be a *syn* elimination. The fact that the eliminations proceed through the *anti* configuration of the transition state rather than the *syn* configuration is accounted for in terms of steric hindrance. By requiring the leaving group and the base to be on opposite sides of the alkane, this mechanism minimizes the steric hindrance between the substituents on the adjacent carbons. Though the account of the stereospecificity of these eliminations is given a mechanistic explanation, the analysis of why the reaction proceeds by the mechanism in question again depends upon the recognition of structural features. In this case, the robust concepts that allow one to convert structural information into relative energy differences are applied directly to the proposed structures of the transition states of the possible alternative pathways.

PHILOSOPHICALLY INTERESTING ASPECTS OF EXPLANATION IN ORGANIC CHEMISTRY

Laws have traditionally played a prominent role in accounts of scientific explanation. However, in the classic sense of precise and general quantitative relationships, laws do not play a central role in the model of explanation articulated in this paper. The only laws that might seem to be at play in these explanations are the laws of thermodynamics and the laws of quantum mechanics. While it is true that both sorts of laws play some role in the explanations, these laws are not sufficient, on their own, to provide even the direct answers to the kinds of questions considered earlier. Because most of the direct answers depend on some account of the kinetics of transformations, the laws of thermodynamics and/or quantum mechanics must be supplemented by some account that allows one to transform information about the paths of reactions into information about their rates. This role is typically played by some version of Transition State Theory. Furthermore, though some of the concepts that facilitate the conversion of structural features into energy differences are given explanations in terms of molecular orbital theory or valence bond theory (hyperconjugation for instance), these theories are themselves only loosely based, at least in the context of organic chemistry, on the laws of quantum mechanics. If this general observation were correct, then it would be a mistake to insist that organic chemists explain by deriving propositions from laws. To insist upon explanations of that sort would be to refuse to recognize the actual, non-law-intensive explanations seen in the scientific practice.

Just as striking as the relatively minor role that strict quantitative laws play in these explanations is the important role that the robustly applicable structural concepts play. These concepts are relatively few in number, but they are applicable across a broad range of chemical structures. They allow one to assess qualitative energy differences on the basis of easily identifiable features of an organic molecule or transformation. By isolating concepts such as leaving group ability and stabilization due to hyperconjugation one is able to project energy differences that can be measured in some context (acid/base reactions or hydrogenation of alkenes) onto molecules where those energy differences have not or cannot be directly measured. Without these portable concepts, each class of reactions would have to be approached anew, without one's being able to apply the lessons learned by studying the last class of reactions. It is these concepts that make organic chemistry comprehensible. If one were unable to rationalize or explain the transformations undergone by organic molecules with these robust concepts, then organic chemistry would be a much more descriptive science—and less of a theoretical science—than it actually is. The heavy dependence of explanations of the sort that have been considered on the identification of structural features of molecules and the consequent tendencies to raise or lower potential energy in various circumstances suggests that the notion of a capacity might play an important role in a more complete account of the reasoning that occurs in organic chemistry. The notion of a capacity and its role in scientific explanation has received significant attention in recent works in the philosophy of science.[12,13] This preliminary work suggests that organic chemistry might provide a fertile ground for considering such issues.

The mechanisms of chemical transformations also played a central role in the explanations considered. Mechanisms were essential to the conversion of structural information into information about the rates of reactions. Questions about the rates of reactions are generally answered by appealing to facts about the energy of the transition state of the reaction. These energies could be evaluated and compared only if one knew enough about the mechanism of the reaction either to directly compare the transition states of various pathways on the basis of their structure or to project energy differences in the transition state from relative energy differences in the reactants or products. The mechanism of a reaction seems to be closely related to the causal process by which a transformation occurs; thus, to a certain extent, this account of explanation reinforces the central role that causal processes have been alleged to play in scientific explanation.

There are several places where one might recognize a role for theories in this account of explanation. First of all, transition state theory and its variants were seen to play an essential role even in the direct answers to the questions considered. These theories allow the conversion of information about the energy along the path of a transformation into information about the rate of that transformation. The possibility of making some such conversion is absolutely essential to the use of facts about the IEA in answers to questions about the rates or product distributions of reactions. This sort of theory seems fundamental to the background understanding of transformations that underwrites explanations of this sort. In addition, some of the structural concepts such as hyperconjugation or resonance receive a justification in terms of theories of chemical bonding. This use of theories was not explored in any detail in this paper, but it does seem less essential than the role of transition state theory. Even in the absence of a convincing justification, these structural concepts can be applied so long as they can be convincingly demonstrated by experiments in related cases. Recall that the capacity to accommodate a negative charge was correlated with the strength of the conjugate acid even if the strength of the acid could not be convincingly explained in terms of more fundamental notions (say atomic size and electronegativity). Lastly, though again this role was only hinted at in this paper, theory played an important role in projecting how the transition state of a reaction would respond to changes in reaction conditions. Consequently, theoretical assumptions such as the Hammond Postulate and its relatives can be expected to play a major role in explanations of how the rate and mechanisms of transformations respond to changes in the substrates or conditions of a reaction.

NOTES AND REFERENCES

1. I have followed van Fraassen's treatment of explanations as answers to *why* questions. I have done this only because it offers a convenient way of focusing my project, not because I wish to endorse van Fraassen's Pragmatic Theory of Explanation. See VAN FRAASSEN, B. 1980. The Scientific Image. Oxford University Press. Oxford.
2. When the question is about a class of transformations that has some feature in common, the facts relevant to the question will be facts true of this class of transformations. All of the questions being considered are about reactions of a certain type, thus, in general, answers to those questions would be facts about transformations of a certain type.
3. HAMMOND, G. 1955. A correlation of reaction rates. J. Am. Chem. Soc. **77**: 334–338.
4. VOLLHARDT, K. & N. SCHORE. 1995. Organic Chemistry, 2nd ed., p. 187. W. H. Freeman. New York.
5. LOWRY, T. & K. RICHARDSON. 1987. Mechanism and Theory in Organic Chemistry, 3rd ed., p. 374. Harper & Row. New York.
6. Roughly speaking, *cis* alkenes have the two largest groups on the same side of the double bond, while *trans* alkenes have the two largest groups on opposite sides of the double bond.
7. VOLLHARDT & SCHORE, Organic Chemistry, p. 395.
8. Hyperconjugation can be given a more rigorous justification in the context of molecular orbital theory, but that would lead us too far afield. It should be noted that the term hyperconjugation is generally applied to cases wherein neighboring orbitals overlap with partially filled orbitals, rather than the completely occupied orbital that are at issue in this case.
9. VOLLHARDT & SCHORE, Organic Chemistry, p. 58.
10. Ibid., p. 97 [Vollhardt's italics].
11. Imagine a case where a moderately bulky base is reacted with a haloalkane that can lead to two different alkenes that are moderately different in energy. If the reaction ended up favoring the more stable alkene, we would explain this result by appealing to the partially formed double bonds in the transition state. On the other hand, if the reaction ended up favoring the less stable alkene, we would explain it in terms of the bulkiness of the base.
12. CARTWRIGHT, N. 1983. How the Laws of Physics Lie. Clarendon Press. Oxford.
13. CARTWRIGHT, N. 1989. Nature's Capacities and their Measurement. Clarendon Press. Oxford.

On the Ordinality of Causes in Complex Autocatalytic Systems

ROBERT E. ULANOWICZ

University of Maryland Center for Environmental Science, Chesapeake Biological Laboratory, Solomons, Maryland 20688-0038, USA

ABSTRACT: The convention in chemical dynamics usually is to identify the reactant and catalytic molecules as the active agencies that combine in mechanical fashion to constitute a reaction process. When macromolecules grow large and complex enough to exhibit some plasticity, however, the subsequent directions in which these molecules change may be guided more by the nexus of chemical reactions in which they participate. In particular, configurations of autocatalytic reactions among plastic macromolecules can come to exert more agency upon the component reactants and mechanisms than *vice versa*, and the nexus of such processes retains its identity longer than do the latter, more transient participants. Whence, the ascendancy of autocatalytic forms as causal agencies provides a natural example of the phenomenon of "emergence" and affords a way out of the conundrums that currently obfuscate the issue of the origin of life.

KEYWORDS: autocatalysis; causality; centripetality; process philosophy; propensities; selection

Chemistry and physics are often lumped together as exemplars of "hard" science. Chemistry departs from physics, however, in that the structures of interacting entities play a more prominent role in chemical dynamics. With structure also arises the possibility for asymmetry, which is essentially foreign to most laws of physics.[1] Furthermore, as compounds increase in complexity, the combinatorics of chemical structures becomes so immense that one can begin to talk of a virtual continuum of chemical forms. That is, complex chemical forms often can be considered to have malleable structures, and this plasticity becomes most evident in large biomolecules, where chemistry makes contact with biology, and material forms actively begin to evolve. On the matter of evolution, the late Karl Popper believed that science will never achieve an "evolutionary theory of knowledge" until our basic notions of causality have been reformulated.[2] For chemistry, as for physics, causality has long been identified as either material or mechanical in nature—all other types being strictly

---

Address for correspondence: Dr. Robert E. Ulanowicz, University of Maryland Center for Environmental Science, Chesapeake Biological Laboratory, Solomons, MD 20688-0038. Voice: 410-326-7266.
ulan@cbl.umces.edu

excluded. This restriction, which dates to the Enlightenment, is at the core of what makes chemistry a "hard" science. One takes on great risk, therefore, by tampering with mechanical closure; but, if Popper is correct, doing so is a prerequisite before one can achieve a clearer picture of the transition from chemistry to biology and endue a rational and natural meaning to the concept of "emergence."

It is, therefore, with some trepidation (but not without ample precedent) that we now consider the proposition that not all causal agencies derive from enduring physical objects. Specifically, we wish to investigate whether it makes any sense to consider objects and events as being elicited by processes. Such is the thrust of the "process philosophy" of Alfred North Whitehead and Charles Saunders Peirce. There are also strong implications of "process as cause" in the field of network thermodynamics (NT), which treats the relationships between generalized processes (called "through variables") that are caused (in conventional manner) by corresponding potential gradients ("across variables"). What is different about NT is the "dual theorem" of Bernardus Tellegen, who demonstrated how it is always possible to construct a "dual" representation of an NT problem wherein the through and across variables exchange roles.[3] In other words, in the quasi-linear realm of NT, a process is just as likely to elicit an object as an object is to drive a process.

Of course, the domain of chemical reactions is usually non-linear, and it is a bit of a stretch to imagine an object's being the result of but a single process. What seems more plausible, however, is that an object might result from the conjunction of several processes. Hence, the proposition that we wish to explore here is that the most prominent cause of some objects is the nexus of processes in which they are imbedded. To be more concise about what might constitute an appropriate configuration of processes, we now consider the example of autocatalysis.

Here we define autocatalysis to be a particular manifestation of a positive feedback loop wherein a positive effect is contributed by every link to its downstream neighbor. Without loss of generality, we will confine our discussion to the serial, circular conjunction of three processes (such as reaction rates) A, B, and C. When these processes involve very simple, virtually immutable chemical forms, then the entire system functions in wholly mechanical fashion. Any increase in A will invoke a corresponding increase in B, which in turn elicits an increase in C, and whence back to A.

Matters become quite different, however, as soon as the elements engaged in the processes become combinatorially complex. Such entities, having many nearby, almost identical, forms are capable of small, contingent changes in structure, and will continue to function in the autocatalytic loop as before, albeit with somewhat more or less effectiveness. It would also be more appropriate under such circumstances to say that the action of process A has a *propensity* to augment the second process B. That is, the response of B to A is not prescribed deterministically. Rather, when process A increases in magnitude, most (but not all) of the time B also will increase. B tends to accelerate C in similar fashion, and C has the same effect upon A.

Autocatalysis gives rise to several system attributes, which, as a whole, distinguish the behavior of the system from one that can be decomposed into simple mechanisms.[4] Most germane is that such autocatalysis now becomes capable of exerting *selection* pressure upon its ever-changing, malleable constituents. To see this, we consider a small spontaneous change in process B. If that change either makes B more sensitive to A or a more effective catalyst of C, then the transition will receive enhanced stimulus from A. Conversely, if the change in B either makes it less sensitive to the effects of A or a weaker catalyst of C, then that perturbation will likely receive diminished support from A. We note three things about such selection: (1) that it acts on the constituent processes or mechanisms as well as on the elements themselves; (2) that it arises within the system, not external to the system; and (3) that it can act in a positive way to select *for* a particular system result (greater autocatalysis), rather than always *against* the persistence of an individual chemical form. The first attribute defeats attempts at reductionism, whilst the latter two distinguish autocatalytic selection from the "natural selection" of evolutionary theory.

It should be noted in particular that any change in B is likely to involve a change in the amounts of material and energy that are required to sustain process B. Whence, corollary to the selection pressure is the tendency to reward and support those changes that serve to bring ever more resources into B. As this circumstance pertains to any and all members of the feedback loop, any autocatalytic cycle becomes the epicenter of a *centripetal* configuration, towards which as many resources as available will converge. Even in the absence of any spatial integument (as required by the related scenario called autopoeisis[5]), the autocatalytic loop itself defines the focus of flows.

Centripetality implies that whenever two or more autocatalyic loops exist in the same system and draw from the same pool of finite resources, *competition* among the foci will ensue. In particular, whenever two loops share pathway segments in common, the result of this competition is likely to be the exclusion or radical diminution of one of the non-overlapping sections. For example, should a new element D happen to appear and to connect with A and C in parallel to their connections with B, then if D is more sensitive to A and/or a better catalyst of C, the ensuing dynamics should favor D over B to the extent that B will either fade into the background or disappear altogether. That is, the selection pressure and centripetality generated by complex autocatalysis can guide the replacement of elements.

Of course, if B can be replaced by D, there is no reason why C cannot be replaced by E and A by F, so that the cycle A,B,C could eventually transform into D,E,F. This possibility implies that the characteristic lifetime of the autocatalytic cycle generally exceeds those of most of its constituents. The incipience of the autocatalyic form before, and especially its persistence beyond, the lifetimes of most of its constituents imparts causal priority to the agency of the configuration of processes. True, the inception of the feedback loop can be interpreted as the consequence of conventional mechanistic causes. Once in existence and generating its own selection pressures, however, those instigating mechanisms become accidental to the selection agency that arises. Any argument seeking to explain the behavior of the whole system entirely as the result of shorter-lived constituents erroneously ignores the ascendant agency of the configuration of processes, which winnows those ephemeral and transitory mechanisms.

Ever since Democritus, the aim of rational explanation has been to portray all processes as the consequence of universal laws that act on eternal and unchanging fundamental atoms. This reductionist agenda has worked reasonably well at atomic and subatomic scales, but once one encounters macromolecules, durability is reversed. At the mesoscales, it is the configurations of processes that are most enduring—in comparison, their constituents appear merely as transients. The most natural direction for causality to act at intermediate scales is from the persistent configurations of processes towards their transient constituents, whose creation the former mediate.

Those who feel uncomfortable with this reversal of roles might become less anxious once it is explained how the new order obviates some of the most pressing philosophical concerns about the mysteries surrounding "emergence" and the origin of life. It should be recalled that before the Enlightenment life was considered almost universal, so that the greatest philosophical conundrum was how to explain the presence of death in the world.[6] Upon the rise of materialism more than 300 years ago the perspective was radically reversed. Today all material is presumed dead, and the greatest challenge has become to explain how life appeared on the scene. If, however, one's perspective were to shift so as to place more emphasis upon processes, both previous viewpoints would appear as unrealistic extremes. Physicists relate how the conjunction of processes that gave rise to the elementary forms of matter as we know them bear marked formal similarity to those involved in the evolution of living systems.[7] No longer must dead matter and living forms be viewed as separated by an unbridgeable chasm, but rather they both can now be viewed as outgrowths of a common form of creative agency.[8]

ACKNOWLEDGMENTS

I wish to express my gratitude to Professor Joseph Earley for his kind invitation to contribute to this volume, even though I was unable to participate in the ISPC colloquium. Partial support was provided by the National Science Foundation's Program on Biocomplexity (Contract No. DEB-9981328.), and the U.S. Geological Survey Program for Across Trophic Levels Systems Simulation (ATLSS, Contract 1445CA09950093).

REFERENCES

1. NOETHER, E. 1983. Gesammelte Abhandlungen. Nathan Jacobson, Ed. Springer Verlag. New York.
2. POPPER, K.R. 1990. A World of Propensities. Thoemmes. Bristol, UK. p. 51.
3. MICKULECKY, D. 1985. Network thermodynamics in biology and ecology: an introduction. In Ecosystem Theory for Biological Oceanography. R.E. Ulanowicz & T. Platt, Eds.: 163–175. Canadian Bulletin of Fisheries and Aquatic Sciences, Vol. 213. Ottawa.
4. ULANOWICZ, R.E. 1997. Ecology, the Ascendent Perspective. Columbia University Press. New York. p. 201.
5. MATURANO, H.R. & F.J. VARELA. 1980. Autopoesis and Cognition: The Realization of the Living. Riedel. Dordrecht, the Netherlands.
6. HAUGHT, J.F. 2001. Science, religion, and the origin of life. Presented at AAAS Seminar, September 13, Washington, DC; JONAS, H. 1966. The Phenomenon of Life. Harper & Rowe. New York. p. 9.
7. CHAISSON, E.J. 2001. Cosmic Evolution. Harvard University Press. Cambridge, MA. p. 275.
8. ULANOWICZ, R.E. 2002. Ecology, a dialogue between the quick and the dead. Emergence **4(1/2):** 34–52.

Chirality and Handedness

The Ruch "Shoe–Potato" Dichotomy in the Right–Left Classification Problem

R. BRUCE KING

Department of Chemistry, The University of Georgia, Athens, Georgia 30602, USA

ABSTRACT: Chirality and handedness are concepts that apply to the structure of molecules. Chirality is defined by the lack of certain features of symmetry, which lead to an object not being superimposable on its mirror image. Handedness is a different phenomenon relating to the ability to classify chiral objects into right-handed and left-handed objects. All handed objects are chiral, but not all chiral objects are handed. In 1968 through 1970, Ruch and coworkers developed a theory of chirality that provided a mathematical basis for the handedness of chiral objects. Handed chiral objects are considered to be analogous to shoes, which are readily classified into right and left shoes regardless of the size, material, style, or other attributes of the shoes in question. Nonhanded chiral objects are considered to be analogous to potatoes, which have no symmetry because of their irregular patterns of "bumps" and "eyes," thereby meeting the lack of symmetry requirements for chirality. There is, however, no unambiguous way to classify a set of potatoes into "left" and "right" potatoes. In the case of molecules, a tetrahedron with four different substituents such as an asymmetric carbon atom in organic chemistry can readily be classified into right and left tetrahedra, as is typical for organic molecules in biological systems. An octahedral molecule with six different substituents, however, exhibits nonhanded chirality. For this reason, the fact that the key building block of natural products is a tetrahedral carbon atom rather than an octahedral atom of some type may be a critical factor in the handedness of many molecules found in biological structures.

KEYWORDS: chirality; Ruch; handedness; symmetry; nonhanded chirality; octahedral molecules

INTRODUCTION

An object is considered to be chiral when it differs from its mirror image. Thus, reflection of a chiral object changes the object into a different object, known as its enantiomer. If an object is not chiral, however, then reflection only changes the position of the object in space. Thus a mirror image of an achiral object is superimposable on the original object by a suitable combination of rotations and translations in space. Chirality can be defined more mathematically by the absence of reflections

---

Address for correspondence: Department of Chemistry, The University of Georgia, Athens, GA 30602. Voice: 706-542-1901; fax: 1-706-542-9545.
rbking@sunchem.chem.uga.edu

and other improper rotations (rotation-reflections and inversions) in the symmetry point group of the object being considered.

The concept of chirality can be used as a basis for the division of sets of objects into three subsets. One of these subsets contains all achiral objects, that is, objects lacking the property of chirality and thus containing reflections and other improper rotations in their symmetry point groups. The other two subsets contain exclusively chiral objects distributed between these two subsets so that no pair of enantiomers is contained in a single subset. Conventionally, one of these subsets of chiral objects is considered to contain right-handed objects and the other subset to contain the corresponding left-handed objects, that is, the mirror images of the right-handed objects.

Such a classification of chiral objects would pose no problem if the concept of chirality were the same as that of handedness. However, Ruch[1-3] first showed that these two concepts are not equivalent, even though the very word "chirality" itself is derived from the Greek word for "hand," namely χειρ. Ruch thus classified chiral objects into two categories. The first category consisted of handed objects (Ruch's "category a"), for which there was a natural and unambiguous way of dividing the objects into two subsets, conventionally called "left" and "right" objects. Such objects were considered to be analogous to shoes,[a] which can readily be separated into left and right shoes despite major differences in color, shape, and size between different shoes. Ruch's second category ("category b") consisted of nonhanded objects, which were considered to be analogous to potatoes. Potatoes meet the criterion of being chiral in that the mirror image of a given potato is not superimposable on the original potato because of the irregular pattern of its "bumps" and "eyes." Nevertheless, there is no unambiguous way of classifying a set of potatoes into two subsets of "left" potatoes and "right" potatoes. Potatoes are thus the prototypical examples of nonhanded chiral objects.

Although the concept of chirality is applicable to any type of object, including shoes and potatoes, the driving force behind the development of the theory of chirality comes from a major objective in chemistry in understanding the structure of molecules. Enantiomeric molecules (i.e., a set of nonequivalent mirror images) represent a special type of stereoisomers, wheras a pair of stereoisomers is defined as a set of two molecules with the same elemental composition and same topology (set of chemical bonds), but with different orientations of the atoms in space. Traditionally in chemistry, the concept of chirality arose as an effort to explain the ability of compounds containing an "asymmetric" carbon atom, that is, a (tetravalent) carbon atom with four different substituents, to rotate polarized light (optical rotation). Optical rotation is the original example of a so-called pseudoscalar measurement, which gives identical absolute values but opposite signs for enantiomers. Other pseudoscalar measurements arising more recently in the history of chemistry include circular dichroism and optical yields (enantiomeric excesses) in asymmetric syntheses.

The asymmetric carbon atom can be modeled using a tetrahedral skeleton with four different labels at the vertices (FIG. 1A) corresponding to the substituents. The tetrahedron is an example of a shoe-like skeleton where an unambiguous right–left classification is always possible when each of the sites have different labels. In con-

---

[a]Even though the term chirality refers to hands rather than feet, shoes rather than gloves are traditionally used for this analogy. Shoes would fit better with terms like "podality" and "footedness," neither of which is seen in the chemical literature in this context.

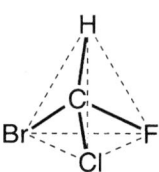
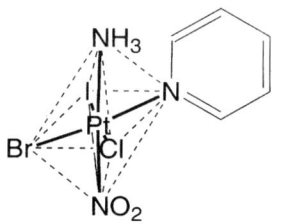

**Tetrahedral
CHFClBr
SHOE
Handed Chiral**

**Octahedral
Pt(NH₃)(NO₂)(NC₅H₅)(Cl)(Br)(I)
POTATO
Non-handed Chiral**

FIGURE 1. (A) An example of a tetrahedral carbon atom with four different substituents. (B) An example of an octahedral platinum atom with six different substituents.

trast the octahedron is an example of a potato-like skeleton, for which an unambiguous right–left classification is no longer possible if all of the sites have different substituents or ligands such as in the inorganic compound Pt(NH₃)(NO₂)(NC₅H₅)(Cl)(Br)(I) (FIG. 1B). This paper will examine later in greater detail the comparison between the tetrahedron and octahedron in this respect.

Ruch recognized the dichotomy of chiral objects into "shoes" and "potatoes" as part of the development of a formal mathematical model of chirality.[2–6] Assimilation of this theory by the international scientific community is made difficult by the advanced mathematics required. This essay is an attempt to extract some broad concepts of nonhanded chirality and the right–left classification problem from this theoretical framework, avoiding as much of the advanced mathematics as possible. The discussion is organized into the following two areas: (1) geometrical aspects of chirality: the concepts of a skeleton, framework groups, and ligand partitions; (2) algebraic aspects of chirality: the description of pseudoscalar phenomena by mathematical functions, the form of which relates to the shoe–potato dichotomy.

THE GEOMETRY OF CHIRALITY

A key concept in the geometry of chirality is that of a skeleton. Thus consider a molecule of the type ML_n where M is a metal or other central atom (e.g., the carbon atom in the case of an asymmetric carbon atom) and the n ligands L (or substituents) may or may not be equivalent. Removal of the n ligands L from ML_n leads to the skeleton.

The symmetry of the skeleton clearly needs to be considered in the theory of chirality. In this connection the concept of framework groups[7] is useful to specify the symmetry of bodies containing a finite number of particles such as the ML_n skeleton and its n sites for the ligands L. Using the concept of framework groups, possible skeletons can be classified into the following four general types according to their symmetries and location of the ligands:

(1) *Linear*: All sites of the skeleton are located in a straight line.
(2) *Planar*: Nonlinear framework groups in which all sites are located in a flat plane.
(3) *Achiral*: Nonplanar framework groups having a symmetry group containing at least one reflection, inversion, or rotation-reflection.
(4) *Chiral*: Nonplanar framework groups having a symmetry group containing no reflections, inversions, or rotation-reflections.

Skeletons with chiral framework groups lead to chiral structures for any distribution of ligands including all ligands identical. Skeletons with linear framework groups can never be chiral even if all ligands are different. The two remaining types of skeletons, namely those with planar and achiral framework groups, are those of interest in the theory of chiral molecules.

An achiral skeleton can lead to a chiral molecule if the distribution of the ligands at its n sites destroys all of its reflection, inversion, and rotation-reflection symmetry. All sites are coplanar in skeletons with planar framework groups. The plane containing all of the sites can be called the *major plane* of the framework group. Any ligand partition of a skeleton with a planar framework group (even that having all ligands different) retains the major plane as a symmetry plane thereby never leading to a chiral structure. However, a chiral structure can be obtained from a skeleton with a planar framework group by destroying the symmetry of the major plane. Since this can be done conceptually by placing a positive charge on one side of the major plane and a negative charge on the other side, this process is conveniently called polarization,[8] From a chemical point of view, polarization can involve polyhapto complexation of a planar skeleton (e.g., the planar hexagon benzene) to a metal atom (e.g., chromium) on one side of the skeleton (FIG. 2).

The symmetry of a skeleton may also be related to its transitivity. Thus a skeleton having all sites equivalent is called a transitive skeleton; otherwise the skeleton is called intransitive. A set of equivalent sites in a skeleton is called an orbit. A transitive skeleton thus has only one orbit consisting of all of its sites. Transitive skeletons play a fundamental role in chirality algebra. FIGURE 3 depicts the transitive skeletons with seven or fewer sites.

The distribution of ligands on the skeleton is important in chirality algebra. The ligand partition can be depicted by a symbol of the type $(a^{ba}[a-1]^{ba-1}[a-2]^{ba-2}\ldots 1)^{b_1}$

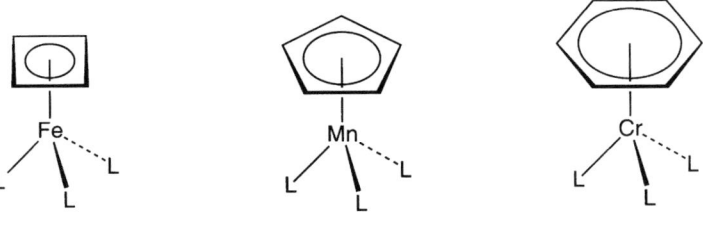

FIGURE 2. Polarization of planar polygonal skeletons by complexation with a metal atom in $C_nH_nML_3$ derivatives (L is commonly CO).

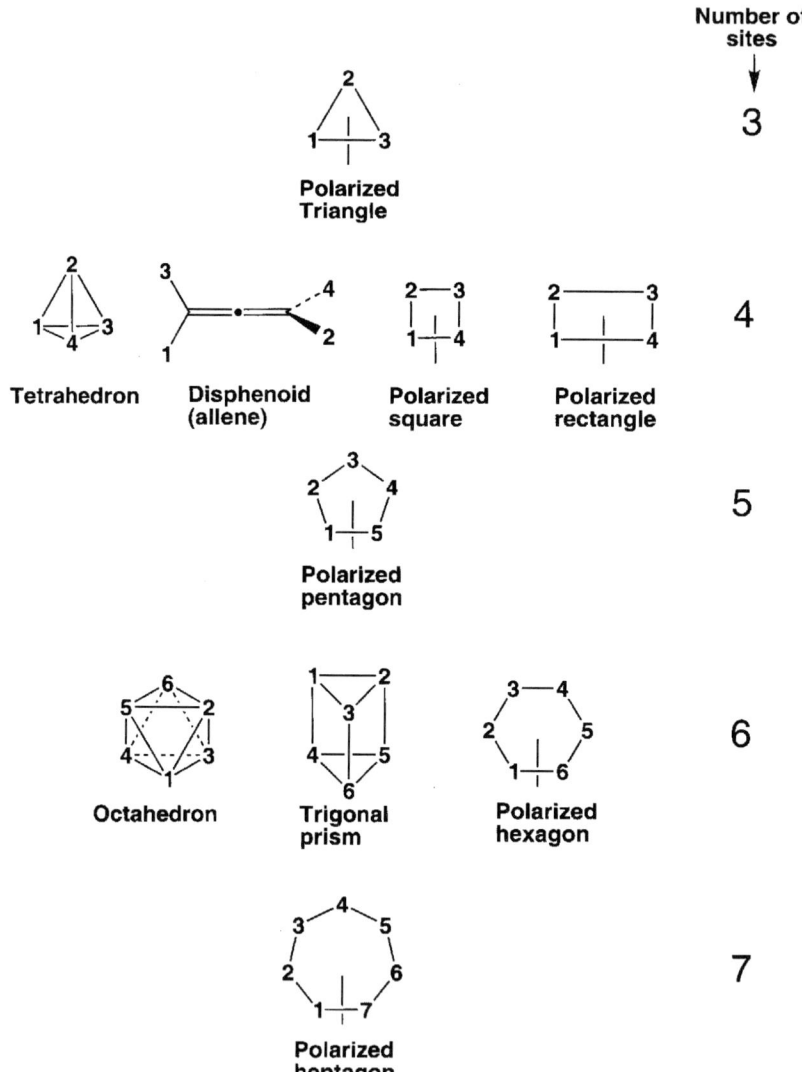

FIGURE 3. The permutationally distinct transitive skeletons with seven or fewer sites. Note that the D_{3d} trigonal antiprism skeleton (not shown) is permutationally equivalent to the D_{3h} trigonal prism.

in which b_k refers to the number of sets of k identical ligands. In addition, the ligand partition can also be depicted by a collection of boxes called Young diagrams, which have the following features:

(1) The rows of boxes represent identical ligands;
(2) The top row is always the longest row and the lengths of the rows decrease monotonically from top to bottom;

(3) The left column is always the longest column, and the heights of the columns decrease monotonically from left to right.

For any given integer n, there is a Young diagram corresponding to all of the possible partitions of n into sums of positive integers.

Young diagrams, and thus the corresponding ligand partitions, can be characterized by the following three parameters:

(1) Order (o): The order of a Young diagram is the maximum number of identical ligands in the ligand partition and is simply the length of the top row. The order also corresponds to the number of columns in the Young diagram.
(2) Index (i): The index of a Young diagram is the number of different ligands in the ligand partition and is simply the height of the left column. The index also corresponds to the number of rows in the Young diagram.
(3) Degree (g): The degrees of Young diagrams provide a basis for assigning an order to the symmetries of ligand partitions required for the theory of chirality. The degree of a Young diagram, g, is less obvious than its order or index but can be calculated by the following equation in which c_k represents the height of column k:

$$g = \frac{1}{2} \sum_{k=\text{order}}^{k=1} c_k(c_k - 1). \tag{1}$$

For Young diagrams having six or more boxes, there are cases in which two or more different Young diagrams with the same number of boxes have the same degree as determined by Eq. (1) so that the ordering of Young diagrams by their degrees is only a partial ordering. In general, Young diagrams having high degrees correspond to relatively asymmetrical ligand partitions, and Young diagrams having low degrees correspond to relatively symmetrical ligand partitions. Thus the degree of a Young diagram may be viewed as a measure of the "asymmetry" of the corresponding ligand partition.

FIGURES 4 and 5 illustrate the Young diagrams for the four sites on a tetrahedral ML_4 skeleton and the six sites on an octahedral ML_6 skeleton, respectively. The Young diagrams are ordered according to their degrees with the lowest degree ligand partition at the top (degree 0 corresponding to all ligands equivalent) and the highest degree ligand partition at the bottom [degree $n(n-1)/2$ for n sites by Eq. (1) corresponding to all ligands different]. Note that for the octahedral skeleton, there are two different ligand partitions represented by different Young diagrams for both the degrees 3 and 6. Thus for an octahedron degrees of 3 and 6 do not uniquely define a ligand partition. Six is the smallest number of sites leading to different ligand partitions having the same degree.

Of particular significance for an achiral skeleton such as the tetrahedron or octahedron is the ligand partition of lowest degree (minimum asymmetry) necessary to destroy all reflection planes, inversion centers, and other rotation-reflections (improper rotations) to give a chiral structure. This ligand partition is called the chiral ligand partition. For the tetrahedron, the chiral ligand partition is that of highest degree (FIG. 4), namely $g = 6$, corresponding to all ligands different and thus the familiar "asymmetric" carbon atom (FIG. 1). For the octahedron there are two chiral

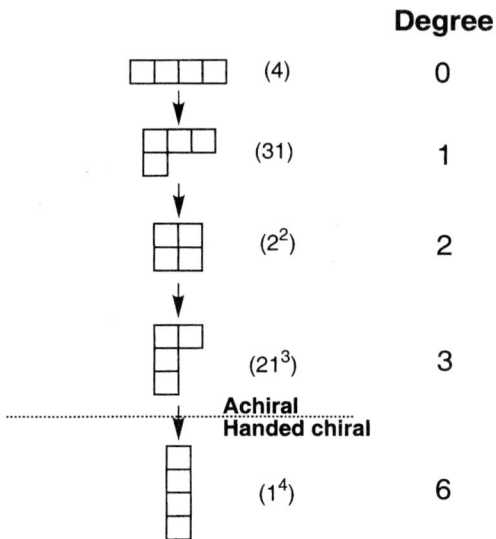

FIGURE 4. Young diagrams representing ligand partitions for four-coordinate tetrahedral ML$_4$ complexes.

ligand partitions, necessarily both of the same degree, namely (31^3 and 2^3), corresponding to chiral isomers of the types *fac*-MA$_3$BCD and *cis*-MA$_2$B$_2$C$_2$, respectively (FIG. 6). Even though the octahedron is a potato-like skeleton with no unambiguous right–left classification available for the most unsymmetrical ligand distribution MABCDEF (e.g., FIG. 1B), chiral octahedra with the substitution patterns *fac*-MA$_3$BCD and *cis*-MA$_2$B$_2$C$_2$ behave like shoes with a right–left classification. Only chiral octahedra with ligand partitions of degrees 7 or greater, for example, ($2^2 1^2$) in MA$_2$B$_2$CD, become nonhanded, that is, like potatoes.

The handedness of achiral skeletons can be recognized by the location of their reflection planes relative to their sites.[9] In this connection, the reflection planes in an achiral skeleton with n sites can be classified into the following types:

(1) Separating planes containing exactly $n-2$ sites;
(2) Nonseparating planes containing fewer than $n-2$ sites.

For an achiral skeleton to be handed, all of its reflection planes must be separating planes. Furthermore, the number of separating planes in a handed achiral skeleton corresponds to the degree of its chiral ligand partition as determined by Eq. (1).

The presence of even a single nonseparating plane in an achiral skeleton eliminates the possibility of an unambiguous right–left classification corresponding to potato-like behavior. Handedness thus allows only reflection planes interchanging only two ligands. In the handed *fac*-MA$_3$BCD and *cis*-MA$_2$B$_2$C$_2$ complexes of the nonhanded octahedral skeletons (FIG. 6), each of the six reflection planes in an octahedron with these particular ligand distributions is seen to interchange only two

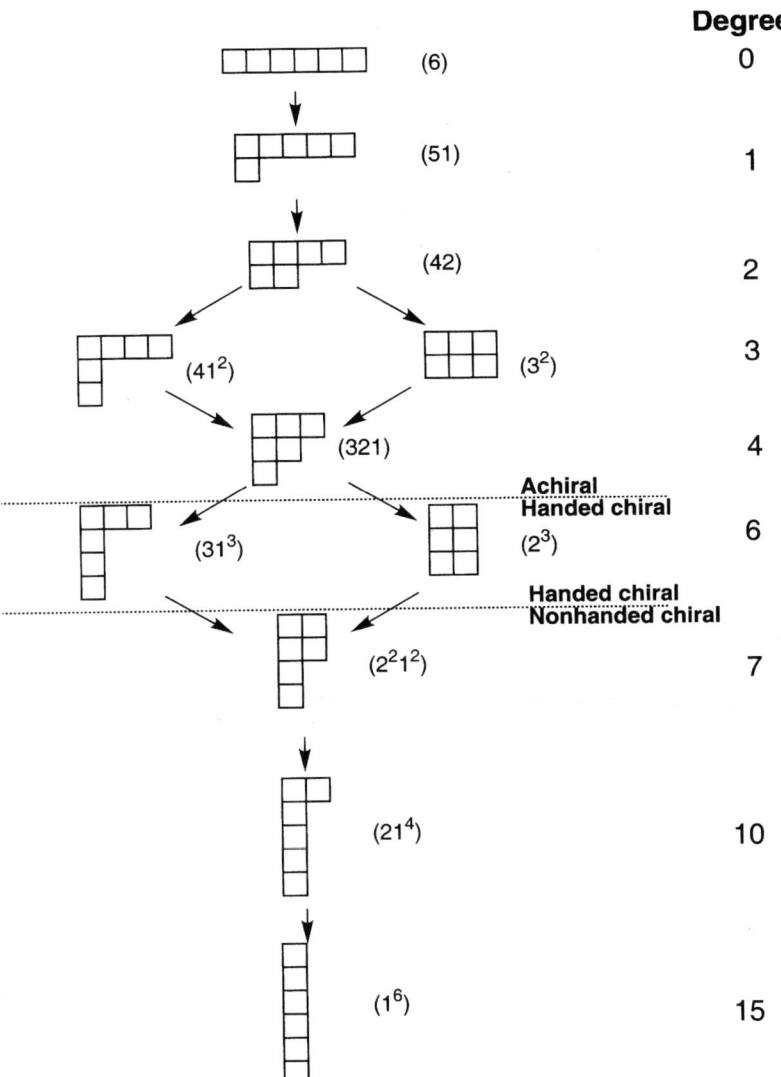

FIGURE 5. Young diagrams representing ligand partitions for six-coordinate octahedral ML_6 complexes.

ligands. However, two of these reflection planes only contain two of the six sites of the octahedron skeleton and are thus nonseparating reflection planes. Thus the reflection planes in a structure provide a trichotomy rather than a dichotomy of chirality phenomena in the following order of increasing ligand partition degree:

(1) Achiral structures containing a reflection plane that leaves the structure invariant so that it is identical with its mirror image;

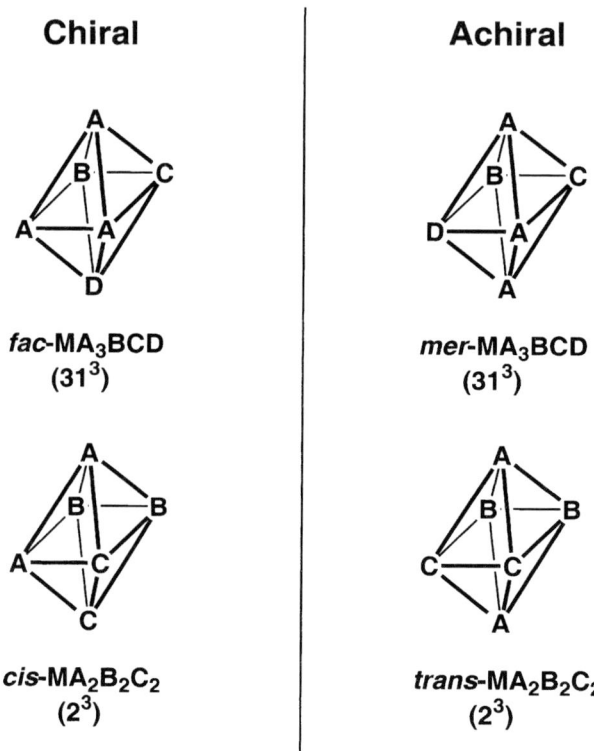

FIGURE 6. (A) Chiral and achiral octahedral MA_3BCD complexes; (B) chiral and achiral octahedral $MA_2B_2C_2$ complexes. The achiral octahedral complexes have a reflection plane through the "horizontal" square.

(2) Handed chiral structures having no reflection planes that leave the structure invariant and having only reflection planes interchanging exactly two ligands;
(3) Nonhanded chiral structures having no reflection planes that leave the structure invariant and having at least one reflection plane interchanging more than two ligands.

This trichotomy is also consistent with the algebra of chirality and the resulting polynomial approximations to chirality functions discussed in the next section.

THE ALGEBRA OF CHIRALITY

The objective of chirality algebra is to derive functions that predict the values of pseudoscalar measurements on chiral molecules such as rotation of plane-polarized light or circular dichroism. Such pseudoscalar measurements have the property of identical absolute values and opposite signs for a pair of enantiomers. Therefore functions that are valid chirality functions must also have such properties. The form of such chirality functions relates to the nature of the chiral skeleton including whether it is handed or nonhanded. The underlying mathematics is relatively complicated and arises from the formal theory of chirality algebra first published by Ruch and Schönhofer.[4,5]

Consider a chiral molecule represented by ML_n where M is a metal or other central atom and the ligands L may or may not be equivalent. Now consider the problem of approximating a pseudoscalar measurement ψ with a function in n variables of the type $f(\lambda_1, \lambda_2, ..., \lambda_n)$, that is, $\psi = f(\lambda_1, \lambda_2, ..., \lambda_n)$, where the n independent variables $\lambda_1, \lambda_2, ..., \lambda_n$ correspond to parameters associated with the ligands at sites $1,...,n$, respectively, that depend only on the ligand, the skeleton, and the pseudoscalar measurement. The set of parameters $\lambda_1, \lambda_2,...,\lambda_n$ can be considered as a point in n-dimensional ligand space. Thus for a given ligand L^1 (e.g., methyl) the associated ligand parameter $\lambda(L^1)$ will be constant for a given skeleton and pseudoscalar measurement. Such ligand parameters can in principle be determined experimentally by fitting a series of pseudoscalar measurements of a given type (e.g., optical rotation of a given spectra line) on a series of compounds having the same skeleton and a restricted set of ligands using such a suitable chirality function. Such a chirality function $f(\lambda_1, \lambda_2,...,\lambda_n)$ is required to have identical absolute values and opposite signs for a pair of enantiomers in accord with the properties of pseudoscalar measurements.

Now consider the problem of finding the simplest type of possible chirality function for a given skeleton having the required transformation properties. This function turns out to be a polynomial $X(s_1,s_2,...,s_n)$ of the same degree as that of the corresponding chiral ligand partition. This lowest degree chirality polynomial $X(s_1,s_2,...,s_n)$ may be regarded as an initial term in a Taylor series approximation of a more accurate chirality function $F(\lambda_1,\lambda_2,...,\lambda_n)$. The accuracy of the approximation of $F(\lambda_1,\lambda_2,...,\lambda_n)$ by its first term $X(s_1,s_2,...,s_n)$ will natural depend on the skeleton and the pseudoscalar property and cannot be predicted *a priori* by theory alone. Thus chirality polynomials are not required to work, that is, provide a close approximation to the experimental measurement of the pseudoscalar property in question. However, chirality polynomials may work so that the investigation of applications to model pseudoscalar properties is a valid pursuit. Note that the ligand parameters $\lambda_1,\lambda_2,...,\lambda_n$ for F and $s_1,s_2,...,s_n$ for its polynomial approximation X are not necessarily the same.

The procedure for obtaining the lowest degree chirality polynomials $X(s_1,s_2,...,s_n)$ involves group representation theory and projection operators beyond the scope of this essay.[4,5] However, for certain special types of skeletons the lowest degree polynomials $X(s_1,s_2,...,s_n)$ take on a particular form. For example, consider an intransitive skeleton with n sites and q orbits. Let X_k be the chirality polynomial for orbit k. Then the chirality polynomial for the intransitive skeleton can be obtained from the following relationship:

$$X(s_1,s_2,...,s_n) = \prod_{k=1}^{k=q} X_k(s_1, s_2, ..., s_n). \qquad (2)$$

Thus the lowest degree chirality polynomial for an intransitive framework group is the product of the chirality polynomials of the individual orbits.

The lowest degree chirality polynomials for handed (shoe-like) skeletons also have a simple form. Thus, consider a handed skeleton with n sites and p planes of symmetry. Any given plane of symmetry of this handed skeleton is a separating plane containing exactly $n - 2$ sites. Consider one of these planes p_k and label the two sites outside this plane a_k and b_k. Reflection through the plane p_k thus interchanges a_k and b_k. Then the corresponding lowest degree chirality polynomial has the form:

$$X(s_1, s_2, \ldots, s_n) = \prod_{k=1}^{\kappa = p} X_k(s_{a_k} - s_{b_k}) \tag{3}$$

Thus the degree of the lowest degree chirality polynomial is equal to the number of symmetry planes.

The lowest degree chiral ligand partition of the tetrahedron (T_d) is the degree 6 (1^4) ligand partition with all ligands nonequivalent (MABCD). The corresponding chirality polynomial is

$$X(T_d : 1^4) = (s_1 - s_2)(s_1 - s_3)(s_1 - s_4)(s_2 - s_3)(s_2 - s_4)(s_3 - s_4) \tag{4}$$

in accord with Eq. (3). For the octahedron with all ligands different (MABCDEF), the full group theoretical algorithm[4,5] gives the following chirality polynomial for the degree 6 chiral ligand partitions (31^3) or (2^3) after some messy algebra:

$$\begin{aligned} X(O_h: 31^3 \text{ or } 2^3) = (s_1 - s_6)(s_2 - s_4)(s_3 - s_5)\{&[(s_1 - s_2)(s_2 - s_3)(s_3 - s_1) + \\ (s_3 - s_6)(s_6 - s_4)(s_4 - s_1) + (s_1 - s_4)(s_4 - s_5)(s_5 - s_1) &+ (s_2 - s_5)(s_5 - s_6)(s_6 - s_2)] + \\ [(s_1 - s_2 + s_6 - s_4)(s_1 - s_5 &+ s_6 - s_3)(s_2 - s_5 + s_4 - s_3)]\} \end{aligned} \tag{5}$$

This rather forbidding chirality polynomial is of degree 6 in accord with the degrees of the corresponding chiral ligand partitions (31^3) and (2^3). However, it is not a product of simple differences between ligand parameters ($s_i - s_j$) like the corresponding chirality polynomial for the tetrahedron (Eq. 4). This is in accord with the nonhanded nature of the octahedron as compared with the tetrahedron. The full chirality polynomial for the octahedron (Eq. 5) would in principle be required to study pseudoscalar phenomena in octahedral complexes with six different substituents such as the platinum complex $Pt(NH_3)(NO_2)(NC_5H_5)(Cl)(Br)(I)$ (FIG. 1B) mentioned in the Introduction. However, such metal complexes are far too rare for generation of a large enough database to provide valid sets of ligand parameters. Furthermore, separating the 15 different pairs of enantiomers from a mixture and resolving individual pairs of enantiomers into pure enantiomers for $Pt(NH_3)(NO_2)(NC_5H_5)(Cl)(Br)(I)$ and related octahedral metal complexes with six different ligands would seem to be a staggering if not impossible task. Therefore the full lowest degree chirality polynomial for an octahedral complex with six different ligands (Eq. 5) is never likely to be tested experimentally.

Considerable simplification of the full degree 6 octahedral chirality polynomial (Eq. 5) can be achieved if handed octahedral complexes of the types *fac*-MA_3BCD and $MA_2B_2C_2$ are considered.[10] First consider the (31^3) chiral ligand partition for octahedral metal complexes of the general type *fac*-MA_3BCD (FIG. 6A), which leads to the equality $s_4 = s_5 = s_6$ for the three equivalent ligands A. Applying this equality to the full chirality polynomial for the octahedron (Eq. 5) leads to

$$X(O_h: 31^3) = 2(s_1-s_2)(s_1-s_3)(s_1-s_4)(s_2-s_3)(s_2-s_4)(s_3-s_4) \quad (6)$$

Note that this chirality polynomial is a product of simple differences of ligand parameters in accord with Eq. (3). Furthermore, this chirality polynomial is seen to be essentially the same as that for the tetrahedron (Eq. 4). This is consistent with the analogy of the (31^3) chiral ligand partition of the octahedron to the (1^4) chiral ligand partition of the tetrahedron. The three equivalent facial ligands A in octahedral *fac*-MA$_3$BCD complexes (FIG. 6A) together thus play the role of one of the four nonequivalent ligands in the tetrahedral MABCD complexes. In fact, the relative orientation of four points including the midpoint of the A$_3$ triangle and each of the unique three ligands B, C, and D in *fac*-MA$_3$BCD is very similar to a tetrahedron.

Now consider the (2^3) chiral ligand partition for the octahedral metal complexes of the general type *cis*-MA$_2$B$_2$C$_2$ (FIG. 6B), which leads to the equalities $s_1 = s_5$, $s_2 = s_3$, and $s_4 = s_6$. Applying these equalities to the full chirality polynomial for the octahedron (Eq. 5) leads to

$$X(O_h: 2^3) = 2(s_1-s_2)^2(s_2-s_4)^2(s_4-s_1)^2 \quad (7)$$

This chirality polynomial is also seen to be a product of simple $(s_i - s_k)$ differences in accord with the handedness of chiral octahedral metal complexes with the (2^3) ligand partition. In the octahedral MA$_2$B$_2$C$_2$ complexes, however, there are only three different kinds of ligands. Since there are only three possible distinct ligand difference terms, each of the terms needs to be squared to give the theoretically required degree of 6.

EPILOGUE

Chirality and handedness are often assumed to be the same thing. Thus, if a molecule is chiral, a right–left classification is assumed to be possible. This misconception has remained for decades after the demonstration of the feasibility of nonhanded chirality in the 1970s for two reasons. First, the mathematics of chirality algebra is difficult (at least for chemists). Furthermore, most chiral molecules, including those of biological importance, are based on "asymmetric" tetrahedral carbon atoms that must be handed so that handedness is the norm for chiral molecules in nature. If biologically significant molecules had structures based on "asymmetric" octahedral atoms of some type rather than "asymmetric" tetrahedral carbon atoms, then nonhanded chiral molecules might be more prevalent, and asymmetry might not even be linked to handedness. However, it is possible that living processes might even require handed chirality and that a coordination polyhedron admitting nonhanded chirality, such as the octahedron, might be unacceptable as a building block for key biologically significant molecules. In this respect the tetrahedron has the rare feature of having only the handed chiral ligand partition (1^4) and no nonhanded chiral ligand partitions; all other ligand partitions for the tetrahedron are achiral. This special feature of the tetrahedron and the propensity of carbon atoms to form four bonds at the vertices of a tetrahedron may contribute to the role of the tetrahedron as the "polyhedron of life" and carbon as the "element of life."

REFERENCES

1. RUCH, E. 1968. Homochiralität als Klassifizierungsprinzip von Molekülen spezieller Molekülklassen. Theor. Chim. Acta **11**: 183–192.
2. RUCH, E. 1972. Algebraic aspects of the chirality phenomenon in chemistry. Acct. Chem. Res. **5**: 49–56.
3. RUCH, E. 1977. Chiral derivatives of achiral molecules: standard classes and the problem of a right–left classification. Angew. Chem. Int. Ed. **16**: 65–72.
4. RUCH, E. & A. SCHÖNHOFER. 1968. Näherungsformeln für spiegelsantimetrische. Moleküleigenschaften. Theor. Chim. Acta **10**: 91–110.
5. RUCH, E. & A. SCHÖNHOFER. 1970. Theorie der chiralitätsfunktionen. Theor. Chim. Acta **19**: 225–287.
6. MEAD, C. 1974. Permutation groups, symmetry, and chirality in molecules. Top. Curr. Chem. **49**: 1–88.
7. POPLE, J.A. 1980. Classification of molecular symmetry by framework groups. J. Am. Chem. Soc. **102**: 4615–4622.
8. KING, R. 1983. Chemical applications of topology and group theory. 13. Chirality and Framework Groups. Theor. Chim. Acta **63**: 103–132.
9. MEINKÖHN, D. 1978. On the investigation of pseudomolecular molecular properties and polynomial approximations according to Ruch and Schönhofer. Theor. Chim. Acta **47**: 67–79.
10. KING, R. 2001. Non-handed chirality in octahedral metal complexes. Chirality **13**: 465–473.

How Harmful Is the First Law?

GEORG JOB AND TIMM LANKAU

Job-Stiftung, Institut für Physikalische Chemie, University of Hamburg, 20146 Hamburg, Germany

ABSTRACT: In his First Law (of thermodynamics), Clausius emphasized the equivalence of heat and work—conservation of energy was mentioned only indirectly. Today, the main emphasis is put on energy conservation, but the equivalence of heat and work has proven not only to be superfluous, but also highly destructive. Because of this emphasis, heat is no longer considered an independent entity. In the guise of the quantity Q heat acts in a strange double role—an entity equivalent to work, but also as something fundamentally different. The place formerly occupied by heat is now filled with an abstract quantity, the entropy S—a phantom without macroscopically relevant properties. By using a phenomenological approach to entropy, it can be shown that entropy S does indeed have easily comprehensible macroscopic properties. This approach can be used to simplify thermodynamic reasoning and to reduce the calculus of thermodynamics to a fraction of its usual extent.

KEYWORDS: entropy; heat; Clausius; quantity of heat; thermodynamics; work

PRELIMINARY REMARKS

The question in the title of this paper raises eyebrows. Today we accept the first law of thermodynamics as a statement of energy conservation. We have learned that conservation of energy is one of the fundamental laws of nature—proven to be valid in both the macrophysical and microphysical worlds. But there is a subtle and inconspicuous difference between the first law of thermodynamics and the principle of conservation of energy—a difference that has far-reaching consequences for the development and understanding of theories of heat, and for related disciplines such as chemical thermodynamics. We are not questioning the principle of the conservation of energy, but its special formulation as part of the First Law of Thermodynamics—with the *equivalence of heat and work* as its central idea since 1850. The statement of the First Law in terms of the equivalence of heat and work establishes a close link between two formerly completely independent concepts and treats the link as an underlying truth without serious examination. Classical thermodynamics is based on this foundation—its success is seen as proof beyond doubt that this was the right approach. For theories in natural sciences their success is often counted as the most important criterion for proving the correctness of underlying assumptions. Doubts

---

Address for correspondence: Dr. Georg Job, Job-Stiftung, Institut für Physikalische Chemie, University of Hamburg, 20146 Hamburg, Germany. Voice: +49-40-42838-3423, +49-4181-380392; fax: +49-4181-380393.

Georg.Job@gmx.de

about the meaning of the heat-work equivalence—accepted without question for one and a half centuries—appear hopeless and almost heretical.

Nonetheless, we hope to show that it is worthwhile to re-examine this point. From a philosophical standpoint, one can pose the question whether it is sufficient to show that a scientific theory is free of contradictions and confirmed by experience. Chemistry offers an interesting parallel, the phlogiston theory. That theory can be formulated in such a way that it is free of contradictions and in agreement with experience, yet it is totally unacceptable.

HISTORICAL BACKGROUND

The Period before 1850

The chemist Joseph Black (1728–1799) was one of the first who brought some order to thermodynamics, during the decades before 1800, by differentiating concepts such as temperature, heat quantity, and heat capacity. He thought of heat as a non-producible and indestructible entity—like a chemical element. He also taught that some substances contain more heat than others with the same temperature. When a substance changes from being thermally poor to thermally rich, like ice to water or water to steam, then the missing heat has to be supplied. When the same process runs in reverse, the excess heat has to be removed. It was realized in those days that it is impossible to create work from nothing—no perpetual motion machine could exist—but it was still believed that work could vanish without a trace. If a wheel with the brake applied has to be turned, work has to be done. Since this work is not retrievable, it was believed to be irreversibly lost. For our purposes, the most important features of pre-energetic thermodynamics are briefly summarized in TABLE 1.

Based on this foundation, in 1824, Sadi Carnot (1796–1832) developed his theory of heat engines by analogy to a water mill. In 1834, Benoit Clapeyron (1799–1864) introduced his equation for the increase in the pressure of steam as a function of temperature. In 1848 William Thomson (1824–1907) created the first universal definition of temperature that was independent of any thermometric material.

The Change

The year 1850 brought a decisive change. This change did not occur suddenly. It had a long history during which more and more facts became known that contradicted the predominant thinking. A simple example—namely, heat generation during abrasion—can clarify this change. This effect was attributed to a heat release from the material ground up during abrasion, which was thought to have a smaller heat capacity than the bulk material. In contradiction to this model, James Prescott Joule (1818–1889) proved in 1843 that there is a fixed relationship between the work expended and the heat generated,

$$W = \alpha \cdot Q \tag{1}$$

independent of the process of generation and of the amount of material ground up, if any. This finding indicates that heat is really generated and conversely, work does

TABLE 1. Main points of the historical development of thermodynamics

	Producible	Destroyable	
Thermodynamics before 1850			
Heat	No	No	Conservative
Work	No	Yes	Semi-conservative
Experimental findings lead to abandonment of pre-energetic thermodynamics			
Heat	No → yes	No	Semi-conservative
Work	No	Yes → no	Conservative
Thermodynamics after 1850			
Heat → entropy	Yes	No	Semi-conservative
Energy = work+heat	No	No	Conservative

not vanish, but that instead a well-defined equivalent entity is formed. Therefore, the old interpretation was wrong on two accounts (TABLE 1).

Joule believed that his research had validated the hypothesis, advocated by him and some of his contemporaries, that heat and work are two forms of the same unchangeable entity occurring in nature. He called it *power, motive force,* or *vis viva* —we refer to it as *energy* today. The thought that the amount of work necessary for the generation of a specific amount of heat could depend on temperature did not occur to him. In that case, the factor α in Equation (1) would not be constant. His simple experimental setup did not allow him to address this problem.

It was known at that time that heat generated at a high temperature could be used to perform more work in a heat engine than heat generated at a lower temperature. Apparently, heat is more valuable at a high temperature than the same amount of heat at a lower temperature—the idea that it takes more work to generate a given heat quantity at a high temperature than at a lower one was therefore close at hand.

Joule's experiments provided evidence for heat generation but not for heat loss. It was known that heat generation depends on the consumption of work. Rudolf Clausius (1822–1888) thought it was justified to assume the inverse also to be true —that a heat engine consumes heat in producing work.[1] With this argument he justified a basic assumption about the nature of heat that later became generally known as the First Law of Thermodynamics. It is worthwhile to mention here that, in the same scientific essay that announces what came to be known as the First Law, Clausius introduced an additional quantity that later came to be called "entropy." Entropy was assigned the same properties as those suggested for heat by the experiments (TABLE 1).

Period after 1850

The old theory of heat had numerous early successes, but was not sustainable in the presence of contradictory new observations. Clausius showed that not all of the old discoveries had to be discarded in order to reconcile the theory with experience. He showed that a few basic assumptions about the nature of heat are sufficient to reach this goal. These are in summary:

- Heat and work are interchangeable in a fixed relationship (the First Law).
- Heat does not flow by itself from low-temperature objects to high-temperature objects.

Please note that the First Law makes a statement about heat while energy conservation is only implied. This version of the First Law allows work to be destroyed in a process where heat plays no role. Conversely, it is possible to formulate the energy principle for example, as Max Born[2] did—without referring to the heat concept. The assumption, commonly made today, that the First Law is equivalent to the principle of the conservation of energy, obscures this distinction: a small distinction but one that is important for us. From his second basic assumption Clausius derived a quantity, later called entropy, which is only a function of the system's state and thus independent of the process used to reach this state. This new quantity was defined by the following integral,

$$S(p, T) = S_0 + \int_{p_0, T_0}^{p, T} \frac{dQ_{\text{rev}}}{T} \qquad (2)$$

Clausius showed that this quantity has a remarkable property:

- The entropy of a closed system increases but never decreases (Second Law).

W. Thomson, who resisted accepting Joule's interpretation of heat for a long time, was persuaded to develop the theory of heat further, starting with the foundation suggested by Clausius. Also Hermann von Helmholtz (1821–1894), Josiah Gibbs (1839–1903), and others expanded the theoretical framework in various directions. For our purposes, the most salient features of "classical thermodynamics" after 1850 are roughly summarized in TABLE 1.

The changes made covered more than that necessary to accommodate the experimental findings:

- Heat was combined with work into a new quantity called "energy" and thus lost its role as an individual entity.
- The newly created, and purely mathematical, quantity *entropy* takes the place of heat.

PROPERTIES OF ENTROPY

Phenomenological Approach to Entropy

Classical thermodynamics is categorized as a phenomenological theory. It has the special advantage of yielding important results without making use of the existence of atoms. In the 20th century, the phenomenological theory was extended to irreversible processes, and thus not limited to equilibrium conditions. Entropy was thought of as something distributed in matter, like a thermal fluid that is producible but cannot be destroyed. It was likened, in some ways, to the electrical charge or a diffusing substance. These comparisons were prompted by the similarities in the differential equations for entropy, electrical charge, and material flows. It is noteworthy that, in this case, entropy is treated, at least formally, as an object with physical properties, not purely a mathematical invention.

The list of properties is still too small to present a picture of entropy that is of practical use. In describing a person, we usually compile a list of "phenomenological" attributes that allow us to evaluate his or her appearance, place of residence, skills, and so on. The conjunction of these attributes is basically what makes up a person; his or her name is only an abbreviation for this list. Data such as fingerprints, blood type, or passport number are important to find and identify a person, but they are no substitute for the "phenomenological" attributes. The "wanted poster" for a person is an example of a concise list of such attributes.

We will show that it is possible to develop a wanted poster or fact file for entropy. This description will focus on phenomenologically relevant attributes that are sufficient to define entropy as a measurable physical quantity—while omitting the usual purely mathematical traits using statistical methods for describing atomic behavior. For our approach, it is not required to quantify temperature. As such, it makes it easy to define temperature on the absolute scale by means of entropy.

Attributes of Entropy

Two important mathematical features of entropy are that it is an extensive quantity and a function of the system state. By comparing the entropy of each of two equal systems in the same state, we find that they have the same value. If the entropy is greater or smaller, then the state of the system is different. The entropy contained in a body is not without effect; it changes the condition of that body. One objective of this section is to describe such observable changes more precisely. Along the same lines, we can translate other abstract statements about entropy into simpler language.

(1) *Entropy as an extensive state function:* Each object contains more or less entropy depending on its state. Objects of the same kind and under the same conditions contain equal amounts of entropy. The entropy of a system of objects is equal to the sum of the entropies of its parts.

(2) *Generation:* Entropy is generated in a variety of ways; for example by rubbing, passage of an electric current, or a chemical reaction.

(3) *The Second Law:* Entropy cannot penetrate thermally insulated walls. Therefore, the amount of entropy in a thermally insulated body cannot decrease but can only increase.

(4) *Heating:* The main effect of an entropy increase is that the body becomes warm. For objects of the same kind, the one with the most entropy is the warmest. A body without entropy is absolutely cold. Entropy flows spontaneously only from hotter to colder bodies. In a body entropy spreads out uniformly until all parts are equally warm.

(5) *Side effects*: An entropy increase causes numerous side effects, including changes in volume, shape, state of aggregation, and magnetism. As an example, let us consider the general effects on a substance of continuously increasing entropy—it does not matter whether the entropy source is internal or external. (*a*) Volume increases for most cases. One could say that entropy requires space. Volume decreases when ice melts to produce water. This is one of only a few exceptions. (*b*) When a solid reaches its melting

point, the additional entropy cumulates in the resulting melt and therefore the body does not become any warmer. The melted substance contains more entropy than the original solid. (*c*) When a liquid reaches its boiling point, the resulting vapor absorbs the additional entropy and therefore the liquid does not become any warmer. The entropy increases further when the steam vapor with the surrounding air.

(6) *Reciprocal effects:* A substance that expands with an entropy input—almost all substances do—becomes warmer upon compression and colder upon expansion according to the Le Chatalier principle.

Working with Entropy

The properties described above make it practical to work with entropy. Entropy can be transferred, poured out, cumulated, distributed... and therefore treated like a physical object.

1. *Controlled exchange:* Let us consider two touching and equally warm bodies. Compressing one body makes it warmer and it gives off entropy to the other one. Expanding the same body makes it colder and then entropy flows in the opposite direction. Entropy can be "squeezed out" and "sucked up" like water by a sponge.

2. *Transfer:* With the aid of an "entropy sponge" one can transfer entropy from a cold to a warmer object or vice versa. By repeating the same procedure, any desired amount of entropy can be transferred.

3. *Preservation of entropy:* In order to guarantee that entropy does not increase during an operation, one has to insist that all operational steps are reversible. This condition is easily stated, but very difficult to satisfy in practice, but methods to circumvent this restriction are known.

4. *Measuring entropy:* The volume loss of ice while melting is proportional to the amount of entropy added. This fact can be used to construct a simple entropy-measuring instrument: a bottle equipped with an inserted capillary tube and filled with an ice-water mixture (ice calorimeter). When care is taken that no entropy is generated or lost during the measurement, then the amount of entropy added is proportional to the water level change of the capillary tube.

TEMPERATURE AND WORK

Definition of Temperature

Entropy always flows in the direction of decreasing temperatures. In other words, temperature difference causes an entropy flow that seeks to equalize those temperatures. One can say that temperature creates a kind of thermal tension for moving entropy, similar to the effect of pressure on a fluid. Already in the 18th century, Johann Lambert (1728–1777) presented this explanation for the expansion of heat.

It takes work to overcome this tension and force entropy into a body. The higher the tension, or the hotter the body, the more work is required. Energy conservation demands that the energy increase in a body is independent of the source; it can be internal or external, as long as the total energy is conserved.

This thought process suggests that the expended energy W can be used as a measurement of temperature. This can be readily accomplished by entropy generation. Since an increase in entropy drives up the temperature, the created entropy quantity S_e has to remain small. The smaller this quantity, the less energy is required. Therefore, W by itself does not provide a proper temperature measurement, but the proportionality factor (T) between W and S_e:

$$W = T \cdot S_e \quad (\text{for } S_e \to 0) \tag{3}$$

Work for Entropy Transfer

If entropy is transferred from a colder to a warmer object—by means of our test aid—then two work contributions have to be considered. First, the work required to force the entropy quantity S into the warmer object against the thermal tension T_2, $T_2 \cdot S$, and secondly, the work output from the colder object, $T_1 \cdot S$. The difference of both contributions is equal to the work W that has to be performed:

$$W = (T_2 - T_1) S \tag{4}$$

It is assumed that no additional entropy is generated by friction or any other means. That would require more work.

If entropy is transferred from a higher to a lower temperature, $T_2 < T_1$, then W becomes negative and work is delivered as in a heat engine. Entropy generation during this operation lowers the work output and therefore, the efficiency.

Entropy Generation in an Entropy Flow

Entropy can be easily generated by passing an electrical current through a resistance. If we know the charge q, the driving voltage $\Delta\varphi$, and the existing temperature T, then we can compute the produced entropy S_e from the energy balance, where $q \cdot \Delta\varphi$ represents the energy from the potential drop and $T \cdot S_e$ denotes the energy required for entropy generation:

$$T \cdot S_e = q \cdot \Delta\varphi \tag{5}$$

It is not commonly known that entropy is also generated by driving it through a thermal resistance. The produced quantity S_e can be computed from a similar energy balance:

$$T \cdot S_e = S \cdot \Delta T \tag{6}$$

In the above equation $S \cdot \Delta T$ represents the energy released by the entropy flow across the temperature drop ΔT, and T is the temperature of the colder side through which both the initial S and newly generated entropy S_e flow. This kind of entropy production can be shown experimentally. For the sake of simplicity, we use the following idealized process (FIG. 1).

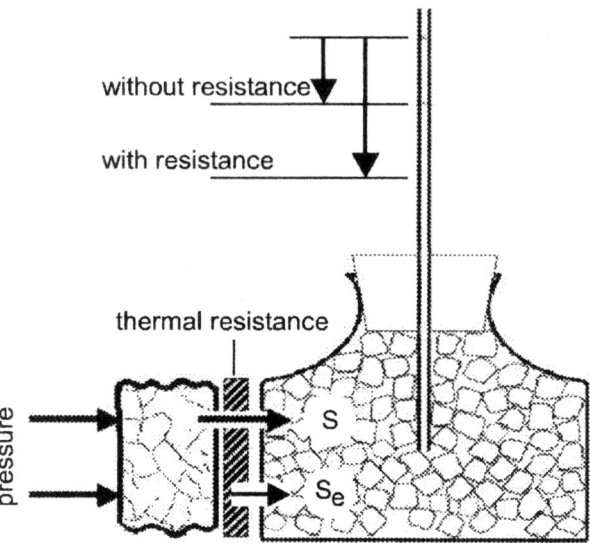

FIGURE 1. Entropy flow through a thermal resistance generates entropy.

1. *Entropy flow without resistance:* By squeezing the "sponge," it remains cold because entropy S is transferred to the bottle. The ice melts and the water level in the capillary tube falls.

2. *Entropy flow through a resistance:* By squeezing the "sponge" as before, it warms up because entropy escapes very slowly. Both S and S_e (generated in the resistance) trickle over to the bottle, the ice melts, and the water level in the capillary tube drops but lower than before! Although the "sponge" released the same amount of entropy for both cases, the bottle now indicates more entropy.

THE FIRST LAW AND ITS CONSEQUENCES

Conservation of Work

Clausius makes a basic assumption about the nature of heat in his First Law—the statement of the equivalence of heat and work. It is not necessary to make this assumption in formulating the principle of energy conservation. It suffices to realize that all work, gained or expended, has an equivalent and the sum of these equivalents remains constant in all processes. Work-equivalents like *potential, kinetic, electrical, internal energy*... can be interpreted as a measure of stored work. We have seen that the energy $W = T \cdot S_z$ can be interpreted as work against the thermal tension T while the entropy S_z is pressed into a body. This energy, which we could call *thermal work* for convenience, flows into a body along with the entropy S_z and was interpreted by Clausius as *heat Q*. If we avoid the exclusion of the quantity Q as "non-work" and treat it as work, then the energy principle can be simply stated as conservation of work:

Work cannot be created nor destroyed.

The quantity Q is a strange mixture of energy and entropy attributes. If we transfer the energy Q to a body, its entropy increases while the entropy of its surroundings decreases. The body becomes warmer and its surroundings colder. However, this phenomenon is not described as entropy flow but as a peculiarity of the energy Q that produces these specific effects. In other words, heating is not considered as caused by an entropy flow, but as a consequence of the unique features of the energy Q that other forms of energy do not possess. Heat is thus "fundamentally" different from other kinds of work. The entropy increase is treated and calculated as a secondary result from this special form of energy transfer. These peculiarities of the quantity Q require unique terminology, calculations, conclusions and a number of supporting mathematical quantities that are not found in other fields of physics. They make the thermodynamics structure strange and confusing.

Entropy as Heat

Naturally, entropy influences everyday activities. Entropy is something that flows into a pot of soup to warm its contents, is lost from a cup of coffee as it cools down, is produced by an electric stove, a microwave oven, and an oil-burning heater. It is delivered via hot water, distributed by radiators, contained by insulated walls in a room and stored by woolen clothing on a body. In short, it is what an unbiased layman calls "heat."

The commonly held concept of "heat" is quite broad and mixes physical aspects like temperature, heat quantity, and thermal energy together with physiological and psychological ones. Therefore, it is not feasible to simply equate entropy and heat based on everyday perceptions without providing precise descriptions. The narrower concept of "quantity of heat" has a better chance. The direct metrication (a philosophical procedure for connecting qualitative concepts like length, time, mass ... with their metrical counterparts l, t, m ...) of the notion "quantity of heat" yields a quantity in full conformance with entropy—as long as the same units and null references are established for both.[3]

H. Callendar[4] already claimed in 1911 that entropy is nothing more than the reconstruction of the heat quantity used by Carnot, with the exception that entropy can be produced while Carnot's heat cannot. From this viewpoint Equation (3) corresponds to Joule's relationship (1), except that the factor α there is not constant, but depends on temperature. Due to a cunning definition of temperature, it is identical with the absolute temperature T.

Consequences

All these observations are incompatible with the equivalence of heat and work assumed by the First Law. Not only that, but the standard interpretation of the First Law precludes a simple interpretation of entropy, by assigning its attributes to the energy quantity Q. As a result, entropy is reduced to a lifeless ghost without concrete physical properties, while Q assumes a schizophrenic double role that it cannot fulfill.

The quantities Q and S are in a relationship similar to that of phlogiston and oxygen at the beginning of modern chemistry. According to phlogiston theory, removing all the phlogiston from a metal yields a heavy, earthy calx (the metal oxide, from the modern point of view), which can readily be transformed again into the (obviously lighter) metal by adding phlogiston.

$$\text{metal} \rightleftharpoons \text{calx} + \text{phlogiston} \quad (\text{today: metal} + \text{oxygen} \rightleftharpoons \text{oxide}) \tag{7}$$

The entity that was later called oxygen is described in the phlogiston theory by its negative—a phantom with fantastic properties, only known in its bound state, but intangible itself, with a negative weight. Before 1800, Lavoisier finally put an end to this "spook." More than 50 years later, Clausius created a similar phantom with his definition of entropy.

H. Fuchs[5] demonstrated the creation of such a phantom in his satirical article "A surrealistic tale of electricity" by demanding electricity and work to be equivalent, analogous to the First Law. Starting from this point, mathematics transforms *charge* into an abstract entity similar to entropy.

Future Outlook

In this text we have tried to show that the treatment of heat as a form of energy by the First Law is more of a disaster than a benefit for thermodynamics. A few modest changes in the assumptions of pre-energetic thermodynamics, as shown in TABLE 1, would have been sufficient to make it consistent with experience. Many difficulties that resulted from this historical arrangement could have been avoided. Within the limited scope of this paper we cannot fully address such questions as how well the new approach holds up under scrutiny, how useful it is for practical applications, and how compatible it is with statistical interpretations. Our investigations over several decades have not encountered any serious difficulties. On the contrary, they show that:

(1) long derivations can often be reduced to one or two lines;
(2) many results are intuitively predictable without any calculations; and
(3) the approach is compatible with microphysical models.

The First Law marks a consequential turning point in the history of thermodynamics, where this science forfeited its heavenly naturalness, perhaps forever—analogous to the Fall of Man.

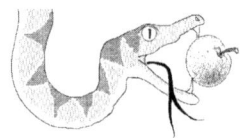

REFERENCES

1. CLAUSIUS, R. 1850, reprint 1921. Über die bewegende Kraft der Wärme.... Ostwald's Klassiker der exakten Wissenschaften, Nr. 99, Leipzig.

2. BORN, M. 1921. Kritische Betrachtungen zur traditionellen Darstellung der Thermodynamik. Physik. Zeitschr. **22:** 218–286.
3. JOB, G. 1972. Neudarstellung der Wärmelehre—Die Entropie als Wärme. Akadem. Verlagsgesellschaft. Frankfurt.
4. CALLENDAR, H.L. 1911. The Caloric Theory of Heat and Carnot's Principle. Proc. Phys. Soc. London **23:** 153–189.
5. FUCHS, H. 1986. A surrealistic tale of electricity. Am. J. Phys. **54:** 907–909.

Physical Explanation of the Periodic Table

V.N. OSTROVSKY

V. Fock Institute of Physics, University of St. Petersburg, 198904 St. Petersburg, Russia

ABSTRACT: The Periodic Table of the elements, the most important generalization in chemistry, is often considered as a representative special case in the study of the relation between chemistry and physics. Its quantum interpretation was initiated, but not completed, by Niels Bohr. In this paper, post-Bohr conceptual developments are discussed from historical and epistemological points of view. The difference between high-precision numerical calculations for individual atoms and the theory of the periodic system as a whole is emphasized. Periodic laws met in Nature are not restricted to the chemical Periodic Table. A comparative study of these laws makes it possible to single out essential features that define the particular pattern of periodicity. It is shown that the periodic system of neutral ground state atoms now has a firm nonempirical quantum-theoretical basis. Alternative approaches, based on group theory and other mathematical schemes, are briefly discussed. It is argued that, while quantum theory is capable of fully accurate calculations for relatively simple atoms or molecular objects, the complexity of polyatomic molecules and chemical reactions guarantees the flourishing of chemistry as a separate scientific discipline.

KEYWORDS: Periodic Table of the elements; *Aufbau* principle; magic numbers; complementarity; reduction

The Periodic Table of the elements is known as "the most important generalization in chemistry."[1] At the same time, it plays a basic role in the branch of physics known as atomic and molecular physics,[a] thus representing, so to speak, the common heritage of physics and chemistry.

The periodic law unifies and systematizes a vast amount of chemical and physical information. Only some of this information can be put in quantitative form, and in physical terms—the rest is qualitative knowledge about chemical compounds, reactions, etc. Advances in physics since the beginning of the 20th century have cast new light on the phenomenon of periodicity. The objective of the present article is to characterize the current status of the explanation of the Periodic Table provided by physics. It is well known historically that the major breakthrough was achieved in

---

Address for correspondence: V.N. Ostrovsky, V. Fock Institute of Physics, University of St. Petersburg, 198904 St Petersburg, Russia. Voice: 7-812-4287569; fax: 7-812-4287240.

Valentin.Ostrovsky@pobox.spbu.ru

[a]The place of this subfield in the whole of modern physics can be characterized by the fact that what is probably the most influential contemporary physical periodical, *Physical Review*, devotes roughly 20% of its cumulative journal volume (section *A*) to the subfield.

Ann. N.Y. Acad. Sci. 988: 182–192 (2003). © 2003 New York Academy of Sciences.

papers by Niels Bohr (see his collected works[2]). However, these famous studies did not exhaust the subject. Active research has continued from that time until now. Often, the progress achieved remains little known to the broad audience of chemists and physicists, preoccupied as they are by fascinating new problems within their narrow subfields. However, the fundamental significance of the periodic law makes broader dissemination of its modern physical representation worthwhile.

In particular, philosophical discussions should be based on updated information about the status of the physical explanation of the Periodic Table. The fact that substantial features of that explanation were dynamically amended in the last thirty years should not be missed. This is particularly important since analysis of the periodic law is considered, quite correctly, as an appropriate case study for pursuing the general issue of the relationship between physics and chemistry.[3–8]

It is important to state, from the very beginning, limitations of the present discussion. First, its objective is to provide a *modern* view—pioneering contributions made by Niels Bohr, well analyzed in the literature, are beyond the scope of this paper. Our first essential citation comes from 1930, and the bulk of the others are from the 1970s and later. Secondly, we concentrate on the Periodic Table of *ground state neutral atoms*—the most simple manifestation of periodicity for quantum-theoretical analysis. (A few comments on more general problems are also included). The word *neutral* is important here, since for positive atomic ions, periodicity patterns are different.[9,10,b] Some regularities for excited atomic states can be understood along the same lines as those for ground states—in cases when the orbital of only one electron is changed.[11] For two- (and more-) electron excitations, the basic principles of classification are, in general, very different. The third limitation is that we consider only the overall structure of the Periodic Table, and leave its more peculiar features aside. (To see what is meant by such features, refer to the paper by Pyykkö[12] or the book by Greenwood and Earnshaw.[13] It is satisfying to see that physical[12] and chemical[13] views on this subject are well correlated). The fourth point is that we concentrate on conceptual and cognitive aspects in the present paper. A reader interested in physical and mathematical detail can consult my recent review papers[14–16] and the bibliographies therein.

We start with the problem of explanation. The phenomenon of periodicity is conventionally illustrated in textbooks by plotting some property of a chemical element as a function of its ordinal number (Z) in the Table—for example, plots where ionization potentials of atoms, I(Z), (i.e., the energies necessary to ionize the atoms) are displayed. Such plots exhibit the well-known non-monotonic quasi-periodic pattern.

We use the term quasi-periodic to stress that the phenomenon of interest here is distinct from mathematical periodicity, where a periodic function is characterized by a single period length. In the Periodic Table, period length is subject to variation, increasing along the Table. The periodicity pattern is thus not fully characterized by a single number, as it might be in mathematics, but by a *set* of period lengths. For the Periodic Table of the elements, this set most often is described by the sequence 2, 8, 8, 18, 18, 32, 32, ..., although some alternative subdivisions have been discussed in the chemical literature.

---

[b]This is the simplest example of multiple facets of the periodic laws in Nature, discussed below.

Values of I(Z) are traditionally supposed to be taken from experiment, although modern methods of quantum mechanics allow physicists to compute this atomic property with an accuracy comparable to the experimental uncertainty. If we replace experimental data by theoretical figures, we obtain essentially the same plot. Does this success mean that we have achieved progress toward a physical *explanation* of the phenomenon of periodicity? Evidently not, although the essential coincidence of the two plots confirms, once more, the validity of quantum mechanics. *Explanation*, as distinct from purely quantitative agreement, requires some overall understanding of the phenomenon—even if accuracy in treating details is sacrificed in favor of revealing basic trends. It seems that explanation and interpretation on the one side, and numerical precision on the other side, are in a *complementary* relationship, as I have previously argued.[15]

Accurate numerical characteristics probably provide the most objective image of Nature, but a researcher (a subject) always also seeks appealing and comprehensive explanations. The complementarity concept originated from Bohr's works on physics.[c] Later he applied it more broadly—for instance, to psychology, biology and anthropology.[17] The epistemological significance of this concept stems from the fact that it concerns a very general pattern of relations between subject and object. Within the complementary pair "numerical calculations" and "explanation," numerical calculations seek to reproduce the physical object with the highest possible precision, whereas explanation is oriented towards the subject. In particular, this means that a specific explanation appeals only to some community of researchers—and explanation might have different meanings for various communities.

This reasoning shows that explanation necessarily means accepting a level of approximation worse than is achieved with the highest numerical precision. If one's objective is to obtain the best numerical results, then approximations are something to avoid—or at least to minimize as much as possible. Fewer approximations provide better numerical output. However, if one seeks explanations, then refusing to accept approximations may hopelessly destroy the entire explanatory framework. Indeed, bare exact equations usually provide very limited insight for a complex system—they are not fertile as explanations. Qualitative explanatory concepts are normally born as a result of approximations.

After this general introduction, the reader should not be surprised to learn that physical explanation of the Periodic Table is essentially based on approximate concepts—such as one-electron quantum numbers, one-electron wave functions (orbitals), electron configurations, effective one-electron potentials, etc. The applicability of these approximations is well established, and tested by the entire development of quantum mechanics and atomic physics.

The physical analysis of the Periodic Table is based on three principles, formulated in all textbooks.

- Many-electron atoms can be characterized by a set of principal and orbital quantum numbers $\{n,\ell\}$ that label occupied one-electron wave functions ("orbitals").

---

[c]The canonical example of complementarity in physics is provided by the Heisenberg Uncertainty Relation, which shows that a particle's coordinates can be fixed only at the cost of enhancing uncertainty in its momentum, and *vice versa*.

- *The Pauli Exclusion Principle:* No two electrons in an atom may have the same four quantum numbers, n, ℓ, m, m_s (the azimuthal and spin quantum numbers—m and m_s—do not play a significant role in the subsequent discussion).
- *The Aufbau Principle:* One-electron orbitals are filled (occupied) sequentially in the order of *increasing energy.*

These principles result in *periodicity,* that is, non-monotonic, quasi–periodic dependence of the system (atom) properties on the number of constituent particles (electrons). Note that, in order to apply the *Aufbau Principle* in practice, one needs to know the sequence (ordering) of one-particle (one-electron) energy levels.

An important issue known to physicists, but almost never mentioned in chemical textbooks, is that these three principles apply to any physical system that is composed of fermions, that is, particles with half-integer spin. The most common elementary particles (electrons, protons, neutrons) are all fermions. This implies that many different periodic laws are realized in Nature for different multi-fermion systems.[d] In addition to the well-known "chemical" Periodic Table, the list of such periodic laws, recently reviewed,[15] includes modifications of the Periodic Table for ionized, compressed, or excited atoms, but also periodic laws for atomic nuclei, clusters of atoms or molecules, so-called artificial or designer atoms (quantum dots), and microscopic particles in traps. Each such system is characterized by a particular periodicity pattern which, as discussed above, is described by a set of period lengths. An alternative characterization is a set of "magic numbers"—numbers of particles in especially stable (i.e., most tightly bound) systems. (In chemistry, the magic numbers correspond to the ordinal numbers of inert gases.) Clearly, the difference of adjacent magic numbers gives period length—both characterization schemes are straightforwardly interchangeable.

In distinction to fermions, the Pauli Exclusion Principle is not operative for bosons (particles with integer spin), and hence quasi-periodicity does not appear for a system of bosons. Instead, one finds the famous phenomenon of Bose–Einstein condensation—all particles occupy a single, lowest-energy, one-particle orbital.[e] Properties of such a system depend smoothly and monotonically on the number of particles. One can argue that fermions play a crucial role in the diversity of our world. The world would be much duller place if its principal constituent particles, electrons and nucleons (protons and neutrons), were bosons.

When science studies some object or phenomenon, it is generally instructive to put it in a broader context of similar objects or phenomena. Then one can gain important insights by looking for what is common and what is specific within such a group. Various periodic laws describe systems composed of a large number (from several to several hundred) of microscopic particles. A specific periodicity pattern follows from the particular features of the system. For instance, an atom is composed of light electrons and a massive nucleus; inter-particle interactions are long range, described by Coulomb's Law. These interactions correspond to attraction between

---

[d]Sometimes the term "Periodic Table" is applied to any attempt at systematization and codification (see, for instance, Ref. 18). We do not think that such a loose terminology is heuristically productive.

[e]The realization of Bose–Einstein condensation for an ultra-cold dilute gas in a trap was subject of the 2001 Nobel Prize in Physics.

electrons and the nucleus, and to electron–electron repulsion. In an atomic nucleus, all particles (nucleons) are, to a good approximation, equally massive; all interactions are of short range and attractive.

In an approximate quantum description, a specific periodicity pattern is governed by the properties of an effective *one-particle potential,* $V_{eff}(\mathbf{r})$, where the vector $\mathbf{r}$ defines the particle's coordinate in space. It should be stressed that this potential appears as a result of the self-consistent motion of all the constituent particles. It does not coincide with any of the potentials that describe interaction between individual particles. Finding $V_{eff}(\mathbf{r})$ is a non-trivial task for quantum theory. In the case of an atom, the tempting intuitive choice is to suggest that $V_{eff}(\mathbf{r})$ simply corresponds to (slightly distorted) Coulomb attraction to the center. This leads to the (n, ℓ) ordering of one-electron energy levels, which, however, generates an incorrect periodicity pattern (see below). The Hartree–Fock method and Thomas–Fermi statistical theory are techniques often used in quantum mechanics to find $V_{eff}(\mathbf{r})$. The latter provides less accurate results, but expresses $V_{eff}(\mathbf{r})$ in a simple and universal form. From the previous discussion, one can anticipate that Thomas–Fermi theory would be appropriate for an analysis of periodicity.

As soon as the effective potential is known, one has to examine what type of energy level ordering it yields. The latter is crucial in the practical application of the *Aufbau* Principle, as emphasized above. This, in turn, governs a set of period lengths that distinguishes the periodic law for the particular system of fermions. It can be seen that analysis of a particular type of periodicity as a distinct phenomenon of Nature implies establishing (i) the form of the effective potential $V_{eff}(\mathbf{r})$ and (ii) the specific periodicity pattern (set of period lengths).

Historically, the discovery of the Chemical Periodicity (by Mendeleev in 1869) and its original quantum interpretation (by Bohr in 1922) were achieved earlier than for other forms of periodicity. In 1955, when Göppert-Mayer and Jensen (recipients of the 1963 Nobel Prize in physics) wrote a book[19] on the then recently discovered periodicity effect (shell structure) in atomic nuclei, they traced its analogy to the shell structure of atoms, which at that time was already familiar, broadly known, and well understood. To the general public, chemical periodicity remains the best (and often the only) known example of periodic behavior. The reason for this seems to be that its manifestations are more on a "human" scale.

Chemistry and biochemistry govern the everyday life of human beings. Nuclear phenomena, such as radioactivity, only rarely interfere (although they could be very important on geological and cosmological scales). The other examples of periodicity—clusters, designer atoms or particles in traps—are produced artificially. Their role in Nature or in (future) technology is yet to be understood. Also, chemical periodicity is manifested in an immeasurable number of chemical compounds, whereas other examples of periodicity the compounds are hardly formed.[f] This circumstance justifies the emergence of chemistry as a separate natural science.

The first step in a deeper analysis of chemical periodicity was an *empirical* recognition that its pattern is described by the following (n + ℓ, n) rule—purposely divided into two parts.

---

[f]Nuclear reactions are sometimes interpreted in terms of transient nuclear quasi-molecules.

- As the nuclear charge Z increases in neutral atoms, one-electron orbitals are filled in the order of increasing value of the sum $N = n + \ell$.
- For fixed N, orbitals are filled in the order of increasing n.

The rule is not satisfied ideally, being subject to some twenty exceptions. However these exceptions are of minor significance[g] and do not destroy the general pattern, as clearly seen in the paper by Demkov and Ostrovsky.[20] The rule is to be compared with its alternative, the (n, ℓ) rule mentioned previously[h]—sometimes considered as a base for an "ideal periodic system." The comparison shows that the latter rule completely fails in the domain of higher ordinal numbers, Z.

Priority issues related to the $(n + \ell, n)$ rule look complicated.[15,16] Briefly, that rule was first published (and applied to prediction of electronic configurations for heavy trans-uranium elements) by Karapetoff[22] in 1930, and only later (in 1936) by Madelung[23]—although Coudsmith[9] testified in 1964 that he received a private communication from Madelung about the rule as early as 1926. Several rediscoveries followed in 1945–1951. Klechkovskii[25] wrote a series of papers and a book[26] analyzing various aspects of the rule in detail (see Refs. 15 and 16 for bibliography). In the 1960s the rule became quite widely known, but remained empirical. In 1969, the prominent quantum chemist Per-Olav Löwdin wrote with regret[24] that:

> It is perhaps remarkable that, in axiomatic quantum theory, the simple energy rule $(n + \ell, n)$ has not yet been derived from first principles.

At the same time, he considered the rule as very important, notwithstanding its empirical background:

> It is... a startling fact that the simple energy rule $(n + \ell, n)$ has not entered any major chemistry textbook, as far as I know, and it is still this rule that gives the first explanation of the occurrence of the transition metals, the rare earth metals, and the over-all structure of the electronic shells of atoms.

This attitude has not changed much even until now. Many chemistry books especially devoted to the Periodic Table do not mention the rule at all. *Chemistry of the Elements*[13] presents a rare exception, and even this book devotes only three lines of text to the rule. Sometimes, it is argued that the rule is contained *implicitly* in any modern periodic chart that displays electronic configurations. However, *explicit* formulation in terms of the sum of quantum numbers $(n + \ell)$ was important in providing a basis for subsequent developments.

In 1971, two years after Löwdin's paper, the status of the $(n + \ell, n)$ rule as only empirical changed. Demkov and Ostrovsky[20] suggested a simple analytical expression[i] for the effective one-electron potential $V_{eff}(\mathbf{r})$ in an atom that allowed them to derive both parts of the rule. This was not some *ad hoc* potential, but was actually the same potential suggested in 1954, seventeen years before, by T. Tietz,[27] as a convenient analytical approximation in the Thomas–Fermi theory of the atom. Between 1954 and 1968, Tietz published eight papers applying this potential to various prob-

---

[g]Carroll and Lehrman[21] argue that the exceptions to $(n + \ell, n)$ rule "are relatively unimportant for chemists."

[h]The orbitals are filled in order of increasing principal quantum number n and, for a given n, in order of increasing ℓ.

[i]In atomic units this reads: $V_{eff}(\mathbf{r}) = Z/[r(1 + r/R)^2]$, $R = \alpha^{-1} Z^{-1/3} [3\pi/(8\sqrt{2})]^{2/3}$, $\alpha \approx \frac{1}{2}$, see formula (2) or (5) in the paper cited, or formula (6) in Ref. 15.

lems of atomic physics (see bibliography in Refs. 16 and 20). At that time, however, the direct relevance of the potential to the $(n + \ell, n)$ rule was not recognized. Most importantly, with this potential the sum of quantum numbers $(n + \ell)$ appeared intrinsically in the analytical solution of the Schrödinger equation. This does not occur in any other model or approximation.[j] The classical trajectories of the valence electron in an atom generated by this potential are also of interest. They appear to be strikingly different from the universally known elliptical trajectories of an electron in a Coulomb field. The trajectories close after not one (as for elliptical orbits) but *two* rotations around the center of force. The names *chemical orbits* or *double necklace orbits* were suggested for these trajectories by the eminent theoretical physicist John Wheeler,[28] who also realized that the $(n + \ell, n)$ rule gives a general pattern of classical trajectories, although he did not find the actual potential. Interestingly, in the potential suggested by Demkov and Ostrovsky,[20] trajectories also possess a fascinating focusing property. The role of this feature in atomic structure is not yet understood. Nevertheless it may be asserted that the periodicity pattern characteristic to the Periodic Table of the elements now has a firm, non-empirical, quantum-theoretical foundation.

We briefly mention another direction of research that was started in 1971. This was a group-theoretical analysis of the Periodic Table. Group theory, in the present context, can be considered as a mathematical framework suitable for the *classification* of objects. There are also other mathematical schemes with this capability, but group theory has proved to be the one that was adequate to account for various *symmetries* intrinsic to particular applications of quantum mechanics. It seems that the abstract group-theoretical approach, as currently developed, amounts to a *translation* of empirical information on the periodicity pattern for atoms into a specialized mathematical language—but no other output is produced. Probably this approach would have explanatory power only within a community that speaks this language. At the same time, group-theoretical symmetry analysis of the effective one-electron potential[20] $V_{eff}(\mathbf{r})$ produced more fruit, allowing an interpretation[11] of the phenomenon of *secondary* periodicity in the Periodic Table. Limitations on the length of this paper do not allow a more detailed discussion (see Refs. 14–16).

In recent years, alternative mathematical schemes (such as the theory of fuzzy sets) have been applied to the Periodic Table (see discussion in Ref. 16). The potential advantage of these mathematical schemes is that they are not restricted to atomic structure. In principle, they could absorb the wealth of chemical and physical information that is available. Only further developments can show how productive these ideas can actually be.

If we want to go beyond the Periodic Table of ground state neutral atoms, then some general comments on the capability of quantum mechanics to explain and reproduce chemistry are in order. We start with a discussion of the famous citation from the 1929 paper by great master of theoretical physics, Paul Dirac[29]:

> The underlying physical laws necessary for the mathematical theory of a large part of physics and the whole of chemistry are thus completely known, and the difficulty is only that the exact application of these laws leads to equations much too complicated to be soluble.

---

[j]For instance, Enrico Fermi evaluated, within his statistical theory, ordinal numbers Z_ℓ for which the first electron with a given orbital number ℓ appears in the Periodic Table; however, the sum $(n + \ell)$ did not emerge in a natural way in his approach.

This quotation might be understood as a declaration of an achieved reduction of chemistry to physics—as such, it created considerable concern among philosophers and chemists (see, for instance, Jaap van Brakel's detailed discussion in a recent book[8]). Dirac's quotation reflects his fascination with new scientific perspectives opened by the powerful, and then recently developed, quantum-theoretical framework. The universality of his claim is understandable psychologically for a person (Dirac) who was one of the principal actors in a "heroic" era of quantum mechanics. As an epistemological message, it can be compared to the similarly famous view on mechanical determinism by Laplace in the nineteenth century. The latter reflects the fascination of a great researcher with the theoretical scheme of classical mechanics—developed to apparent universality and perfection.

Presently, we are not concerned with the subjective side of both of these provocative statements; that is to say, with whether those researchers understood their declarations literally, or merely used them to provoke subsequent discussion. The objective picture in both cases looks similar. Further development of science has revealed new problems—both within the theoretical scheme itself, and in the relation of that scheme to Nature. In the case of Laplace, the situation looks clearer, simply because a longer historical trail shows the real status of mechanical determinism. Even if we put aside uncertainties induced by quantum physics and remain solely within classical mechanics, we see that any deterministic declaration (taken literally) underestimates the *qualitative significance* of the great complexity of the task of prediction. Complexity follows from the enormous number of coupled dynamical equations to be solved, and also from the necessity to specify the initial conditions with unthinkable precision—in view of the chaotic character of non-linear dynamics. As a resolution (perhaps only partial) of this paradigm, alternative approaches in the form of statistical mechanics and, later, of the modern science of chaos were developed. The deterministic manifesto, although it proved to be incorrect, played a positive role by consolidating thought.

Equations "much too complicated to be soluble" create a problem that is not automatically solved by the advent of more powerful computers. According to Hegel's dialectics, when gradual quantitative changes cumulate, a qualitative leap occurs. In turn, the qualitatively new situation stimulates new scientific approaches. In case of Laplacian determinism, such a new approach was provided by statistical mechanics. Probably, in case of the quantum description of the micro-world, such an alternative approach is given by the science of chemistry—which came into existence long before the formulation of quantum ideas.

Equations "much too complicated to be soluble" do not allow theory to make *predictions*. The capability to predict is of major importance when the validity of any new theory is to be established. Dirac's quotation expresses his belief that, on the atomic scale, all fundamental laws of nature are already known—but it also warns us that the predictive power of a theory drops as more and more complicated (e.g., atomic and molecular) objects are considered.

One can ask what the grounds for Dirac's belief might be. Of course, it is often reasonable to make extrapolations, based on the comparison of theoretical predictions with experiment for simple systems—where equations can be solved with sufficient accuracy. Such a comparison is a process which continues now, and will be continued in the future. Within experimental errors and the uncertainties of theoretical calculations, no discrepancy has yet been found between quantum theory (more exactly, quantum electrodynamics) and experiment. This should not preclude our fully appreciating the fact that for complicated systems, especially polyatomic molecules, fully *ab initio* calculations will not be achieved in the foreseeable future. Since such molecules are of great importance and interest both fundamentally and for applications, the existence of chemistry as a separate scientific discipline is assured.

Sometimes, philosophical criticism of reductionist ideas concentrates on revealing approximations made by physicists when treating relatively simple systems. However, approximations are intrinsic to physics—one who can be satisfied only by exact and rigorous results should turn to mathematics, which is not a natural science. To criticize physics for using approximations is beyond the particular context of relationship between physics and chemistry. In fact, this would mean questioning entire physical approach to understanding Nature. In view of the many successes achieved by physics, such criticism does not appear productive. (Incidentally, chemistry cannot boast that it uses fewer approximations than physics.)

Criticism of this type is usually based upon the standpoint of some "ideal" science—which exists only in imagination of the critic. It seems that it is much more interesting to analyze how and why the real science of physics works. Of course, free and vivid critique of any concrete approximation is one of cornerstones of physics as a living natural science. However, such a scrutiny is a job for professionals, bearing in mind the growing sophistication of theoretical and experimental tools developed by physics.

As stated above, the really key point for a separation between the two sciences (just as between chemistry and biology) is the difference in the level of *complexity* of the objects of study—and, as a corollary, the difference between approaches, techniques, mathematics employed, ways of formulating results,[k] etc.

It seems that some sense could be found in drawing some kind of border between the subjects of chemical and physical studies, making all appropriate reservations about its approximate and probably time-dependent character, and about inevitable subjectivity—and allowing for mutual penetration. It seems that, currently, all the atoms are subjects of atomic and molecular physics. In particular, as discussed above, physics provides a non-empirical explanation of the Periodic Table of ground state neutral atoms. Diatomic molecules also mostly belong to physics.[l] The simplest triatomic molecules with a small number of electrons (as, for instance, molecular ion H_3^+, which is important for astrophysical applications[31]) can also be treated by purely physical methods. It should be stressed that atomic and molecular physics, although being strongly influenced by computational physics, does not (by far) reduce to it—as one might infer from Dirac's manifesto. The more complicated triatomics seem to belong to chemistry, and even more so the polyatomics. Triatomic systems provide the simplest possible case to study chemical reactions. This presents a demanding test for the quantum theory. It would be of interest to enter this area in more specific detail.

---

[k]See discussion by Van Brakel[8] on how many laws are there in chemistry.
[l]The underlying mathematical fact is that the simplest one-electron diatomic, H_2^+, could be treated exactly within the Born–Oppenheimer (fixed-nuclei) approximation, although the solution is quite complicated. See, for instance, the book by Komarov *et al.*[30] For triatomics similar solutions are absent.

In many cases, electronic transitions do not play a significant role for low-energy reactions of interest to chemistry. This situation is easiest for a theoretical treatment since the standard Born–Oppenheimer approximation is applicable, and reaction can be considered as motion on a single multidimensional potential surface. The calculation of this surface for each particular triatomic is a task for quantum chemistry. The dynamical problem of treating motion along the potential surface belongs to a significant extent to physics, where it is known as "collision with rearrangement." Many efficient theoretical methods have been developed to treat molecular dynamics. The bottleneck for the entire problem seems to be the potential surfaces —which are known with low accuracy. The result is that the whole characterization of chemical reactivity suffers from serious uncertainty, and is often dependent on the choice between several potential surfaces that may be available in the literature (see, for example, the concrete case study in the paper by Tolstikhin *et al.*[32]). The computation of chemical reactions requires knowledge of potential surfaces for a broad range of atomic-nuclear configurations. This is a considerably more difficult problem than calculations for configurations in the vicinity of nuclear equilibrium positions that are needed in the theory of molecular vibrations. The complexity factor is drastically enhanced when reactive processes are considered in systems with more than three atoms.

ACKNOWLEDGMENTS

I am thankful for support from the International Society for the Philosophy of Chemistry that allowed me to participate in the 2002 ISPC Symposium, and extend my gratitude to E.R. Scerri and J.E. Earley for their warm encouragement. Critical reading of the manuscript and much useful advice by R. Hefferlin are also highly appreciated.

REFERENCES

1. HAKALA, R. 1952. The Periodic Law in mathematical form. J. Phys. Chem. **56:** 178–181.
2. BOHR, N. 1977. Collected Works, vol. 4 [Periodic System (1920–1923)]. North-Holland. Amsterdam.
3. HETTEMA, H. & T.A. KUIPERS. 1988. The Periodic Table: its formalism, status, and relation to atomic theory. Erkenntnis **28:** 387–408.
4. SCERRI, E.R. 1997. Has the Periodic Table been successfully axiomatized? Erkenntnis **47:** 229–243.
5. SCERRI, E.R. 1997. The Periodic Table and the electron. Am. Sci. **85:** 546–553.
6. SCERRI, E.R. 1998. The evolution of the Periodic System. Sci. Am. **279:** 56–61.
7. SCERRI, E.R. 1998. How good is the quantum mechanical explanation of the periodic system? J. Chem. Ed. **75:** 1384–1385.
8. VAN BRAKEL, J. 2000. Philosophy of Chemistry. Leuven University Press. Leuven, Belgium.
9. GOUDSMITH, S.A. & P.I. RICHARDS. 1964. The order of electron shells in ionized atoms, Proc. Natl. Acad. Sci. USA **51:** 664–671.
10. KATRIEL, J. & C.K. JORGENSEN. 1982. Possible broken supersymmetry behind the Periodic Table. Chem. Phys. Lett. **87:** 315–319.
11. OSTROVSKY, V.N. 1981. Dynamic symmetry of atomic potential, J. Phys. B **14:** 4425–4439.

12. PYYKKÖ, P. 1991. Relativistic effects on periodic trends. *In* The Effects of Relativity in Atoms, Molecules, and the Solid State. S. Wilson, I.P. Grant & B.L. Gyorffy, Eds.: 1–13. Plenum. New York.
13. GREENWOOD, N.N. & A. EARNSHAW. 1984. Chemistry of the Elements. Pergamon. Oxford.
14. OSTROVSKY, V.N. 1996. Group theory and Periodic System of Elements. *In* Latin-American School of Physics XXX ELAF. O. Castaños, R. Lópes-Peña, Jorge G. Hirsch & K.B. Wolf, Eds.: 191–216. AIP Conference Proceedings 365.
15. OSTROVSKY, V.N. 2001. What and how physics contributes to understanding the Periodic Law? Found. Chem **3:** 145–182.
16. OSTROVSKY, V.N. 2003. Modern quantum look at the Periodic Table of Elements. *In* Fundamental World of Quantum Chemistry: A Tribute Volume to the Memory of Per-Olov Löwdin. E.J. Brändas & E.S. Kryachko, Eds. Kluwer. Dordrecht. Vol. 2, Chapt. 23. In press.
17. BOHR, N. 1999. Collected Works, Vol. 10 [Complementarity beyond Physics]. D. Favrholdt, Ed. Elsevier. Amsterdam.
18. BASHFORD, J.D. & P.D. JARVIS. 2000. The genetic code as a Periodic Table: algebraic aspects. BioSystems **57:** 147–161.
19. GOEPPERT-MAYER, M. & J.H.D. JENSEN. 1955. Elementary Theory of Nuclear Shell Structure. Wiley. New York.
20. DEMKOV, YU.N. & V.N. OSTROVSKY. 1971. *(n+ℓ)* filling rule in the Periodic System and focusing potentials. Zh. Eksp. Teor. Fiz. **62:** 125–132, [Errata **63:** 2376, 1972]. [English translation: Sov. Phys.—JETP **35:** 66–69, 1972].
21. CARROLL, B. & A. LEHRMAN. 1948. The electron configuration of the ground state of elements. J. Chem. Ed. **25:** 662–666.
22. KARAPETOFF, V. 1930. A chart of consecutive sets of electronic orbits within atoms of chemical elements, J. Franklin Inst. **210:** 609–614.
23. MADELUNG, E. 1936. Die Mathematischen Hilfsmittel des Physikers, 3rd ed. Springer. Berlin. p. 359.
24. LÖWDIN, P.O. 1969. Some comments on the Periodic System of Elements. Int. J. Quant. Chem. (Symp.) **IIIS:** 331–334.
25. KLECHKOVSKII, V.M. 1951. *(n+ℓ)* gruppy v posledovatel'nom zapolnenii elektronnykh konfiguratzii atomov [(n+ℓ) groups in successive filling of atomic electron configurations.] Doklady Akademii Nauk **80:** 603–606.
26. KLECHKOVSKII, V.M. 1968. Raspredelenie Atomnyh Elektronov i Pravilo Posledovatel'nogo Zapolnenya *(n+ℓ)*-Grupp [The Distribution of Atomic Electrons and the Rule of Successive Filling of (n+ ℓ) Groups]. Atomizdat. Moscow.
27. TIETZ, T. 1954. Approximate analytic solution of the Thomas-Fermi equation for atoms. J. Chem. Phys. **27:** 2094.
28. WHEELER, J.A. 1971. From Mendeleev's Atom to the Collapsing Star. *In* Atti dei Convegno Mendeleeviano. M. Verde, Ed.: 189–233. Academia delle Scienze di Torino. Torino.
29. DIRAC, P.A.M. 1929. Quantum mechanics of many-electron systems. Proc. Roy. Soc. London **A123:** 714–733.
30. KOMAROV, I.V., L.I. PONOMAREV & S. YU. SLAVYANOV. 1976. Sferoidal'nye i Kulonovskie Sferoidal'nye Funktzii [Spheroidal and Coulomb Spheroidal Functions]. Nauka. Moscow.
31. KOKOOULINE, V., C.H. GREENE & B.D. EZRY. 2001. Mechanisms for destruction of H_3^+ ions by electron collision. Nature **412:** 891–894.
32. TOLSTIKHIN, O.I., V.N. OSTROVSKY & H. NAKAMURA. 2001. Cumulative reaction probability and reaction eigenprobabilities from time-independent quantum scattering theory. Phys. Rev. A **63:** 891–894 042707 (1–18).

Gauge Theory and Chemical Structure

JAMES MATTINGLY

Philosophy Department, Georgetown University, Washington, DC 20057, USA

ABSTRACT: The possibility of chemical structure in the context of quantized matter is examined by way of Richard Bader's *Atoms in Molecules*. I critically examine his notion of "electronic charge density"—showing that he cannot really mean "density of charge"—and I argue that the appropriate concept is expectation value of charge. This still allows him to define chemical structure, but it makes problematic his appeals to the explanatory power of structure. This is because, as Rosenfeld and Bohr showed, the expectation value of charge cannot be taken as the electronic field experienced by other charges. I suggest that we can recover the efficacy of structure by thinking of chemistry as a gauge theory. Current consensus in the study of gauge theories indicates that gauge potentials represent a new type of property; while no member of the family of functions comprising the gauge potential is real, the potential itself is causally potent. I illustrate this in the case of electrodynamics, where the vector potential can causally influence charges in the absence of electric or magnetic fields. I show how chemical structure can be considered to be a gauge field. Following Bader, I take it to be a family of geometric configurations, no one of which is possessed by a given molecule. I claim that current research in gauge theory licenses the attribution of causal potency to this notion of structure, despite its lack of reality. I thus begin the process of freeing the explanatory resources of gauge theory from physics alone.

KEYWORDS: gauge theory; chemical structure; Bader; scientific explanation

INTRODUCTION

Back in the good old days, we are told, electrodynamics was philosophically much simpler to understand than it is now. There were electric fields and magnetic fields, and when these were specified, so were all possible electrodynamical observables. Thus the picture of what was real for electrodynamics was just that: the electric and magnetic fields. Of course things hadn't always been so simple. Indeed in the good even older days, the whole idea of the field concept was itself new, and untried and, frankly, suspicious. Yet with the reception and widespread acceptance of Maxwell's formalization of Faraday's researches (and of course his own and others'), the field concept came more and more to dominate physical theorizing, until the whole practice of thinking about field action was enshrined by Einstein in the principle of local action. Now we are told that, rather than good old property-valued

---

Address for correspondence: Dr. James Mattingly, Philosophy Department, Georgetown University, Washington, DC 20057. Voice: 202-687-2592; fax: 202-687-7493.
jmm67@georgetown.edu

fields, the world is populated by operator-valued distributions—and this is true for all quantum fields, including electrodynamics. If this were not enough of a strain on our metaphysical imaginative faculty, it turns out that even good old electricity and magnetism have suffered a reversal of fortune. To wit, there's an extra field there that has properties completely unlike those of earlier fields. This is all to say that our researches in quantum physics have led to a revision of how we ought to understand what was really going on in electrodynamics even at the classical level.[1]

A similar problem has arisen for students of chemical structure. It seemed, around the end of the 19th century, that the dream of Newton, Hooke, Boyle, and others that we resolve the action of chemicals into the behavior of very small bodies with hooks and other mechanical contrivances had finally succeeded, in a manner of speaking. There were indications that the molecular theory could explain the behavior of various chemicals by the interactions among them caused by their electric fields, and that what determined the details of these fields was the shape of the individual molecules. This shape was, moreover, taken to be a constant for a given species of molecule. So, roughly speaking and with hedging all around, the strong solvency properties of H_2O (as well as the fact that it expands at its freezing point) are to be understood as arising from the peculiar "V" shape of the water molecule. This shape gives water a strong dipole moment, and this in turn allows water to "pry" other species of molecule apart from each other. But with the advent of quantum mechanics, and thus quantum chemistry, the very notion of chemical structure has been again called into question. Since atoms have no particular locations in space, it is argued, then molecules can have no particular geometry. Somewhat more carefully, one would point out that, insofar as the relative momenta of the atoms in some molecule are well-defined, the relative positions of these atoms are not. But for the naive story I just told about the efficacy of molecular structure in explaining molecular behavior, we need the shape to be stable and so we need the relative momenta of the various atoms to vanish. Without getting caught up in the modern debate over chemical structure that has been going on more or less since the beginning of quantum mechanics, I do want to suggest that the principles of gauge theory may be able to make plausible the idea that chemical structure really is ill-defined, and yet at the same time it really is causally effective in chemical interactions.

A little context may be in order, and this context will, I hope, make sense of the connection between gauge theory and chemical theories. My plan for producing this context and illuminating the connection between chemical structure and gauge theory is as follows: I begin with a brief introduction to the notion of gauge theory. The main thrust of this section is the idea that, in current views of gauge theory, the properties of gauge fields are causally active despite our inability to say what kind of thing these properties are. The second part of the paper concerns the issue of chemical structure. Indeed it is quite narrowly focused on one particular proposal to recover the concept of chemical structure in the context of quantum mechanics—Richard Bader's *Atoms in Molecules*.[2] The book is an ambitious attempt to show that chemical structure is meaningful for quantum-mechanical molecules despite the fact that chemical shape is not. Moreover, Bader attempts to show how his notion of chemical structure can recover the explanatory power that chemical shape was supposed to have. A discussion of Bader's program occupies the second and third sections. Finally in the last section I propose that Bader's proposal, though philosophically unsatisfying at first blush, may be made more satisfying (or at least brought into line with some popular views of proper explanatory structure) by recasting it in analogy with gauge theory.

GAUGE THEORY AND NOVEL PROPERTIES

To illustrate the notion of gauge theory, I'll appeal just to the example of electrodynamics. This will keep the discussion manageable, but will still allow its interesting and novel explanatory strategy to emerge. Naturally this discussion is too compressed for a full analysis of gauge theory. In particular, it is not uncontroversial to suggest that the appeal to novel properties is generic of gauge theories and not just those theories with Aharonov–Bohm type effects. I think it is correct, but I won't argue for it here.[3]

The vector potential in electrodynamics, which is the field that forced a change in our understanding of the ontology of electrodynamics, is not really a new field. What is new is the conviction, widely held by physicists and philosophers of physics, that we must take it seriously as a physically significant quantity. The potted history goes like this: from early in the development of electrodynamics as a mathematical field theory, it was known that the magnetic field could be used to define another field, the vector potential, via $\nabla \times \mathbf{A} = \mathbf{B}$, where $\mathbf{B}$ is the magnetic field. But this equation does not uniquely specify $\mathbf{A}$, for one can always add the gradient of an arbitrary scalar function to $\mathbf{A}$ to obtain a new vector potential $\mathbf{A}'$ that satisfies the defining condition. (A complementary point obtains with respect to the scalar potential ϕ. The electric field can be given as $\mathbf{E} = \nabla \phi$, but ϕ is not uniquely defined thereby. One can always add some constant to a given scalar potential to produce a different potential that is, nonetheless, physically equivalent.) This is what is called a gauge transformation, and $\mathbf{A}$ is called a gauge field or gauge potential. There is a significant number of different ways of expressing the mathematical concept of gauge, and some of these are explained in the literature I cite below. But for the purpose I have in mind, this description of $\mathbf{A}$ as a gauge potential should do nicely. The crucial feature I want to highlight is that no particular choice of $\mathbf{A}$ is *the* vector potential. The vector potential is in some sense all $\mathbf{A}$s and in some sense none. The equations of motion do not single out a unique $\mathbf{A}$, so we have no grounds for saying that one is privileged over the others. We are simply stuck with an ill-defined electromagnetic property.[4] Until the second half of the 20th century the ill-definedness of $\mathbf{A}$ could be safely ignored.

It was thought because of the gauge freedom in $\mathbf{A}$ that the vector potential had no metaphysical (or even physical for that matter) significance. Nothing distinguishes the "right" $\mathbf{A}$ from all the others, and $\mathbf{A}$ is dispensable in all of our electrodynamical explanations. Then rather than maintaining that despite all appearances to the contrary there really is just one real $\mathbf{A}$, why not simply abandon it? The appropriate stance seems to be to acknowledge the irreality of both $\mathbf{A}$ and ϕ, and to regard them as merely mathematical fictions. Assessments of the appropriateness of this stance changed with the publication of Aharonov and Bohm's famous paper in 1959.[5]

In that paper, Aharonov and Bohm proposed a thought experiment that would show the "reality" of both $\mathbf{A}$ and ϕ. For the former they suggested scattering a quantum-mechanical electron around a solenoid. Because the solenoid was so long, there would be no appreciable magnetic field in the region of motion of the electron even when there was a current flowing in the solenoid and hence a magnetic field inside. And yet, when the current was flowing there would be a measurable effect: the interference pattern of the electron would be shifted from that produced when the current was zero. Thus, since **A** was non-zero in the region of travel, Aharonov and Bohm concluded that it was real. (There have been a number of experimental confirmations of this well-known effect. Indeed researchers now regularly use "Aharonov-Bohm flux rings" to probe quantum-mechanical mesoscopic effects.)

Before I say what lesson was drawn from this experiment, I want to explain why there is one lesson that certainly should not be drawn: that gauge fields are real only when they are quantum-mechanical. I make this explanation is more detail elsewhere,[6] but in its simplest form, the argument goes like this: the electromagnetic field used to observe the efficacy of **A** was itself—in the context of the derivation as well—a classical field. Thus it is the classical **A** that was observed, not a quantum-mechanical **A**. So whatever we decide about how to understand **A**, and gauge fields generally, will have to apply to our understanding of classical field theories as well.

To return to the Aharonov–Bohm effect: now that the reality of **A** has been established, there are a number of approaches one could take to its interpretation. The simplest, I suppose, would be simply to say "oops, I guess one of the fields *was* the real one," and then move on to more pressing issues. There are important considerations though that apparently disallow this option. Looming large among them are various combinations of worries about non-determinism, non-locality, and epistemological underdetermination. I will not consider these here in any detail.[7] Instead I will simply repeat the consensus view of those physicists and philosophers of physics who have written on this: one cannot choose the naive option. There, however, the consensus very nearly ends. For there are a number of conflicting proposals setting out the right way to understand **A**. But what all these approaches have in common is their willingness to accept the causal efficacy of **A**—while denying the efficacy of any one member of the family of functions that makes up **A**. For, it is argued, **A** is clearly what causes the shift in interference patterns: there is no **E** field there; there is no **B** field there; the shift is functionally related to **A**, which is non-zero there; so it has to be **A**. Thus in one way or another we have a new kind of entity that appears to be a family of vector fields—again, none of which exists—which itself does exist and exerts an influence on the paths of electrons. What could be stranger than that?[8] We conclude from this that on the current best view of the physics and philosophy of physics community that gauge fields are causally efficacious; that the "unreality" of a quantity is no argument against its role in causal explanations, for the sense of unreality may be no more than an effect of the need to choose a new kind of property.

The most important reason for considering gauge theory in this context is to see whether the techniques of physics can be imported into analyses of chemistry without begging the question of the autonomy (or dependency) of chemistry from (or on) physics. So while I think there is an important sense in which gauge theoretical techniques can be applied to chemistry—especially in the context of debates over chemical structure—this work is speculative. I will not be presenting a mathematical embedding of chemistry into the category of gauge theories. I will not even be showing the mathematical possibility of re-characterizing certain chemical properties as gauge fields. Instead I will be pursuing an analogy between chemical properties and gauge theories, and observing that the way theorists who reason about the connection between gauge theory and properties in physics has a resonance with some ways of reasoning about chemical properties. The conclusion I draw is that further work on this issue is warranted in order to see whether this analogy can be fruitfully supported with more detailed characterization of chemical properties as, say, constrained Hamiltonian systems.

MOLECULAR STRUCTURE: THE CHARGE DENSITY

As I pointed out above, concern with (and suspicion about) the notion of structure in chemistry is not new. But I can here offer neither a comprehensive overview of the history of structure debates in chemistry nor an analysis of the present consensus or lack thereof about what constitutes chemical structure.[10] Instead I will be focusing exclusively on Bader's account of how we ought to understand and define chemical structure. But note that I am not particularly concerned with whether Bader's proposal holds up to critical scrutiny. And I will not be offering much such scrutiny myself. I am, however, very interested in the pattern of explanation he employs, and the way this pattern connects with patterns of explanation employed by those who study gauge theories. It is this that I wish to explore here. So while the remarks I make will be specific to his account, I believe that the lessons I draw will be more general.

I begin this part of the paper with a brief account of Richard Bader's paradigm for attributing structure to molecules described by quantum-mechanical state functions. After a brief clarification in this section of one of his candidate physical properties, I suggest, in the next section, how this paradigm might be seen as an application of gauge theoretical ideas to the question of molecular structure. I will indicate below certain similarities between the resources to which Bader appeals in his account of structure and the general account of gauge theories given above. The key point is that appeals to underlying but unrealized features of the situation in chemistry resemble—both in a technical sense and in what might be called their philosophical flavor—the physicists' appeal to dynamical variables that never take on particular values.

Bader begins his account of chemical structure with a litany of the problems that cannot be solved by appeals to geometric analyses of structure: we are unable, he says

> ...to assign a single geometric structure, average or otherwise, to rotation- or inversion-related isomers, to a molecule in an excited vibrational state with geometrical parameters very different from those for the same molecule in its ground state, or to a molecule in a "floppy" state wherein the nuclear excursions cover a wide range of geometrical parameters. (p. 54)

He identifies the cause of these inabilities as follows:

> In reality, these are shortcomings of attempts to impose the classical idea of geometry on a quantum system. The nuclei, like the electrons, cannot be localized in space and instead are described by a corresponding distribution function. (p. 54)

Before detailing Bader's solution to the conundrum of applying geometry to quantum systems, I need to back up a little and outline the most important resource he has developed for this purpose. This is what Bader calls "the charge density." Now this is a funny quantity. As he assures us in various places, "the charge density, $\rho(\mathbf{r})$, is a physical quantity which has a definite value at each point in space." (p. 13)

Since this quantity plays such a central role in his analysis, it is crucial that we see what precisely it is. The charge density, or "electronic" charge density as he sometimes calls it, is the probability density of the electron wave function multiplied by the unit of electronic charge. (chapt. 1, especially pp. 6–7) In what sense is this a charge density? In fact it simply isn't, except at best in some collapse theories of quantum mechanics where $\psi^*\psi$ is taken to be density of stuff. But Bader can't be using one of these versions since he needs his electrons to be fully non-localized. That is, he needs for what follows to define a charge distribution that is continuous and twice continuously differentiable (except at the location of the nuclei). But on collapse, the electron wave function localizes to a delta function distribution. Of course this localization spreads almost immediately to a distribution throughout space, but this doesn't help in Bader's quest for an in-principle notion of structure. More crucially Bader will need to be able to show that the properties of the electronic charge distribution attain local maxima only at the location of nuclei. But on any collapse model the electron's wave function will be peaked sharply about its localization position. Thus it is clear that Bader is using a no-collapse model of quantum mechanics. (Note that it won't do to suggest that there is some kind of continuous collapse going on here, induced by the coupling of the vibrational or rotational modes of the molecule, or by some other mechanism. Bader's argument for why we cannot use classical geometry is a *principled* objection. No matter how small the magnitude of the probability density far away from the classical nuclear location, that density is non-zero. Bader is under no illusions about this, but is not always explicit about what is entailed by the fact that the "nuclei...cannot be localized in space.")

I will not comment on the various complications introduced into Bader's account by his lack of attention to the quantum-mechanical measurement problem (which is what the above issue amounts to). Instead I will suggest that he would do just as well to consider what his charge density really is: the expectation value for the charge at each point of space. What's the difference? Well first, it isn't an average. For while the expectation value of the charge at some given point may be non-zero, on a no-collapse view of quantum mechanics there is no answer to the question "what's the average charge here?" To have an "average charge" at a point, one needs the charge at that point to be well-defined at each moment. But on a no-collapse view of quantum mechanics, there is no answer to the question "what's the charge here now?" Even on a collapse view, where these questions do have answers, the answer is not Bader's "charge density." It is perfectly consistent to suppose a non-zero expectation value for the charge at a point while also supposing that no charge is ever at that point. I submit that since Bader's "charge density" as defined is really just the expectation value of the charge, he should not obscure matters by calling it the average charge.[11] On the other hand, the expectation value won't do as a quantum-mechanical quantity which couples to, say, the nuclear charge of some atom. This is essentially the point to Rosenfeld and Bohr's 1933 paper, where they show that expectation values of the charge cannot, for experimental reasons, play the role of the electric field. Instead, they show, the electric field in quantum mechanics must itself be a true quantum-mechanical field.[12]

Now because Bader's account of the charge density is incorrect, we cannot say that the behavior of the molecule is specified by the action of this scalar field. So what do we have instead? I'll return to this question after outlining the rest of Bader's proposal. Here I will merely suggest that the results he obtains employing the notion of equivalence classes of geometries suggest that we take seriously the idea that the topology of the "electronic charge density" is the real dynamical quantity of interest, and that we follow Bader in rejecting geometric configuration as primary. To this I now turn.

MOLECULAR STRUCTURE: EQUIVALENT GEOMETRIES

We have, as Bader shows us, a notion of structure given in the pre-quantum understanding of the atomic constituents of molecules. This is simply that molecular structure is the molecular geometry given by the arrangement of constituent atoms. The principle distinction between the quantum and pre-quantum cases is that in the latter we can make coherent the notion of a fixed geometry. But in the former case, recall, a fixed geometry of unsharply localized electronic and nuclear constituents is incoherent. To get around this problem, Bader will use instead the idea of equivalence classes of geometries as the structure of the molecule. An equivalence relation over possible nuclear configurations provides Bader with the notion of structure he requires to carry out his program of demonstrating "that the existence of atoms with definable properties and the associated concepts of the molecular structure hypothesis are a consequence of the quantum description of matter."(p. 2)

The new notion of structure is complicated in execution, but quite simple in idea. The rough and ready outline goes like this: pick, as exemplar, a classical geometry of the atoms involved in the molecule of interest. Consider, in "geometry space" the neighborhood of the exemplar geometry. For all geometric configurations in the region defined by the molecular bond structure of the exemplar molecule, we say that these geometric configurations have the same molecular structure.(p. 54ff) These then are equivalent geometries. But notice that we cannot mean by this the following hypothetical suggestion: "If the actual geometry of the molecule is in the equivalence class for most of some period of time, we assign it the structure associated to that class. If, on the other hand, it spends most of its time exemplifying the geometries of another class, we assign it the structure of that class." For the whole problem is that molecules just don't have a given geometry. On the other hand, if we allow that Bader has succeeded in his descriptive task, we *can* understand that the geometric possibilities of the molecule are constrained to lie inside a given equivalence class of geometries.

Bader takes it that a conception of structure developed along these lines will allow for a solution to the "central problem of molecular structure—to predict the discontinuous changes in structure that are caused by a continuous change in nuclear coordinates." (p. 88) Here Bader makes it clear that for him the issue is not primarily about whether we *can* attribute structure to molecules in the way outlined, since he takes himself to have succeeded already in that. Rather the issue is how, given that they do have structure, we can account for the observed evolution of molecular structures. As a philosophical issue, however, the connection between attribution of structure and its causal agency is not so clear.

Bader immediately goes on blandly to observe:

> It is important to remember that, while the structural aspects of the theory of atoms in molecules are described by the dynamical changes in the topology of the charge densi-

ty, the theory is rooted in quantum mechanics. It is the atom and its properties which are defined by quantum mechanics. The bond paths and the structure they define just mirror and summarize in a convenient way what the atoms are doing, performing the same role here as does the assignment of a set of bonds in the molecular structure hypothesis. (p. 89)

But the bond paths he mentions are summaries of the properties of the "charge density," as are the sets of bonds. For us it is important to remember this, for it is here that Bader's reasoning is most transparent. Molecular structure is irreal. It is a kind of "mirror" of the behavior of atoms. But the atoms don't have any particular behavior that is being mirrored. That is where we started this investigation. Instead the atoms have a connection to the expectation value of the charge distribution. But this is itself irreal—irreal in the sense that it cannot be taken to be the way various charges interact with electronic fields. As I said above, this is precisely the point that Rosenfeld and Bohr make. That said, the expectation value of charge is still well-defined. The problem is how to understand that it (or rather the structure it is used to define) has any efficacy. Assuming, along with Bader, that this notion of equivalence classes of geometry is sufficient to his purposes, however, seems to commit us to the idea that irreal properties are causally potent. For this reason, I suggest that the resources of gauge theory are sufficient to underwrite an explanation for the causal powers of chemical structure.

WHY CHEMICAL STRUCTURE IS A GAUGE PROPERTY

First to summarize the problem: The idea that there is some definite meaning that attaches to the topology of molecules but not to their geometries is a little puzzling. For again, as Bader tells us, the reason that molecules have no geometry is that the nuclei of their constituent atoms, being quantum-mechanical, have no definite location. (p. 54) And yet a similar problem obtains for the topological features of the molecules as well. There is no definite sense to the claim that a nucleus is on *this* side of some critical surface of structure change as opposed to *that* side. One could, I suppose, invoke some theory of quantum-mechanical measurement at this point and say that the nucleus is localized by the observer, or the environment, or whatever. But again, and more strongly, if there is any spread at all in the wave-function of the nucleus, then there is spread to arbitrary distances. And then there cannot be a sharp demarcation from one equivalence class of structure to another. So we must explain in a different way how we understand structure as well as how we understand change of structure.

It is here that I think the analogy between chemical theories and gauge theories may be fruitfully articulated. For we know that the attribution of structure is crucial to much of chemical practice. Some of the analogical features are the following: (1) The equivalence classes Bader defines may be seen as constraint manifolds, with change of structure given by the breaking of constraints by the introduction of outside forces—a "lifting off of" the manifold; (2) Attributing a particular geometrical configuration to a molecule is like picking out a single point of the constraint manifold, and tracking the choice would be analogous to specifying a gauge trajectory—but this apparently violates determinism and is disallowed; we needn't (indeed we cannot) believe that a given structure is reflective of just one single configuration of the molecule, just as a particular choice of gauge need not be associated to a given gauge potential—structure is crucial, but no given geometry is real.

Then to summarize the "solution": We have, in chemistry, a situation that is similar in certain respects to the situation in gauge theories. On the face of things it seems impossible to attribute structure to quantum-mechanical molecules. And yet we require structure to underwrite much of our explanatory practice in chemistry. Fortunately we can make sense of this explanatory practice without imposing an arbitrary notion of structure on chemistry, without, that is, appealing to extrachemical physical resources, by noting how a similar practice in physics is underwritten using gauge-theoretic techniques—techniques that are not themselves particular to physics but should be considered general principles of physical theorizing.

What remains to be done is clear, but not easy. One would need to show, at a minimum, that the analogy I have suggested between chemical structure and gauge theory can be extended in the following way: One would have to show that we can recast analyses of chemical structure (and here I refer specifically but not necessarily exclusively to Bader's account) into the mathematical language of gauge theory. There are a number of different candidates for that language, but the likeliest choice for use in chemistry is that of constrained Hamiltonian systems. The idea would be to represent the various different geometries in each equivalence class as points in a submanifold of configuration space, find an appropriate sense of dynamical evolution for geometric states, and then characterize the tendency of molecules to remain within a single equivalence class as the operation of a weak constraint on the system. (We assign weak constraints because we know that molecules can, without outside interference, move between geometric equivalence classes.) We then would try to represent the interactions of molecules in terms of the constraints they jointly obey. It is too early to report much in the way of progress on this front, but the work is ongoing.

In advance of carrying out the project of the previous paragraph, I still think there is something that can be said in terms of the usefulness of the project for the philosophy of chemistry. What we see here is that possibilities exist for the peaceful coexistence of chemistry and physics as "fundamental" theories. How so? "If anything," a critic might say, "this program reduces even more chemistry to physics, by its importation of constrained Hamiltonian systems, principle fiber bundles—the whole panoply of physics tools." And yet that's all that's being imported: the relatively new mathematical tools associated with gauge theory. On the other hand, the application of these tools shows how to speak meaningfully about the autonomy of systems that are clearly built up out of other systems. If these tools are applied in the way I have suggested, then one need neither pose nor answer the question: "Is chemistry reducible to physics?" The irrelevance of *that* question will be manifest.

NOTES AND REFERENCES

1. The idea of giving a retrospective reevaluation of what *was* true for older theories is an extremely interesting topic in its own right, but it would take us too far afield to consider it now. Some of this is done in MATTINGLY, J. 2001. Singularities and scalar fields: matter theory and general relativity. Philos. Sci. **68**: S395–S406.
2. BADER, RICHARD F.W. 1990. Atoms in Molecules—A Quantum Theory. Clarendon Press. Oxford, UK.
3. MATTINGLY, JAMES. 2003. Why there are no gauge fields. Unpublished manuscript.

4. I speak all along in this paper as though I believe the story I'm telling about gauge theory. I don't. But the proper connection between what I do believe and the story I'm telling here will have to wait. What I express is the overwhelming consensus in present discussions of the nature of gauge theories.
5. AHARONOV, Y. & DAVID BOHM. 1959. Significance of potentials in the quantum theory. Phys. Rev. **115:** 485–491.
6. MATTINGLY, "Why there are no gauge fields."
7. I carry out a critical appraisal in the work cited in note 6.
8. These and other issues have received extensive treatment in the literature. See for instance: BELOT, G. 1998. Understanding electromagnetism. Br. J. Philos. Sci. **49:** 531–555; HEALEY, R. 1997. Nonlocality and the Aharonov–Bohm effect. Philos. Sci. **64:** 18–40; LEEDS, S. 1999. Gauges: Aharonov, Bohm, Yang, Healey. Philos. Sci. **66:** 606–627; and for an early and still crucial treatment, WU, T.T. & C.N. YANG. 1975. Concept of nonintegrable phase factors and global formulation of gauge fields. Phys. Rev. **D12:** 3845–3846.
9. See the first three references in note 4 for representative views.
10. However, these issues are addressed by other contributions to this volume, and references therein.
11. For those who are inclined to demand that "actually there" be given an operational definition, you are already opposed to saying that the expectation value is the average charge. Average charge is itself a nonoperational kind of quantity.
12. ROSENFELD, L. 1963. On quantization of fields. Nucl. Phys. **40:** 353–356.

Symmetry in Basic Physical Laws

ELMAR KÜHL[a] AND TIMM LANKAU[b]

[a]*Berufsgenossenschaftliches Institut für Arbeitsschutz—BIA, 53754 Sankt Augustin, Germany*

[b]*Job-Stiftung, Institut für Physikalische Chemie, University of Hamburg, 20146 Hamburg, Germany*

ABSTRACT: Recent physical experiments suggest that chirality is not a purely chemical phenomenon, but seems to be an inherent feature of basic physical laws. A fundamental physical principle (the CPT theorem) connects asymmetry in space with the "arrow of time," and shows that such an arrow exists even on the level of mesons.

KEYWORDS: symmetry; CPT theory; parity; chirality; arrow of time; antimatter; particle decay

INTRODUCTION

Asymmetry has been a major concern in chemistry since 1848 and Pasteur's pioneering work on crystallization. Chirality has been regarded as a typically chemical phenomenon, since it has commonly been believed that the basic laws of physics are totally symmetric.[1] This statement is true for all classical and quantum-mechanical phenomena, but recent theoretical and experimental research have shown that it is not applicable to the weak nuclear force (e.g., β-, K^0- or B^0-decay) (TABLE 1). In 2002, Y. Karyotakis and G. Hamel de Monchenault[2] published numerical results for symmetry violation in B-meson decay.

THE CPT THEOREM[3]

C, P, and T refer to different symmetry operations. The parity operation ($\hat{P}$) changes the sign of all spatial coordinates. This operation is essentially a point reflection of the particle, placed at the origin of a coordinate system. If the mirror image of a system and the system itself are physically indistinguishable, the system is called "parity invariant." The $\hat{C}$ operator (charge conjugation) transforms a particle into its anti-particle, and the $\hat{T}$ operator reverses time. A time-invariant process, such as a planet orbiting a star, is physically reasonable with time running either backwards and forwards; a time-variant process looks different if time runs backwards. An illustrative example for a time-variant process could be a ball dropping

---

Address for correspondence: Elmar Kühl, Berufsgenossenschaftliches Institut für Arbeitsschutz - BIA, 53754 Sankt Augustin, Germany. Voice: +49 (0)2241 231-2862; fax: +49 (0) 2241 231-2234.

Elmar.Kuehl@hvbg.de

TABLE 1. Symmetry in basic physical processes

Interaction		Symmetry Conserved (+) or Violated (−)			
Gravity		+	+	+	+
Electromagnetism		+	+	+	+
Strong nuclear force		+	+	+	+
Weak nuclear force	β-decay	−	−	+	+
	K^0-decay	−	−	−	+
	B^0-decay	−	−	−	+

from a certain location (place A) on a table to the floor. If time runs now backwards, and the ball jumps onto the table, but comes to rest at a different place A', the process would be considered time-variant. Time variance does not say anything about whether it is possible for a process to run backwards in time; it just demands that the process would look different if time did run backwards.

The CPT theorem states that all interactions are invariant under the successive operation of $\hat{C}$, $\hat{P}$, and $\hat{T}$ and (that is, $\hat{C}\hat{P}\hat{T} = \hat{1}$). This theorem can be derived from very basic assumptions. An anti-particle can be described as a particle moving backwards in time ($\hat{C}\hat{P} = \hat{T}^{-1}$). The theorem demands that a process that violates CP symmetry has to be time-variant, in order to maintain CPT invariance.

BETA DECAY[4]

Soon after Lee and Yang[5] published their theoretical work predicting lack of parity conservation in the emission of particles (electrons) from nuclei, C.S. Wu presented experimental proof[6] confirming that prediction. The momentum vector of the electron ejected in β-decay can either point into the same direction as the spin of the atomic nucleus from which it is ejected or in the opposite direction. If $\hat{P}$ (simplified as a mirror reflection) is applied to the problem, the alignment of both vectors is inverted and so is the handedness of β-decay. If β-decay is a parity-invariant process, electrons have to leave the nuclei in equal amounts in both directions, aligned with the nuclear spin and the opposite direction; otherwise the process and its mirror image are different. By aligning the nuclear spins of a sample of ^{60}Co in an adiabatic demagnetization experiment, Wu was able to show that electrons from β-decay leave the nuclei preferentially in the direction that is opposite to the nuclear spin (left-handed) so that β-decay violates parity and is therefore a chiral process.

It was later shown that β-decay also violates charge conjugation (**C**) and that the combination of $\hat{C}$ and $\hat{P}$ (**CP**) is once again a good symmetry operation. According to the CPT theorem β-decay should therefore be invariant against time inversion.

KAON DECAY

After confirmation of parity violation in β-decay by Wu (1957), particle physicists were convinced that at least the joint transformation of parity and charge conjugation should be symmetric for all systems. Contradicting this assumption, Christenson *et* al.[7] reported, in 1964, the observation of a small CP-violation in the decay of the neutral kaon. Neutral kaons are mesons containing an anti-strange quark and a down quark ($\bar{s}$ d). The kaon K^0 and its antiparticle ($\bar{K}^0$) decay by weak interactions to the same final states, and form a pair, which oscillates between two different states:

$$|K_S^0\rangle = \frac{1}{\sqrt{2}}(|K^0\rangle + |\bar{K}^0\rangle) \text{ and } |K_L^0\rangle = \frac{1}{\sqrt{2}}(|K^0\rangle - |\bar{K}^0\rangle)$$

decays into 2 pions with a characteristic time constant $\tau_S = 0.9 \cdot 10^{-10}$ s, while K_L^0 decays into 3 pions ($\tau_L \approx 500\,\tau_S$). The final states are distinguished by their CP quantum numbers.

$$K_S^0 \to 2\pi \;:\; \mathbf{CP}|K_S^0\rangle = +1|K_S^0\rangle \text{ and } \mathbf{CP}|2\pi\rangle = +1|2\pi\rangle$$
$$K_L^0 \to 3\pi \;:\; \mathbf{CP}|K_L^0\rangle = -1|K_L^0\rangle \text{ and } \mathbf{CP}|3\pi\rangle = -1|3\pi\rangle$$

Although the kaon oscillates between two states, after a sufficient amount of time ($t \ll \tau_s$) we expect all kaons to be in the K_L^0 state, and only decays from that state into three pions should be observable. But Christenson et al. occasionally observed a decay producing two pions! Therefore, a CP-violating decay process has to exist for K_L^0. The CP-violation in kaon decay is a tiny effect—only one of about 500 long-lived kaons decays to two pions, which is equal to an asymmetry factor A of approximately 0.002. But it gave the first experimental suggestion that matter and anti-matter may behave differently.

B-MESON DECAY

The standard model of particle physics would be consistent with a much bigger asymmetry in B-meson decay than that observed in kaon decay (values for A close to unity have been predicted), but another theory predicts an asymmetry of similar size. So analysis of B-meson decay can be used to test both theories. The B-meson ($\bar{b}$ d) forms a particle–antiparticle pair (B) similar to the kaon. Particle physicists are interested in one particular type of decay of the B-meson, since the amount of CP-violation responsible for the asymmetry can be deduced directly from it:

$$B \to J/\Psi + K_S^0$$

(J/Ψ is a single particle, the vector meson $c\bar{c}$.) The standard model of particle physics uses the Cabibbo-Kobayshi-Maskawa matrix V_{CKM} to describe processes involving the weak interaction. V_{CKM} is a 3×3 unitary matrix whose nine components measure the strength of transitions of quarks from down-type quarks (with charge $-1/3$) to up-type quarks (with charge $+2/3$). Because of the unitarity of the matrix, only 4 of the 9 components can be treated as independent parameters. One of these parameters is known as the complex Kobayashi-Maskawa phase. This phase is the only possible way to incorporate CP-violation into the standard model: A CP-transformation changes V_{CKM} into the complex conjugate matrix V_{CKM}^*—a process that would not

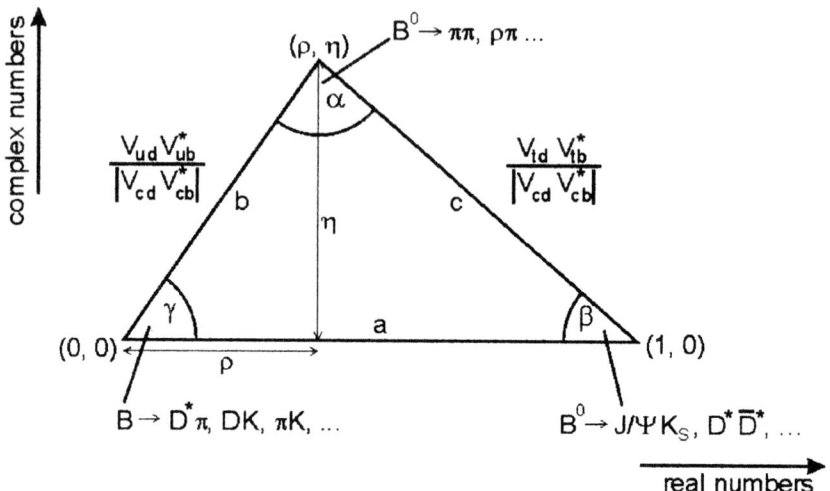

FIGURE 1. The components of V_{CKM} are connected by the unitary relation $V_{ik}^* V_{ij} = \delta_{kj}$. With $k = 3$ and $j = 1$ the unitarian relation transforms to: $V_{ub}^* V_{ud} + V_{cb}^* V_{cd} + V_{tb}^* V_{td} = 0$. This equation can be represented with the unitary triangle in the complex plane. The figure shows the normalized unitary triangle ($V_{cb}^* V_{cd} = 1$) after a rotation to align $V_{cb}^* V_{cd}$ with the real axis. Now the apex of the triangle has the coordinates ρ and η, where the value of η can be used to quantify the asymmetry in time.

change a matrix with real coefficients. A good approximation of the V_{CKM} is the Wolfenstein parametrization[8] with four independent parameters (λ, A, ρ, and η).

$$V_{CKM} = \begin{bmatrix} V_{ud} & V_{us} & V_{ut} \\ V_{cl} & V_{cs} & V_{cb} \\ V_{td} & V_{ts} & V_{tb} \end{bmatrix} = \begin{bmatrix} 1 - \lambda^2/2 & \lambda & A\lambda^3(\rho - i \cdot \eta) \\ -\lambda & 1 - \lambda^2/2 & A\lambda^2 \\ A\lambda^3(1 - \rho - i \cdot \eta) & A\lambda^2 & 1 \end{bmatrix}$$

Using the Wolfenstein parameterization, one of the unitary relations can be geometrically represented in form of a triangle in the complex plane—the so called unitary triangle (FIG. 1).

CP-violation requires that the area of that triangle be non-zero. The size of two of the sides of the triangle can be deduced from various known B-meson decay rates. The interpretation of these measurements is limited by theoretical uncertainties, but the measurement of the angle β—from time-dependent CP-violating asymmetries in the decay of the B-meson—is free from such uncertainties. With the knowledge of angle β, the parameter responsible for CP-violation can be calculated. During 2002, the BABAR experiment at the Stanford Linear Accelerator Center (SLAC) and the BELLE experiment at the KEK Laboratory (Japan) have reported[2] the first observations of CP-violations in the B-meson system yielding the value $\sin 2\beta = 0.78 \pm 0.08$.

CONCLUSIONS

The reported value for CP-asymmetry ($\sin 2\beta = 0.78$) validates the standard model of particle physics and rules out the idea of super weak interactions. The equation for the asymmetry factor $A(t)$, which is proportional to $\sin 2\beta$, has also a time-dependent factor. $A(t)$ oscillates therefore on a psec timescale following so the $B^0 \bar{B}^0$ mixing with a maximum value[2] for $|A_{max}| \approx 0.4$ and offering another explanation for the observed matter anti-matter imbalance in the universe. So far, chemistry has focused only on the properties of matter, but physicists have recently been able to produce anti-hydrogen in greater quantities.[10] From the chemist's point of view, the most important consequence of that achievement may be the chance to compare the chemistries of hydrogen and anti-hydrogen. It is generally believed that both forms of matter will behave in the same way. Yet, the K- and the B-meson experiments suggest that matter and anti-matter can behave differently, raising the question as to whether such differences might influence the chemistries of hydrogen and anti-hydrogen.

Further, if the CPT theorem is correct, CP-asymmetry demands time variance, which also has been experimentally validated[11] in appropriate systems. The "arrow of time"[12] is commonly believed to manifest itself as entropy only in many-particle systems. The K^0-experiments suggest that an arrow of time may also exist also on a subatomic level. One possible response to this situation is to wonder whether questions such as irreversibility and chirality—traditionally thought to arise only at the macroscopic (or chemical) level, and to be specifically chemical—could have their origin at the level of mesons and quarks, that is to say, in fundamental physical laws. If so, this would be an example of the reduction of chemistry to physics.

[*Editor's note*: Since the origins of irreversibility and chirality in relatively macroscopic systems can be understood quite satisfactorily without invoking broken symmetry of underlying physical laws (see papers by Prigogine and by King, in this volume), another interpretation, and a more plausible one, is that the types of symmetry breaking that are well-studied in chemical systems also may be involved in microphysical processes that are currently experimentally inaccessible to physicists. In a sense, this might be considered a reduction of physics to chemistry.]

REFERENCES

1. ULANOWICZ, R.E. 2002. On the ordinality of causes in complex autocatalytic systems. **988**: this volume.
2. KARYOTAKIS, Y. & G. HAMEL DE MONCHENAULT. 2002. A violation of CP symmetry in B meson decays. Europhysics News **33**. <http://www.europhysics.com/full/15/article4/article4.html>
3. KOSTELECKY, A. 2002. Background information on Lorentz and CPT violation. <http://media4.physics.indiana.edu/~kostelec/faq.html>
4. MYNENI, K. 1984. Symmetry destroyed: the failure of parity. <http://ccreweb.org/documents/parity/parity.html>
5. LEE, T.D. & C.N. YANG. 1956. Question of parity conservation in weak interactions. Phys. Rev. **104**: 254–258. <http://prola.aps.org/abstract/PR/v104/i1/p254_1>
6. C. S. WU *et al.* 1957. Experimental test of parity conservation in beta decay. Phys. Rev. **105**: 1413–1415. <http://cornell.mirror.aps.org/abstract/PR/v105/i4/p1413_1>
7. CHRISTENSON, J.H. 1964. Evidence for the $2pi$ decay of the $K^0{}_2$-meson. Phys. Rev. Lett. **13**: 138–140. <http://cornell.mirror.aps.org/abstract/PRL/v13/i4/p138_1>

8. WOLFENSTEIN, L. 1983. Parametrization of the Kobayashi-Maskawa matrix. Phys. Rev. Lett. **51**: 1945–1947. <http://cornell.mirror.aps.org/abstract/PRL/v51/i21/p1945_1>
9. KAMIEN, R.D. 2001. Chirality. <http://dept.physics.upenn.edu/~kamien/chiralweb/>
10. JORGENSEN, L.V. 2002. Why is antihydrogen interesting? <http://athena.web.cern.ch/athena>
11. PERRICONE, M. 1998. The time machine. Fermi News. **21**: 1–3. <http://www-fmi.final.gov/MI_Fermi_News/FermiNews98-10-30.pdf>
12. CALLENDER, C. 2001. Thermodynamic asymmetry in time. *In* Stanford Encyclopedia of Philosophy. <http://plato.stanford.edu/entries/time-thermo/>

The Metaphorical Foundations of Chemical Explanation

THEODORE L. BROWN

School of Chemical Sciences, University of Illinois, Urbana-Champaign, Urbana, Illinois 61801, USA

> ABSTRACT: To address the question of whether typical chemical explanations have characteristics that distinguish them from explanations in other sciences, it is important to recognize the deeply metaphorical nature of scientific explanation in general. The theory of conceptual metaphor, based on results from modern cognitive sciences, postulates that largely unconscious thought processes, grounded in embodied experience and in experiences drawn from the social domain, are at the core of scientific reasoning. I argue, using several examples, that models employed by chemists as explanatory instruments are metaphorical in nature and have a character distinguishable from explanatory representations employed in other fields of science.
>
> KEYWORDS: metaphor; conceptual metaphor; chemical explanation; scientific explanation; unconscious thought processes; models; models as metaphors

The various fields of science are distinguished in part by the level of generality at which a discipline or field of science claims to offer explanations. Physics is judged to be more general than chemistry, because the laws of physics account for more than just chemical phenomena. Chemistry in turn is more general than the chemical subfield of biochemistry, which is more general than cellular biology, in the sense that biochemical processes extend beyond cells.

The various scientific disciplines also differ in terms of the complexity associated with the phenomena under investigation. Roughly speaking, complexity tends to run in opposition to generality. Much of chemistry is thought about and communicated in terms of models and theories that have no place in physics. The chemist focuses on emergent properties of systems that are generally too complex to be dealt with using the models and methods of a physicist.[1] Similarly, cellular biology deals with phenomena and explanations that are not encompassed within biochemistry. For the cellular biologist, the cell's complexity of structure, organization, and processes result in emergent properties that cannot be understood and modeled entirely in terms of all the biochemical reactions occurring there. Thus, each area of science possesses

---

Address for correspondence: School of Chemical Sciences, University of Illinois, Urbana-Champaign, 601 South Goodwin Street, Urbana, IL 61801. Voice: 217-333-2992.
tlbrown1@earthlink.net

a characteristic domain of subject matter, albeit loosely defined, and a style of approach, as it were, to its subject matter.

These observations relate to a question that underlies the theme of this volume: is there a characteristic content or style of explanation in each area of science? It lies at the heart of whether one can even usefully talk about a philosophy associated with a scientific discipline. The mere designation of a body of subject matter will hardly suffice to make for interesting inquiry. The answer to this and related questions is to be found in analysis of the representations, or models, that scientists employ. Study of the ways in which scientists observe the world, formulate hypotheses, design experiments, and communicate with one another in their chosen areas of scholarship, provides the basis for discerning the character of distinct areas of science. I will argue that analysis of the metaphorical content of scientific language and thought provides a productive avenue of approach to these kinds of questions.

To fully understand how science works, we must look into questions of how the scientist interacts with the world. In part this involves analysis of the nature of experiment. To understand the origins of scientific creativity, we must know what is involved in the scientist's engagement with the world in the course of "doing science." How does the scientist approach observations, reason about observational data, create and develop new theories, and communicate ideas to others? We begin with the premise that tacit[2] or implicit knowledge[3] and embodied experience play essential roles. These subconscious forms of understanding are manifested in the metaphorical character of scientific reasoning, as reflected in the language employed in reasoning and communicating about science.[4] Analysis of language provides insights into the nature of the cognitive processes employed in reasoning, and into the structured nature of our understanding of the physical world.

THE THEORY OF CONCEPTUAL METAPHOR

During the past few decades, the study of metaphor has moved from primarily literary and philosophical territory to the realms of psychology, linguistics, and other cognitive sciences. As cognitive scientists have learned more about human conceptual systems, the essential roles played by metaphorical thought have become more evident. Many of the entities that we want to think about and talk about, such as love, time, or the meanings of scientific observations, are abstract concepts. To convey ideas about these abstract entities, we call upon language and conceptions that we normally employ in speaking and thinking about more concrete experiences. For example, we conceptualize the abstract idea of "mind" in terms of a spatial metaphor. We "put things out of" our minds, or "things pass through" or minds, and so on.

The spatial metaphor for mind is but one example of the general observation that we talk about abstract concepts by using language drawn from concrete domains. Such use of language is metaphorical, but not in the sense employed in classical theories of language, which concerned themselves with novel constructions, in which words are not used in their ordinary senses. The new view of metaphor asserts that our everyday language is replete with metaphors that we use without being conscious of their metaphorical character. These are termed *conventional metaphors*, or preferably, *conceptual metaphors*, to distinguish them from the novel constructions found in fiction, poetry, and scientific theories. George Lakoff asserts that conceptual metaphors are not simply matters of language[5]:

> [T]he locus of metaphor is not in language at all, but in the way we conceptualize one mental domain in terms of another. The general theory of metaphor is given by characterizing such cross-domain mappings. And in the process, everyday abstract concepts like time, states, change, causation, and purpose also turn out to be metaphorical.

Note that Lakoff here refers to metaphor as a conceptualization process. The linguistic expression that may result from this cross-domain mapping is, as it were, a surface manifestation of a more fundamental and deeper matter of thought. Lakoff and Mark Johnson describe the theory of conceptual metaphor in their influential 1980 book, *Metaphors We Live By*,[6] in which they cite a host of examples drawn from a wide range of human experience. As just one example, consider the ways in which we conceptualize the abstract idea of time. Among the many metaphorical mappings for time, one of the most general is, TIME IS A RESOURCE. A subset of this is, TIME IS MONEY. Here, from Lakoff and Johnson, are some of the manifestations of this conventional metaphor in contemporary English[7]:

- You're wasting my time.
- This gadget will save you hours.
- I don't have the time to give you.
- That flat tire cost me an hour.
- I've invested a lot of time in her.
- He's living on borrowed time.
- I lost a lot of time when I got sick.

The idea of time as a resource, money in this case, is drawn from our everyday social experiences with time. Of more direct significance for science, we also think of time in terms of spatial representations. Future times are in front of the observer, past times are behind the observer, and the passage of time is continuous and one-dimensional. The metaphor TIME PASSING IS MOTION has two special cases. In one, the observer is fixed, and times are entities moving with respect to the observer. Time has a velocity relative to the observer. On the other hand, times may be imagined as fixed locations, and the observer moves with respect to them.

These ways of conceptualizing time show up in everyday language when we say, for example:

- "In the coming months...."
- "I'm looking ahead to summer vacation."
- "The time has long since passed when ..."
- "She's facing the future with optimism."
- "I can't believe how quickly the time has passed!"

In science the conceptualization of time as a spatial entity is reflected in the general mapping, TIME IS LENGTH. For example, data may be collected at specific time intervals over a period of time, and the results displayed as a two-dimensional graph in which the measured quantities are displayed along one axis and time along the other. The length of the time axis represents the time elapsed from some reference starting time.

In summary, we attempt to understand abstract concepts in terms of our most basic physical experiences with the world, accumulated from birth, and in terms of more structured experiences in the world, *experiential gestalts*, that are products of more complex physical and social interactions. As an example of the latter, we speak of protein "folding." Here there is a metaphorical association of something that happens to proteins in solution with the acts of folding macroscopic objects like napkins, travel maps, or card tables. We do not literally see proteins fold; what we have are observations on proteins, such as ultracentrifuge data, optical rotatory dispersion data, or other similar experimental artifacts. The association of these data with folding is metaphorical in nature. As another example, we speak of "chaperone" proteins, drawing upon our understanding of the social role of the chaperone to explain what happens when certain proteins act upon others.[8] When we analyze the thought processes of scientists in designing experiments and making observations, the terms in which they formulate and advance hypotheses and theories and the ways in which science is communicated, we find that metaphor is at the core of what scientists do. The reader is invited to explore this subject further in my book, *Making Truth: Metaphor in Science*.[4]

MODELS IN CHEMISTRY

If we grant the importance of metaphorical thought and representation in science, the question of whether there exists a characteristic form of explanation in chemistry becomes one of whether the metaphors employed are peculiar to chemistry, differing significantly from those employed in other scientific disciplines. The answer to this is clearly yes. While a host of examples might be cited, I will point to just two by way of example.

The role of molecular models in chemistry has a long history. As the range of organized chemical facts expanded in the nineteenth century, the need for ways of representing substances grew ever more urgent. The work of Biot, Pasteur, and others in identifying optical activity in substances, coupled with knowledge of isomers, provided the impetus for van't Hoff and LeBel's model for tetrahedral carbon. At that time, many influential scientists were resisting the use of models to represent unobservable entities. Ernst Mach and Pierre Duhem were among the leading scientists of the time who advocated a positivist outlook, in which only descriptive statements that correspond directly to observation are valid. Although this philosophical movement persisted for some time (it gave rise eventually to the school of logical positivism), the usefulness of models and theories became overwhelmingly evident, and molecular models became commonly accepted metaphorical representations. Three-dimensional models of the ball-and-stick or space-filling variety (FIG. 1) are employed today to represent the structures of simple substances, as well as complex polymeric substances such as proteins.

Metaphors can be thought of as mappings of properties from the known and familiar, the source domain, onto the properties we wish to understand, the target domain. The latter consist of the observations made on a system under study. Molecular models, whether three-dimensional or two-dimensional representations (such as

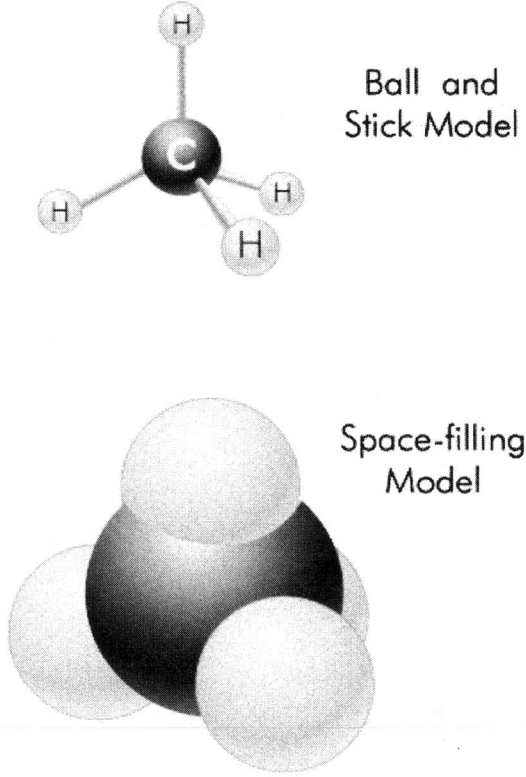

FIGURE 1. Ball-and-stick and space-filling models of methane. (Produced by Janet Sinn-Hanlon, reprinted with permission from T.L. Brown, *Making Truth: Metaphor in Science*, University of Illinois Press, 2003, Figure 2.4, page 23.)

those in FIG. 1 or as viewed on a computer screen), are metaphors in this sense: they convey chemically important aspects of molecular substances, such as shape, volume, topology, and geometry, that map onto chemical, X-ray scattering, or spectroscopic data. Their explanatory power lies in their capacity to capture the interpretations placed upon those experimental data. Such models are primarily chemical in nature. While they convey simple physical ideas of molecules, they are most useful in relation to chemical questions about substances. Thus, they constitute a peculiarly chemical form of explanation. The shorthand structural notations used by organic chemists to denote chemical structures and reactions in two dimensions are even more obviously chemical metaphors. It is difficult to imagine a scientist from any discipline other than chemistry making use of the following metaphorical representation of a chemical process:

The chemist is concerned not only with various representations of chemical structure, but also with models for chemical change. The chemical equation shown above contains information not only about the particular chemical reactants, but also pro-

$CH_3CH=CHCH_3 + H-\ddot{\underset{..}{Br}}: \xrightarrow{slow} CH_3-\underset{\underset{H}{|}}{\overset{\overset{H}{|}}{C}}-\underset{\underset{}{|}}{\overset{\overset{H}{|}}{C}}-CH_3 + :\ddot{\underset{..}{Br}}:^- \xrightarrow{fast} CH_3-\underset{\underset{:\ddot{Br}:}{|}}{\overset{\overset{H}{|}}{C}}-\underset{\underset{H}{|}}{\overset{\overset{H}{|}}{C}}-CH_3$

vides at least a partial model of how the reaction is thought to proceed. A formal theory of chemical reactions grew out of models for gases. In 1872, Boltzmann brought the kinetic molecular theory of gases to maturity, by deriving the Maxwell distribution by a method based on the dynamics of molecular collisions. In this form, the Maxwell-Boltzmann theory of gases is a *physical* theory. In 1889, however, Svante Arrhenius pointed out that the rates of many gas phase reactions increase exponentially, according to the expression

$$k = Ae^{-E_a/RT}$$

where A and E_a are constants characteristic of the reaction. A simple interpretation of the Arrhenius equation is that colliding molecules require a certain minimum energy of relative motion, E_a, to bring about the needed bond breakings and formations that lead from reactants to products. Arrhenius in effect introduced a characteristically chemical consideration into gas behavior. Over time, the empirical Arrhenius model developed into more detailed models, under the general rubric of activated-complex theory, of how chemical reactions occur. In activated-complex theory, molecules are assumed to move on energy surfaces. Colliding molecules move higher in energy on such a surface as the kinetic energy of the collision is converted into the potential energy of bond length and other geometrical changes. At some point on the surface, the colliding molecules attain a configuration, the activated complex, possessing a critical energy and geometrical form, from which they can pass into products. FIGURE 2 shows an illustration of a simple reaction.

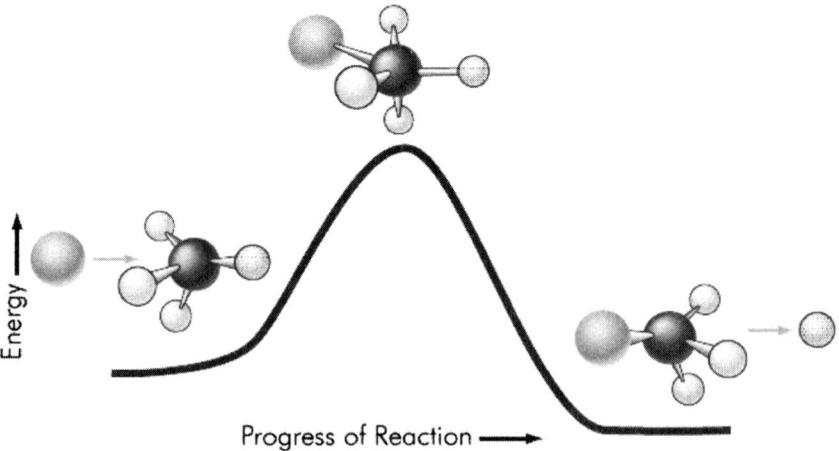

FIGURE 2. A schematic representation of a chemical reaction surface. (Produced by Janet Sinn-Hanlon, reprinted with permission from T.L. Brown, *Making Truth: Metaphor in Science*, University of Illinois Press, 2003, Figure 7.3, page 131.)

FIGURE 2 is a complex metaphorical representation. It shows on the one hand the proposed changes in molecular structure that occur in the course of reaction, employing the metaphorical properties of model representations. In addition, however, it represents energy changes in the process as movement over a surface. Notice that the *change in state* from reactants to products is portrayed as a *movement* from one place to another. This movement is not, of course, a literal movement from one place to another; the dimension along which the reaction "moves" is an abstraction called "progress of reaction." Notice also that the vertical movement on the surface is purely metaphorical. It corresponds to a basic conceptual metaphor, MORE IS HIGHER, one drawn upon heavily in scientific representations.

METAPHORICAL EXPLANATION AND REDUCTIONISM

The foregoing examples illustrate the thesis that the conceptual foundations of chemistry and other sciences are thoroughly metaphorical. Presented with new aspects of the world, we humans understand them in terms of deeply engrained bodily and social experiences that already form the framework for dealing with life on a day-to-day basis. The fact that metaphor is so inextricably a part of the fabric of science also means that it plays many roles. Scientists employ metaphorical reasoning to interpret observational data, in creating models to account for new observations and to reinterpret older data. Metaphors, once created and put to use in these ways, serve in communication among scientists, and between scientists and the public. Such communication depends upon explanatory language that conveys the essential ideas with clarity. A good metaphor can also be persuasive. Selling a new idea and receiving credit for it are important for the scientist's goals of achieving recognition.

Metaphors form the key ideas that go toward defining distinct areas of research. The examples given above illustrate the argument that observations we regard as chemical in character are explained in terms of metaphors characteristic of chemistry, distinct from those employed in other areas of science. But explanations generated in one area of science may and often do have currency in others. Metaphor is the vehicle by which ideas and models from one scientific discipline are transferred to another. For example, the idea of a "code" has found its way from information theory into molecular biology; the idea of a "spin glass," important in the study of disordered systems in physics and metallurgy, has been applied in models for protein folding. Metaphors that conceptualize the mind as a computer are clearly traceable historically to the development of digital computers. Obversely, neural nets and genetic algorithms, both fruitful developments in computer sciences, are grounded in metaphors drawn from the biological sciences.

Changes in the tools available to the scientist are often reflected in the content, and even the character, of explanation. A prime example is the move towards reductionism; as computational and other analytical tools become more powerful, there is a movement towards accounting for observations in ever more fundamental terms. Thus, cellular biology is increasingly dealt with in biochemical terms. Within chemistry, computational tools now make it possible to calculate properties, including reactivities, through the use of complex programs that involve a multitude of underlying assumptions that may not be visible to the user. Such a program might, for example, compute the energy of an activated complex in a reaction, pointing towards one particular possible pathway as optimal. Because such programs are grounded in quantum mechanical theory, it might be argued that reductionism is at work: the explanation of a particular chemical reaction has become more characteristically physical than chemical. But data are not explanatory in themselves. For the chemist to make effective use of powerful computational resources, there must still be an underlying metaphorical model of what is happening in the conventional chemical sense. Seen this way, the computational tools serve to draw more accurately the curve in FIGURE 2, not to define an altogether different way of thinking of the reaction. Of course, the latter could occur; new research might suggest that the models implicit in FIGURE 2 should be replaced by others. We would expect, however, that any new model, whatever its form, would be grounded in the same kinds of metaphorical representations of energy, reaction rate, and reaction progress, all of which have strongly chemical characteristics.

NOTES AND REFERENCES

1. This is not meant to deny that there is significant overlap of physics with chemistry at the chemistry–physics interface.
2. POLANYI, M. 1958. Personal Knowledge: Toward a Postcritical Philosophy. University of Chicago Press. Chicago; POLANYI, M. 1966. The Creative Imagination. Chem. Eng. News, 25 April, p. 88.
3. REBER, A.S., R. ALLEN & P. REBER. 1999. Implicit versus explicit learning. *In* The Nature of Cognition. R.J. Sternberg, Ed. MIT Press. Cambridge, MA.
4. BROWN, T.L. 2003. Making Truth: Metaphor in Science. University of Illinois Press. Urbana, IL.
5. LAKOFF, G. 1993. The contemporary theory of metaphor. *In* Metaphor and Thought. 2nd edit. A. Ortony, Ed. Cambridge University Press. Cambridge, UK. p. 203.
6. LAKOFF, G. & M. JOHNSON. 1980. Metaphors We Live By. University of Chicago Press. Chicago, IL; LAKOFF, G. & M. JOHNSON. 1999. Philosophy in the Flesh: The Embodied Mind and Its Challenge to Western Thought. Basic Books. New York.
7. LAKOFF & JOHNSON, Chapter 2.
8. In the first use of the chaperone concept, Laskey and co-workers specifically referred to its metaphorical origins: LASKEY, R.A., B.M. HONDA, A.D. MILLS & J.T. FINCH. 1978. Nucleosomes are assembled by an acidic protein which binds histones and transfers them to DNA. Nature **275**: 420.

"Causes" in Chemical Explanations

JANET D. STEMWEDEL

*Department of Philosophy, San José State University,
San José, California 95192-0096, USA*

ABSTRACT: Chemists talk about causes and make frequent use of the verb "to cause." They use the verb "to cause" for its generality in identifying causal chains, whether all the details of those chains are established or some remain unclear. Most of these causal chains bottom out in attractions and repulsions. Attractions and repulsions are causal relations that are chemically basic; chemistry has no further story to tell about why attraction and repulsion are causal. If there is some further story to tell about the causal status of attraction and repulsion, it is left to physics to tell it, but it is not obvious that the absence of such a story would prevent chemists from using attractions and repulsions in their causal explanations.

KEYWORDS: causation; explanation; autonomy of chemistry; black box

INTRODUCTION

Although it may come as a surprise to some philosophers of science, chemistry talks about causes. The causal status of the mid-level relations that chemists most frequently discuss depends on further stories in terms of lower-level causal relations that are chemically basic, chiefly attractions and repulsions. Causings in chemistry are *chains* connecting front-ends and tail-ends of phenomena—"before" and "after" segments. The frequency of chains reflects the pressure to analyze chemical phenomena in terms of a very small set of basic causal relations. Moreover, although attraction and repulsion are relations that are unanalyzed by chemistry, they are not in principle *unanalyzable*. Chemistry accepts that the causal status of attraction and repulsion may depend on a further story physics provides, a story in terms of causal relations that are truly unanalyzable primitives — if, in fact, physics provides such a story. Nonetheless, the importance of attraction and repulsion in grounding causal explanations in chemistry might well prompt chemists to assert their autonomy if developments in physics threatened this explanatory resource.

In this paper, I examine the ways chemists use the verb "to cause" in their explanations. Here, I focus on the particular uses of this general way of identifying or describing causal connections. It is a further question whether chemical explanations, and chemical practice in general, requires that chemistry talk about causes, or whether the concept of causation is something chemists could dispense with. I have taken

---

Address for correspondence: Janet D. Stemwedel, Department of Philosophy, San José State University, One Washington Square, San José, CA 95192-0096. Voice: 408-924-4468; fax: 408-924-4527.

jstemwed@email.sjsu.edu

up this issue elsewhere,[1] arguing that if causation is excised from chemistry, chemists may be robbed of tools they use to distinguish good explanations from bad ones. For the present discussion, however, I focus on the use chemists actually make of "cause."

BROAD FEATURES OF CAUSAL TALK IN CHEMISTRY

Despite Bertrand Russell's famous 1917 claim that "in advanced sciences...the word 'cause' never occurs,"[2,a] chemists do use the verb "to cause" in its various forms.[b] They use this verb not only in textbooks, where one might wonder if it is just a colorful or metaphorical way of speaking, employed to engage students in the material. The verb "to cause" is also used in chemical journals, where chemists speak to other full-fledged chemists and might be expected to be careful to say just what they mean. And occurrences of this verb are by no means rare. For example, in a single 11-page article (by De Luca et al.[3]), "causes" and "caused" appear no fewer than seven times. Of 11 journal articles I examined, six contain forms of "to cause," another one contains the verb "affects," and two identify "effects."

How do chemists use "to cause"? There are two sorts of usage we might expect here: either "causes" captures a paradigmatically causal relation, or it identifies a chain each of whose links is a paradigmatically causal relation. A paradigm causal relation is a process that is primitively causal. Its status as causal doesn't depend on there being any further causal story. (I take it the causal capacities in Cartwright's account of causation[4] are supposed to mark just this sort of primitive causal linkage.) Once a causal relation can be analyzed in terms of a causal chain, the relation is no longer a primitive one; the paradigm causal relations that form the links of the chain become the source of the causal status. Nonetheless, while it is not a paradigm, the causal relation may still be an important and useful one. (Many authors with differing accounts of the core concept of causation have examined causal chains and the role they play in our explanations. Harré[5] gives a particularly interesting discussion of the role of hypotheses citing causal mechanisms.) As it happens, chemists frequently use "causes" to identify chains. "Causes" is used less often to identify a relation that is paradigmatically causal because chemists seek to understand the processes they study in terms of a very small number of chemically basic causal relations.

---

[a]It is worth noting that, while Russell was most concerned with physics and its alleged ability to do without causes, he further claimed that there are *no such things* as causes. Thus, presumably, any chemistry that qualified as an "advanced science" would dispense with causal talk as well.

[b]With the exception of one textbook primarily aimed at secondary school students, the textbooks from which I draw my examples are all used in college-level instruction. Some are introductory texts, others are texts used in upper-division courses. The journal articles from which I draw examples were collected from journals in the Current Periodicals section of the Swain Library of Chemistry and Chemical Engineering at Stanford University during a single visit early in 1998. I selected articles only from "big name" journals in the field, and I did so based on how interesting and comprehensible I found their abstracts. I do not claim that this unscientific sample of the chemical literature is a perfect representation of chemical discourse as a whole, but I assert that the specimens I examine are not anomalies.

In chemistry, there are many important kinds of causal relations. Such relations deal with transformations of matter, specifically of molecules and aggregates of molecules. Chemists describe the front-ends and tail-ends in terms of structural features of a certain sort: atoms, bonds, bond lengths, bond angles, molecular motions and collisions, electron distributions, etc. Finding the causal links between such features helps chemists understand, explain, manipulate, and control chemical systems.

Among the causal relations chemists identify is a very small set of relations that behave like paradigm causal relations in chemical discourse. Earlier in the history of chemistry, many more sorts of chemical relations were primitively causal. But chemistry has taken it as a project to figure these relations out in terms of lower-level structure. Like all scientists, chemists operate under the assumption that today's paradigm causal relation may become tomorrow's chain.

Chemistry sees itself as interlocking with physics in a particular way. This means that the most basic types of causal relations in chemistry are provisionally taken to depend on the underlying stories physics provides to make these relations causal. As such, they are not paradigm causal relations. Rather, they are *quasi*-paradigm causal relations. The chemical judgment that a relation is causal is grounded in there being an underlying story in terms of quasi-paradigmatic causal links; the causal status of these links is then passed on to physics.

USES OF "TO CAUSE"

Chemists use forms of the verb "to cause" for their *generality* in identifying causal links. "Cause" can be used for the identification of new causal relations for which there is not yet a particular causal verb; for chains where no particular causal verb captures the whole process; for chains that are not yet fully worked out; or for chains whose particular details are not relevant. The most common usage makes "cause" a placeholder for a chain.

The role "cause" plays holding the place of a chain is clearest in the instances where the steps in the chain are then spelled out. A ubiquitous causal claim in introductory chemical textbooks is that placing a solid substance like sodium chloride in water causes it to dissolve. As one introductory textbook explains this causing,[6] the following steps take place:

(1) The positive ends of water molecules are attracted to the negative ions on the surface of the solid; the negative ends of water molecules are attracted to the positive ions on the surface of the solid.
(2) Attraction between the water molecules and the ions pulls the ions out of the crystal arrangement.
(3) The ions that have been pulled out of the solid attract more water molecules, until they are surrounded by water molecules.

In other words, "causes" here gives the causal upshot of a (relatively short) causal chain in which attractions between ions and water molecules pull the ions out of the solid and into the solution.

"Causes" can also stand for a chain whose steps will be elaborated at more length. The same textbook poses the question, "What **causes** differences in the reactivities of the metals—or in other properties that vary from element to element?"[7] The initial answer it offers is, "Many properties of elements are determined largely by the number of electrons in their atoms and how these electrons are arranged."[8] But the detailed answer is explored over the course of several chapters that follow. The preliminary answer also suggests that "causes" here will group together more than one type of chain responsible for bringing about differences in elemental properties. "Causes" generalizes these types of chains. At least one of the chains has to do with the number and arrangement of electrons in atoms, but there might be other causal chains here as well, chains that deal with different structural details.

Another way "causes" is used is to represent a chain whose links are not identified but are left to the reader to spell out. Here is an example from a study to characterize the binding interaction between ice and an eel antifreeze protein (AFP). The authors write: "[T]he exposed ice surface will grow with a curvature between the bound AFPs. This curvature yields a larger ratio of surface area to volume, which in turn **causes** a local depression of the freezing point of that surface relative to the bulk solvent freezing point."[3] The chain left to the reader to identify is the one that connects a larger ratio of surface area to volume with a local depression of the freezing point of the ice surface. At first glance, it seems odd to cite the surface area to volume ratio as a cause of something, but some further information about the freezing of water helps set out the links in the chain by which this happens. Freezing (that is, formation of a solid) of water (a polar molecule) requires that the negative ends of each H_2O molecule line up with the positive ends of the adjacent H_2O molecules, and vice versa. Surface H_2O molecules cannot form attractions on all sides like H_2O molecules in the interior of the solid can. Greater curvature makes the greatest degree of positive-end to negative-end alignment harder to achieve without lowering the temperature, because of the size and geometry of the water molecules and the speed with which they move at this temperature.[c] So here, the surface area to volume ratio is a property of the ice curvature, and the curvature is itself a property of the arrangement of the H_2O molecules between the bound AFPs. This property causes the freezing point depression (that is, the freezing will occur at a lower temperature) given the properties of H_2O molecules. "Causes" here stands for the chain in which increasing the surface area to volume ratio leads to more H_2O molecules at the surface and fewer in the interior, which leads to fewer attractions between the H_2O molecules, which results in less H_2O in solid form at this temperature (that is, a freezing point depression). The background information drawn on here is familiar to any chemist—indeed, it is the sort of information introductory textbooks and chemistry courses spend a great deal of time conveying.

Why, in a case like this, do the authors leave it to the reader to fill in the details of the causal chain? Textbook authors sometimes do this to give the student practice identifying such chains. A journal article, however, would probably not have as its main aim presenting exercises for the reader. It is possible that the chain in this case is already well known to chemists who study antifreeze proteins (the main audience for the paper). Since these details are so well known, the authors don't need to rehearse them again, but rather use "causes" as shorthand for this chain. On the other hand, there may be rhetorical value in leaving the details of this chain to the reader. Going through the process of building a description of the causal chain oneself can

---

[c]Temperature is a measure of the mean kinetic energy of the molecules in a system, so the higher the temperature, the faster (on average) the molecules of the system are moving.

make what seems at first to be a surprising result more convincing. "Surface area to volume ratio causes *what*? Let me see that.... Why yes, of course, it does!"

The verb "to cause" can also stand for causal chains by black boxing them. The term "black box" connotes something whose interior is obscure or inaccessible. Sometimes the obscurity arises from ignorance—important details are not yet known. Other time, details that are known are not illuminated in a discussion because they are irrelevant.

Sometimes the details of a chain are hard to fill in because they require expertise from another field. Examples of this sort of usage are especially common when chemists gesture to physical details of chemical systems. In a classic textbook on quantum mechanics in chemistry, Pauling and Wilson write, "[T]he relativistic change in mass of the electron **causes** a splitting of energy levels correlated with the observed fine structure of hydrogen-like spectra."[9] Relativity is not a standard piece of background knowledge for chemists, so the details of *how* a relativistic change in electron mass brings about splitting of energy levels are not ones they could easily fill in. For such details, they would defer to physics. As it turns out, however, the details inside the black box are not judged by chemists to be essential to the explanatory task at hand. If they were, chemists would have to make issues like relativistic effects on mass part of their own area of expertise. Instead, they can rely on the judgment of physics that there is a genuine causal chain in the black box. What is important for chemists to know is that there is a splitting of energy levels in a hydrogen atom, and that it is caused—somehow—by properties of the electron.

Another example is seen in an investigation of ways to model xenon clusters. The authors write: "[T]he charge, through its migration, is trapped between two atoms because of the polarization field that it **causes**. That is to say, a dimer forms."[10] Here, *something* about the migrating charge sets up a polarization field, although here the authors do not tell us what. The reader might be able to draw on his background knowledge here. Certainly, she could consult a physicist or a textbook on electricity and magnetism to fill in the missing details about how moving charges set up such fields. But it is not the point of this article to explain how moving charge brings about a polarization field. Rather, what is important is the *consequence* such a field has in this system—namely, that it traps charge between two atoms. The black box here isolates a genuinely causal link whose details are not essential for the task of understanding dimer formation. The details are known—by *someone*, if not by the authors or the reader—but they don't need to be filled in here.

A different kind of black boxing happens when provisional causal claims are advanced. Here, the verb "to cause" captures a connection that looks causal, but whose details remain to be worked out. "Causes" stands for the incompletely characterized chain that we think will turn out to be a causal one. What is in the black box is the chain connecting the suspected cause and effect (if, in fact, they are causally connected). Filling in more details of the chain should settle the question of causation and remove the box.

Examples of this sort of usage are especially common in journal articles, since these are more likely than textbooks to present work on the frontiers of chemical knowledge, trying shed light on phenomena that are not yet well understood. For example, in a study of muscle thin-filament activity, the authors state: "Addition of Ca^{2+} **causes** an approximately 25° azimuthal movement of tropomyosin from the outer to the inner domain of actin."[11] Addition of Ca^{2+} in these circumstances appears to be necessary and sufficient for the movement of tropomyosin, giving us reason to hope there is a causal link here. But it is not yet clear *how* exactly Ca^{2+} brings about this movement of tropomyosin. Part of the point of the study is to propose a model for filament activation, but this model leaves the role of Ca^{2+} somewhat vague. The authors model tropomyosin behavior under conditions of high and low calcium—in other words, Ca^{2+} concentration is a parameter in the model. But nothing in the model takes up the issue of *how* Ca^{2+} brings about the shift of the tropomyosin—whether by binding to actin or tropomyosin and causing a conformational change, or by undergoing a chemical reaction, or by some other means. If there is a genuine causal chain here, some of its details remain unknown.

While identifying the chains linking putative cause and effect is the main project of many research projects, for some projects certain of these chains are not of the utmost importance. Sometimes provisional causal claims are made where the details of the chain are not as important for the task at hand as the fact that there *is* a causal link. In a journal article examining a transmembrane transporter, one set of authors claim, "Binding of the cation **causes** a conformational change in the sugar binding site, which results in an increase in affinity for sugar."[12] How exactly does cation binding bring about the change in the shape of the sugar binding site? Knowing all the details of this would confirm the hunch that cation binding is a cause here. But in some studies, the important thing to know is that the shape *is* changed in a particular way, rather than the fine details of the cation-protein interaction that bring this change of shape about. In an article about interferon production, on which the compound reserpine has an apparent effect, the authors write, "Reserpine ... **causes** catecholamine depletion and blocks their re-uptake."[13] The details of the link between reserpine and catecholamine depletion are not spelled out here. It is possible that this is because the connection is not fully understood. But even if it is, knowing reserpine's *effects* in a biochemical system (where the level of catecholamine may be quite important for the planned experiments) could be more useful to these authors than knowing the details of *how* reserpine brings these effects about. As was the case with the black boxing of chains that depend on the expertise of another field, such a black box pushes off the problem of the lower-level links that make a relation causal. So long as the relation *is* causal these lower-level links are not essential to the explanatory project at hand.

Besides being used as shorthand for causal chains, "to cause" can also identify a relation that is a paradigmatic causing to chemists, a link which is primitively causal. In such a case, the causation is obvious, something that can be "seen." Such usage is not as common as using "causes" in place of a chain. But this is because sciences such as chemistry generally strive to take what was once a straight-shot causal linkage and eventually break it down into a chain at a lower level.

Certainly, chemists use "causes" to identify modifications whose consequences are obvious. Sometimes, the obviousness of a newly observed process moves chemists to *identify* the relation as causal. Consider, for example, the claim, "To evaluate how the binding of myosin itself **affects** tropomyosin position, filaments were saturated with myosin subfragment-1 (S-1) in the absence of ATP ... S-1 binding **causes** a further shift of about 10° from that in the high-Ca^{2+} state such that tropomyosin now lies entirely over the inner domain of actin."[14] The authors of this study were able to use electron microscopy to actually visualize this interaction. They could see the binding of S-1 to the actin filament—see where the tropomyosin is initially, see where S-1 binds to actin and how much space it takes up, and see how this binding pushes the tropomyosin on an additional 10°. One object displacing another is a very commonsensical sort of causing. The authors provide a detailed description of what they see in the visualization of this process: "In displacing tropomyosin, S-1 binds to and crosses the face of actin subdomain-1, it covers the subdomain-1-3 junction and approaches subdomain-3."[15] The details of S-1 binding here are likely important to the effect that is brought about. But these additional details are not the links of a chain. Rather, they sharpen the description of the cause and effect, helping a reader who does not have a micrograph in front of her to "see" the causation, too.

But not every obvious causing is a single-link paradigmatic causing. Sometimes, the process that initially looked paradigmatically causal is later re-described in terms of lower-level interactions. Or sometimes a causing is obvious *as a consequence* of lower-level paradigm (or quasi-paradigm) causal relations rather than *because it exemplifies* such a paradigm relation. Return to the article on models of xenon clusters, in which trapping charge between two atoms forms a dimer. The authors go on to assert, "[T]he formation of the dimer would **cause** dislocations in the structure of the cluster."[16] The use of "cause" in this claim captures consequences that are obvious to a chemist. Here, localizing charge between two atoms in the cluster changes the shape of the cluster, bringing two atoms closer together and others farther away. The change in relative position of these atoms *is* a dislocation in the structure of the cluster.[d] At first blush, "cause" seems to identify a paradigmatic link between localizing charge between two atoms and bringing the two atoms closer together. But, what makes this effect obvious to a chemist is the chemist's internalized grasp of the behavior of positively and negatively charged bodies. Further details can be provided about how the positive charges in the xenon nuclei and the attractions these have with the negatively charged electrons bring about the dislocations in the cluster. "Cause" here still captures an alteration with obvious consequences, but there is a (short) chain by which this causing takes place. Indeed, the tropomyosin displacement above can also be re-described in terms of molecules, where the incoming electron-dense regions of one repel the electron-dense regions of the other.

One more usage of "caused" in this literature is worth noting. "Caused" can be used to identify effects that are *not* caused, features that are not present in the tail-end of a happening connected to a particular front-end. For example, in the study of eel antifreeze protein the authors state, "[T]he added bulk of the Cys residues **caused** little perturbation of the protein backbone... and no displacement of the ice-binding atoms."[17] Such a locution identifies the absence of particular effects that might have been expected. One might think the substitution of a bulky amino acid into this protein would cause such perturbation and displacement, but it does not. This suggests that the process by which the effect is brought about is less straightforward, and that the steps of the chain involved here are not yet fully understood. However, this claim of what isn't caused may eliminate certain chains that might have been proposed, simply because these chains would result in an effect that does not in fact occur.

To summarize briefly, chemists use "cause" as a general way to identify relations in which the link between the identified front-end and the identified tail-end has the

---

[d] A good spatial analogy for this sort of relationship might be a rug with buckets of dirt swept under it. Stomping on it to flatten the pile of dirt under one part of the rug will shift that dirt under another part of the rug.

right kind of structural character. Chemists could use "cause" to identify relationships that are paradigmatically causal, but more frequently they use "cause" to identify chains of chemically basic causings. Some are chains where all the relevant details are known by the chemists. Some are chains where all the details are known, although some are known on the authority of experts from another field. Still others are provisional claims where significant parts of the chain remain to be worked out. The predominance of chains is a consequence of the push to understand a system by understanding its parts, where causings that were once primitive end up being understood in terms of a chain of lower-level connections. So, "caused" in the claim "A caused B" may start out identifying a paradigmatic causing and end up being a placeholder for the relations in each of the links of a causal chain.

Why do chemists use "cause" in these instances rather than using particular causal verbs? They use it for its generality. In some instances, there is no particular causal verb that would capture the entire chain for which "cause" stands. In others, where more than one process type may be involved, "cause" is a generalization of the multiple particular causal verbs that might apply. Where "cause" is used as a black box, chemists may have yet to establish what causal verbs properly apply to a situation, or are identifying a causing where the particular causal verb would not do much to convey meaningful information. Finally, although we have not seen this in our examples, a chemist might elect to use "cause" in the identification of a new paradigmatically causal relation for which no particular causal verb has yet been coined.

THE ROLE OF PHYSICS IN CHEMICAL CAUSINGS

Chemists frequently use "attract" and "repel" in their discussions. For example:

The...valence shell electrons on the *central atom* **repel** one another.[18]

[A] lone electron pair from the water **is attracted** to the positive charge on the central C.[19]

The electrons in a bond **repel** each other, but they are prevented from getting too far apart because of their **attraction** to the two nuclei in the bonded atom.[20]

We might expect that chemists would defer to physics to provide precise technical meanings of the verbs "to attract" and "to repel." However, chemists appear not to do this.

Nowhere in chemistry textbooks are chemists given official accounts from physicists as to what these verbs "properly" mean. Instead, these verbs are taken to be basic and to have intuitive meanings. In other words, "to attract" and "to repel" maintain their commonsense meanings in chemical discourse.[e] For all practical purposes in chemistry, attraction and repulsion are unanalyzed.

---

[e]This does not mean that it is common sense that *decides* what attractions and repulsions are, since the appearance of these types of causings in common sense is a result of the influence of scientific information on common sense. However, chemists do not need to learn meanings of "attract" and "repel" that are any different from what the grade-school student with minimal exposure to such scientific concepts learns.

Do chemists use "attract" and "repel" to bracket a lower-level story about attraction and repulsion? We might think this is so if chemists have learned such a lower-level story about attractions and repulsions in the course of their training. Most chemists must study some amount of physics in the course of their training, during which they are exposed to an account of attractions and repulsions in terms of movements of charges in fields of electrical force. Chemists-in-training are sure to encounter definitions like the following: "the electric field vector E at some point in space is defined as the electric force F acting on a positive test charge placed at that point divided by the magnitude of the test charge q_0."[21] True enough, chemistry students learn to make calculations involving electric fields. But even such a definition of an electric field does not settle the question of whether attractions and repulsions cause the force, which causes the movement of the test charge, or whether the field causes the attractions and repulsions, which cause the movement of the test charge. Indeed, physics textbooks speak of "the charge distribution responsible for the electric field."[22] This is compatible with attractions and repulsions causing a particular force on a test charge, the net effect of which is captured by the electric field. If physics does settle the question of the causal relation between fields and attractions and repulsions, it does so elsewhere. Chemists occasionally consider complicated systems in terms of the force fields they generate, but they do so in cases where this is the easiest way to determine the net causal upshot of many attractions and repulsions. In chemical explanations of phenomena, it is attractions and repulsions rather than force fields that are basic.

However, chemists know that physics is a field that looks for the fundamental forces. If there is any lower-level story about attractions and repulsions to be told (perhaps in terms of very small particles like quarks), it is up to physics to tell it. If there is such a story now, chemists as a group do not know the details of it. And they don't think they *need* to, provided that the story from physics allows chemists to understand attraction and repulsion in just the way they do.

Many of the causal stories of chemists take attractions and repulsions as their most basic relations. This is where chemical stories about causation usually bottom out. Attraction and repulsion *act* like paradigm causal relations, but they aren't really viewed by chemists as unanalyzable primitives. Rather, the causal status of attractions and repulsions is a matter handed off to physics. Under the assumption that physics does (or will) give an underlying causal story for attraction and repulsion, the lower-level structure of the underlying physical story is what *makes* attraction and repulsion causal. If physics made a surprising discovery that called the causal status of attraction and repulsion into question, chemists would expect that their own usage of these quasi-paradigm causal relations should change.

Exactly how it would change, however, is not obvious. Chemists might be pushed by such a change in physics to adopt different quasi-paradigms as the basic building blocks of their stories, looking to physics to specify good candidates. Or, chemists could decide to reinstate attraction and repulsion as true paradigm causal relations until the physicists have fully worked out their lower-level story. Or, chemists could even decide that going down too far, and trying to ground attraction and repulsion in a lower-level story, misses the level of structure that is causally basic. The importance of attraction and repulsion to causal explanations in chemistry is sufficient that chemistry might well assert its autonomy, parting ways with physics rather than sacrificing chemistry's formidable explanatory power.

REFERENCES

1. STEMWEDEL, J.D. 2002. Explanation, unification, and what chemistry gets from causation. Paper presented at the Eighteenth Biennial Meeting of the Philosophy of Science Association. Milwaukee, WI, November 9, 2002.
2. RUSSELL, B. 1970. On the notion of a cause. *In* Mysticism and Logic. 2nd ed. Barnes and Noble. New York. p. 132.
3. CARTWRIGHT, N. 1989. Nature's Capacities and Their Measurement. Clarendon. Oxford.
4. HARRÉ, R. 1987. Varieties of Realism. Blackwell. Oxford.
5. AMERICAN CHEMICAL SOCIETY. 1993. ChemCom: Chemistry in the Community. Kendal/Hunt. Dubuque, IA. p. 55.
6. Ibid., p. 115. Emphasis mine.
7. Ibid., p. 115.
8. DE LUCA, C.I. *et al.* 1998. The effects of steric mutations on the structure of type III antifreeze protein and its interaction with ice. J. Mol. Biol. **275:** 515–525.
9. PAULING, L. & E.B. WILSON, JR. 1985. Introduction to Quantum Mechanics. Dover. New York. p. 36. Emphasis mine.
10. ATHANASOPOULOS, D.C. & K.E. SCHMIDT. 1998. An isotropic hopping model for singly charged Xe clusters. J. Phys. Chem. A. **102:** 1615–1624. Bold emphasis mine.
11. VIBERT, P., R. CRAIG & W. LEHMAN. 1997. Steric-model for activation of muscle thin filaments. J. Mol. Biol. **266:** 8–14.
12. HIRAYAMA, B.A., D.D.F. LOO & E.M. WRIGHT. 1997. Cation effects on protein conformation and transport in the Na^+/glucose cotransporter. J. Biol. Chem. **272:** 2110–2115.
13. CONTI, B. *et al.* 1997 Induction on interferon-γ inducing factor in the adrenal cortex. J. Biol. Chem. **272:** 2035–2037.
14. VIBERT, CRAIG & LEHMAN, "Steric-model for activation," p. 10.
15. Ibid.
16. ATHANASOPOULOS & SCHMIDT, "An isotropic hopping model," p. 1616. Bold emphasis mine.
17. DE LUCA *et al.*, "The effects of steric mutations," p. 520.
18. WHITTEN, K.W., K.D. GAILEY & R.E. DAVIS. 1992. General Chemistry with Qualitative Analysis. 4th ed. Saunders. New York. p. 299. Bold emphasis mine.
19. DICKERSON, R.E., H.B. GRAY & G.P. HAIGHT, JR. 1979. Chemical Principles. 3rd edit. Benjamin/Cummings. Menlo Park, CA. p. 866. Bold emphasis mine.
20. STREITWIESER, A. & C.H. HEATHCOCK. 1985. Introduction to Organic Chemistry. 3rd ed. Macmillan. New York. p. 180. Bold emphasis mine.
21. SERWAY, R.A. 1987. Physics for Scientists and Engineers. 2nd ed. Vol II. Saunders. New York. p. 512.
22. SERWAY, Physics, p. 513.

John Dalton and the Aesthetics of Molecular Representation

TAMI I. SPECTOR

Department of Chemistry, University of San Francisco, San Francisco, California 94117-1080, USA

> ABSTRACT: This paper examines the negative response to Dalton's symbolism in the context of the representational system of affinity tables. When situated in an iconoclastic scientific culture that rejected overt speculation, affinity tables reflected a functionalist empirical aesthetic, while Dalton's symbols embraced a deductive aesthetic suggestive of alchemical iconography.
>
> KEYWORDS: Dalton; aesthetics; affinity tables; iconoclasm

Typically, twentieth century histories of chemistry cast the origins of the modern atomic theory in reproductions of Dalton's atomic symbols.[1] Indeed, the inclusion of Dalton's symbols in these modern accounts gives the false impression that they were embraced by the chemical community upon their publication, when in reality they were never adopted.[2,3] Despite this inclusion of Dalton's symbols in many modern chemical histories, biographies of Dalton actually tend to primarily focus on the origin and reception of his chemical atomic theory as delineated in the chapter "On Chemical Synthesis" in Volume I of *A New System of Chemical Philosophy*.[4] By contrast, the commentary on the response to Dalton's atomic and molecular symbols has been quite limited, principally noting that he was the first to use a "modern chemical symbolism"[5] and placing his symbolic notation in the context of the relative acceptance of Berzelius' alphabetically based one.[6] Some authors briefly mention the taboo against speculation in the nineteenth-century chemical community and discuss Dalton's notation as a "paper tool" for the development of the chemical atomic theory, while others describe the symbols as prescient signs for our modern understanding of molecular geometry.[7] Quite often Dalton's symbols are diminished by the notion that they were not taken up by other chemists because they were unwieldy to use and difficult to reproduce with the available print technology.[8] In this paper, my concern is to understand the rejection of John Dalton's atomic symbols by his contemporaries through the lens of the visual and aesthetic milieu of the late eighteenth and early nineteenth century.

Address for correspondence: Tami I. Spector, Department of Chemistry, University of San Francisco, 2130 Fulton Street, San Francisco, CA 94117-1080. Voice: 415-422-2927; fax: 415-422-5157.
 spector@usfca.edu

In the nineteenth century, atomic and molecular representations ideally served only as a convenient system for organizing experimental data, and arguments surrounding the development of these symbolic forms were based on the ideal of hypothetical neutrality.[9] For chemical symbols to be accepted and disseminated into the chemical community, it was necessary that they make no pretense of attempting to represent the unseeable universe of atoms and molecules. Conventions of writing in the early nineteenth century chemical community also involved avoiding all speculation, regardless of a particular author's ideas about the underlying forces determining chemical interactions.[10] Thus, as noted by Alan Rocke in *Chemical Atomism in the Nineteenth Century*, Dalton's chemical atomic theory suggested a convincing explanation for the experimental observation that chemical elements combined in precise and reproducible weight proportions to form specific chemical compounds. This explanation rested on the theory that

> There exists for each element a unique "atomic weight," a chemically indivisible unit (i.e., a chemical atom), that enters into combination with similar units of other elements in small integral numbers.[11]

Notwithstanding the reasonableness of this assertion, Dalton undermined the acceptance of his ideas when he constituted his astute "theory of definite proportions" out of hypothetical drawings of interacting circular atoms. Specifically, he justified his concept of particulate atoms using neo-Newtonian reasoning to extrapolate the invisible forces that exist between discrete real-world objects onto diagrammatic atomic particles, conceptualizing liquids and solids as globular particles interacting through "the forces of attraction and repulsion"[12] and pure elastic fluids (that is, gases) as "analogous to that of a square pile of shot."[13]

For the nineteenth-century chemical community, Dalton's hypothetical drawings evoked particular anxieties about overt speculation and more general anxieties about visual images. In this context, the oft-reproduced "arbitrary marks of signs chosen to represent the several chemical elements or ultimate particles"[14] from Dalton's *New System of Chemical Philosophy* appear in actuality to be anything but arbitrary.[15] Indeed, with these signs Dalton challenged the accepted chemical aesthetics of the early nineteenth century represented by the widely disseminated tables of elective affinities. Modern studies on affinity tables have examined their status as "the paradigmatic organizational device" for the eighteenth-century chemical community.[16] My concern is not to replay these accounts, but rather to recast the tables themselves as aesthetic objects that set a context for understanding the visual impact of Dalton's symbols. The aesthetics of affinity tables is bound up with the fact that they were not merely collections of experimental data, but *the* formal representation of chemical theory in the eighteenth century. In the following, I show why affinity tables became the normative representation for eighteenth-century chemists, and how their formal aesthetic properties expedited the acceptance of Lavoisier's nomenclature and the rejection of Dalton's iconic symbols.

In early nineteenth-century Britain, these tables took the form typified in John Webster's 1826 *Manual of Chemistry,* as shown in TABLE 1.[17] This table can be understood most immediately as an empirical listing of elementary substances arranged in accordance with their relative affinities, or in modern parlance, reactivities, for the substance Sulphuric Acid at the head of the table. In actuality, decoding this seemingly simple table is predicated on knowledge of the now out-

TABLE 1. A typical affinity table

Sulphuric Acid
Baryta
Strontia
Potassa
Soda
Lime
Magnesia
Ammonia

moded concept of elective attractions. This key chemical concept of the eighteenth and nineteenth centuries was brought to the forefront of chemical practice by Torbern Bergman's exhaustive 1775 *Dissertation on Elective Attractions*, in which he defined attraction, or affinity, as a special strong force of "contiguous attraction" between small particles akin to the Newtonian force of gravity.[18] Regardless of a chemist's belief in these mechanical principles, such tables of elective attraction were embraced by eighteenth- and early nineteenth-century chemists because they tightly linked what appeared to be a "non-hypothetical" representational system to experiment.

Displacing the observable properties of chemical reactions into a confined written form, affinity tables offer a one-to-one homology between experiment and a synoptic figure that encapsulates the essential attribute of observation without speculation. For eighteenth- and early nineteenth-century chemists, affinity tables emblematized the ideal of empiricism, which was based on the power of observation to reveal, through induction, the truths of the natural world. In their list-like presentation, devoid of syntactical connection, affinity tables have the aesthetic function of revealing the aesthetic properties of a unified chemical theory of nature through their organizational form.[19] Unlike Webster's nomenclature-based table, eighteenth-century affinity tables employed an alchemically traceable symbolism. This further enhanced the aesthetic function of affinity tables by linking their organizational structure with the power of alchemy to metaphysically reveal the "inner nature of the material world."[20]

Throughout the eighteenth century, affinity tables gradually expanded to accommodate more and more experimental observations, yet their structure remained relatively unchanged. Ultimately, this profusion of accumulated data was most impressively summarized by Bergman's tables of elective attractions.[18] Bergman's tables were incredibly comprehensive, but in contrast to their intent, also figuratively and literally symbolic of how unwieldy chemistry had become in the eighteenth century. Coinciding with this profusion, and ultimately with the demise of the belief that affinity tables could reveal the ultimate nature of chemical combinations, was the publication of Lavoisier's *Method of Chemical Nomenclature* (*Méthode de Nomenclature Chimique*) and the *Elements of Chemistry* (*Traité Élémentaire de Chimie*).[21] In this context, these texts, which contain tables of nomenclature, were formally conservative and philosophically progressive, encompassing, in the words of Stephan Weiniger, "a complex interplay of continuities and discontinuities that the semiotic system both illuminated and concealed."[22] Ultimately, Lavoiser fostered acceptance of his "chemical revolution" by draping its content over the conventional structure of affinity tables.

On the surface, the shift from the symbols of affinity tables to the words of nomenclature tables might appear to empty them of any symbolic or, more precisely, alchemical connotations. On closer inspection, however, since Geoffroy published the first affinity table, the distinction between words and symbols had been blurred.[23] For most of the eighteenth century, the iconic, hieroglyphic forms of chemical communication existed side by side with a linguistic chemical system. The connection between word and image in affinity tables is not just proximate, however. Rather, in the graphical representation of the tables, the arbitrary and conventional (that is, nonrepresentational) neo-alchemical symbols and words are displayed as equalities, dissolving the word/image dichotomy.[24] As defined by W.J.T. Mitchell, the relationship between "word and image" can be understood as a "dialectical trope" or, in other words, as a "figurative condensation of a whole set of relations and distinctions," that shift and transform each other "from one conceptual level to another" and shuttle between "relations of contrariety and identity, difference and sameness."[25] In this case, the symbolic forms are literally figures that carry their alchemical traces into the signifying words. This sedimentation of alchemical content into words is made explicit by Lavoisier's intent for his tables of chemical nomenclature to embody linguistically the power of nature, through experimental manipulation.[26] In early nineteenth-century Britain, writers of influential chemical textbooks and dictionaries embraced Lavoisier's "chemical revolution" but also continued to include Bergman's tables of affinities.[17] Often copying from one to another, these partial affinity tables were, most often, transformed from Bergman's symbols into an English translation of Lavoisier's nomenclature. In this form, these fragmentary artifacts no longer serve as guides for experimentation, but as linguistic emblems of British empiricism.

This shift from Bergman's symbolically based tables to Lavoisier's nomenclature tables can also be understood within the larger visual culture of the enlightenment. Barbara Marie Stafford has shown that, in the eighteenth century, the word was privileged by the "invisible patrimony of literate civilization," while the "radical sensory skepticism" as "ushered in by Cartesians, British empiricists, and French sensationalists, identified visualization with the lures of sophistry."[27] This iconoclasm was practiced by early nineteenth-century chemists through overt alchemical distancing. Historians of chemistry have shown that chemists in the eighteenth and early nineteenth centuries were busy asserting their discipline as a valid, independent, and thoroughly modern science.[28] As part of the convention of textbook writing, this professionalism involved producing texts with little or no alchemically traceable symbolism, and prefaces that extolled the virtues of modern chemistry. Textbook authors were also obliged to provide a history of chemistry that included references to alchemy as part of chemistry's ancient past. For example, in his *System of Theoretical and Practical Chemistry* (1807), Fredrick Accum wrote that, upon publication of Johann Becher's *Physica Subterrarea* (1669), "chemistry escaped forever from the trammels of alchemy, and became the rudiments of the science which we presently find."[29] Thus, by excluding its visual remnants and including it as part of the historical narrative, nineteenth-century chemists placed alchemy firmly in chemistry's disciplinary past.

Set in a cultural labyrinth of aesthetic empiricism, alchemical distancing, and virulent anti-iconicism, Dalton's figurative symbols provoked his peers in myriad ways. Dalton's "arbitrary marks of signs" visualized the invisible, blurred the line between hypothetical conjecture and justifiable conceptualization of experimental fact, and exposed the sublimated conceptual underpinnings of chemistry. These symbols were controversial and revolutionary because with them he *overtly* speculated, hypothesized, and contradicted the theoretical hegemony of early nineteenth-century chemistry contained within the structure of affinity tables. These symbols were derided, fully forty-four years after their initial publication, by Michael Faraday as "not the slightest use to chemistry."[30] Dalton's ultimate heresy was to unabashedly challenge chemistry, symbolically transmuting its underlying aesthetic epistemology from the nonvisual to the visual, from the empirical to the constitutive, and from the inductive to the deductive.

REFERENCES AND NOTES

1. For example, see PARTINGTON, J.R. 1989. A Short History of Chemistry. 3rd ed. Dover. New York. pp. 176–177; SALZBERG, H.W. 1991. From Caveman to Chemist: Circumstances and Achievements. American Chemical Society. Washington, DC. p. 216; HUDSON, J. 1992. The History of Chemistry. Chapman & Hall. New York. p. 83; MIERZECKI, R. 1990. The Historical Development of Chemical Concepts. Kluwer Academic Publishers. Dordrecht. p. 119; and CROSLAND, M.P. 1962. Historical Studies in the Language of Chemistry. Harvard University Press. Cambridge, MA. p. 261.
2. Dalton's symbols were first published in DALTON, J. 1808. A New System of Chemical Philosophy. Volume 1. S. Russel for R. Bickerstaff. Manchester, UK.
3. A few of Dalton's contemporaries did comment that his symbols might be pedagogically useful (for example, see URE, A. 1821. A Dictionary of Chemistry. Entry on Equivalents. Thomas and George Underwood. London; and DAUBENY, C. 1831. An Introduction to the Atomic Theory. S. Collingwood for J. Murray. London. pp. 97–98.), but the majority of nineteenth-century authors implicitly rejected the symbols by not including them in their own texts even when other modes of representation were included (for example, see BRANDE, W.T. 1821. A Manual of Chemistry. J. Murray. London; and HENRY, W. 1810. An Epitome of Chemistry, 2nd American from the 5th English Edition. William Andrews. Boston).
4. DALTON, New System, pp. 211–216. For biographies of Dalton, see, for example, ROCKE, A. 1984. Chemical Atomism in the Nineteenth Century from Dalton to Cannizzaro. Ohio State University Press. Columbus, OH; THACKRAY, A. 1972. John Dalton: Critical Assessments of His Life and Science. Harvard University Press. Cambridge, MA; GREENAWAY, F. 1966. John Dalton and the Atom. Cornell University Press. Ithaca, NY; CARDWELL, D.S.L., Ed. 1968. John Dalton and the Progress of Science. Manchester University Press. Manchester; and PATTERSON, E.C. 1970. John Dalton and the Atomic Theory. Doubleday. New York.
5. For example, see PATTERSON, John Dalton and the Atomic Theory, p. 191.
6. For example, see THACKRAY, John Dalton: Critical Assessments, pp. 115–124.
7. For example, see RUSSELL, C.A. 1968. Brezelius and the development of the atomic theory. *In* CARDWELL, Ed., John Dalton, pp. 258–273; and THACKRAY, A. 1970. Atoms and Powers. Harvard University Press. Cambridge, MA. pp. 256–267.
8. For example, see CROSLAND, M.P. 1962. Historical Studies in the Language of Chemistry. Harvard University Press. Cambridge, MA. p. 261; and BROCK, W.H. 1992. The Norton History of Chemistry. W.W. Norton. New York.
9. For a typical example of nineteenth-century commentary on this, see DAUBENY, C. 1850. An Introduction to the Atomic Theory. Oxford University Press. Oxford. pp. 97–109.
10. DUNCAN, A. 1996. Laws and Order in Eighteenth-Century Chemistry. Clarendon Press. Oxford. p. 102.

11. ROCKE, Chemical Atomism in the Nineteenth Century, p. 12.
12. DALTON, New System, p. 143.
13. Ibid., pp. 189–190.
14. Ibid., p. 219.
15. Ibid., plate 4.
16. For example, see DUNCAN, Laws and Order; and ROBERTS, L. 1991. Setting the Table: The Disciplinary Development of Eighteenth-Century Chemistry as Read through the Changing Structure of its Tables. *In* The Literary Structure of Scientific Arguments. P. Dear, Ed. University of Pennsylvania Press. Philadelphia, PA. pp. 99–132.
17. WEBSTER, J.W. 1826. A Manual of Chemistry. Richardson and Lord. Boston. p. 14.
18. BERGMAN, T.O. 1970. Facsimile reproduction of the 1785 Dissertation on Elective Attractions. Frank Cass. London.
19. ZANGWILL, N. 2001. Aesthetic functionalism. *In* Aesthetic Concepts. E. Brady & J. Levinson, Eds. Clarendon Press. Oxford. pp. 123–148.
20. http://www.levity.com/Alchemy/metaphys.html.
21. LAVOISIER, A.L. 1789. Traité élémentaire de Chemie, présenté dans un ordre et d'après les découvertes modernes; avec figures. Cuchet. Paris; LAVOISIER, A.L. 1787. Méthode de Nomenclature Chimique. Cuchet. Paris.
22. WEININGER, S.J. 1998. Contemplating the finger: visuality and the semiotics of chemistry. Hyle **4:** 4.
23. An engraving of Etienne-François Geoffroy's 1718 affinity table can be found in Encyclopedia Britannica. 1771. Vol. II. A. Bell & C. MacFarquhar. Edinburg. p. 106.
24. For example, see BERGMAN, Facsimile Reproduction of the 1785 Dissertation. Chemical Signs Explained.
25. MITCHELL, W.J.T. 1996. Word and image. *In* Critical Terms for Art History. R.S. Nelson & R. Shiff, Eds. University of Chicago Press. Chicago, IL. p. 53.
26. ROBERTS, L. 1991. Setting the table: the disciplinary development of eighteenth-century chemistry as read through the changing structure of its tables. *In* Literary Structure, Dear, Ed., p. 123.
27. STAFFORD, B.M. 1997. Good Looking: Essays on the Virtue of Images. MIT Press. Cambridge, MA. pp. 45–46.
28. For example, see NYE, M.J. 1993. From Chemical Philosophy to Theoretical Chemistry. University of California Press. Berkeley, CA.
29. ACCUM, F. 1807. System of Theoretical and Practical Chemistry. Printed for the author. Vol. 1: Section III. London.
30. FARADAY, M. 1853. Lecturer's Notes for a Speech Given Before the Royal Institution. Longman, Brown, Green & Longmans. London. p. 51.

Writing as Thinking

JEFFREY KOVAC

Department of Chemistry, The University of Tennessee, Knoxville, Tennessee 37996-1600, USA

ABSTRACT: In this article I try to sketch preliminary answers to several questions. (1) What is the role of writing in science? (2) How can rhetorical analysis illuminate the nature of scientific writing? (3) What is the role of models and metaphors in scientific thinking and writing? Since this is a preliminary inquiry, my remarks are brief and suggestive, but I think they point to some profitable directions for further study.

KEYWORDS: writing; rhetoric; metaphor

INTRODUCTION

To understand the role of writing in human thought, it is useful to turn to those who use language most creatively, poets, so I begin with the words of William Stafford.

> The writer is not so much someone who has something to say as someone who has found a process that will bring about new things that he would not have thought of had he not started to say them.[1]

In some ways, this is all I have to say—but unlike the poet, who allows readers to find their own meanings, I will elaborate further.

When I was first learning to be a scientist, I was taught the standard view, or myth, of the role of writing in science. First, one does the research: formulate a problem, plan and execute experiments or theoretical calculations, analyze the data, draw conclusions, and so forth. Then, as a last step, and a necessary evil: write it up. The term "writing up" suggests that all the hard work has been done and that the writing is just a report. Even more, perhaps, it suggests that the scientist is merely observing and recording nature—which of course reinforces the view that science is completely objective.

As I have matured as a scientist, I have discovered that writing is an integral part of science. Writing is part of everything that a practicing scientist does. The writing of formal documents, such as proposals and research articles, is certainly part of it, but, more important, it is in both formal and informal writing that scientists think. As David Mermin says, we do not write up science, we write it.[2]

Address for correspondence: Jeffrey Kovac, Department of Chemistry, The University of Tennessee, Knoxville, Tennessee 37996-1600. Voice: 865-974-3444; fax: 865-974-3454.
jkovac@utk.edu

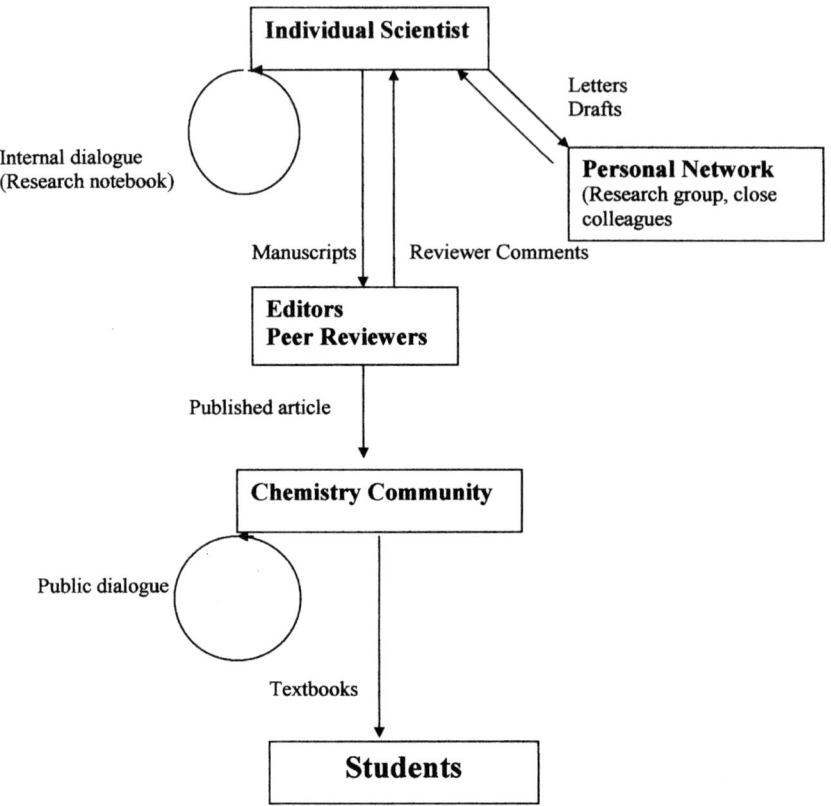

FIGURE 1. A "Feynman diagram" for the relationship between writing and science—the internal scientific process. The *arrows* show the mutual and self-interactions between individuals and groups involved in the development of scientific knowledge. Alongside the arrows are listed the kinds of written documents that mediate these interactions.

THE ROLE OF WRITING IN SCIENCE

To illustrate the essential role of writing in science, I have constructed two "Feynman diagrams," one for the internal process in which scientific knowledge is developed (FIG. 1) and a second for the interactions of science with society (FIG. 2). Feynman invented his diagrams to show the interactions between particles, interactions that are mediated by the exchange of other particles.[3] For example; the electromagnetic interaction corresponds to an exchange of photons. My diagrams show how exchanges of written documents (write-ons?) serve to mediate the interactions between individuals and groups in the scientific process. The division into internal and external processes is somewhat arbitrary, but it makes the diagrams easier to read and understand. Of course, the two are parts of the same overall enterprise.

As individual scientists work, they engage in an internal dialogue, or "self interaction," which is usually carried on in a research notebook—where ideas are devel-

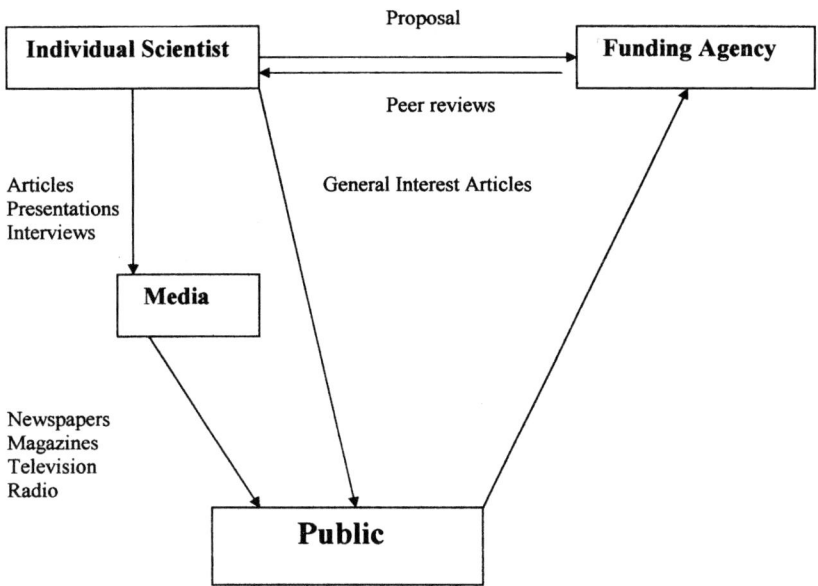

FIGURE 2. A "Feynman diagram" for the relationship between writing and science—science and society. The *arrows* show the mutual and self-interactions between individuals, groups, and organizations involved in the primary relationships between science and society. Alongside the arrows are listed the kinds of written documents that mediate those interactions.

oped and data are recorded and processed. The use of a private journal to aid thinking and creativity is common. In addition, scientists engage in a private dialogue with a personal network of colleagues, including their co-workers in a research group. While some of this dialogue is oral, much is written. In the past, scientists exchanged letters. These days, much of the communication is by e-mail, but the process is the same. Questions are raised and answered; experimental procedures are developed; interpretations of data are proposed and critiqued; conclusions are formulated.

Eventually, the internal and private dialogue results in a manuscript that is submitted to a journal, initiating another exchange. The editor and peer reviewers send back written comments resulting in a revised manuscript. This cycle continues until an acceptable article is produced. The published article generates its own self-interaction or internal dialogue within the relevant scientific community. An interesting or important article stimulates comment, criticism, extensions, and applications—much of this occurs in writing, both in formal published articles and informal exchanges of letters. Finally, the scientific knowledge developed in this public process is distilled into textbooks and pedagogical literature to educate the next generation.

I focus on two aspects of the relationship between science and society: the grant process through which science is funded, at least in the academy, and the relationship between science and the public. In the grant process, individual scientists, or small groups, write formal proposals for funding. This results in an exchange of written documents, because written peer reviews are sent back. In some cases, the proposal is rewritten and resubmitted, starting the cycle again. Most funding agencies also require regular written reports. Scientists communicate with the public either directly through general interest articles or indirectly through the media. There is a feedback mechanism, because public opinion influences the research agenda in various ways.

The two diagrams give an overview of how writing is built into the process of science. All important interactions are mediated by written documents. While the diagrams are probably incomplete, I think they illustrate the most important ways that writing is used in the process of science. Having shown the structure, I now turn to a discussion of the substance.

THE RHETORIC OF SCIENCE

A common myth about scientific writing, reinforced by the use of the phrase "writing up," is that it is not a form of rhetoric. As Roald Hoffmann has so persuasively argued, the scientific article is a human-made text designed to persuade.[4] Initially, scientists write to convince themselves. Chemist and philosopher Michael Polanyi has argued that knowing is an act of skill.[5] This has certainly been my experience in science. Nature can be subtle—understanding and interpreting the data are difficult. Scientists must construct and test careful arguments to persuade themselves that they have found the truth. History shows that it is easy to be deluded.[6] Once a scientist is convinced that the results are both correct and interesting, the next task is to persuade a community of peers.

The form of the scientific article is well defined by tradition. Experimental articles are usually written so as to suggest that science is inductive, whereas theoretical articles are written to suggest that all theory is deductive. As Peter Medewar suggested long ago, the scientific article is a misrepresentation of the actual process of investigation.[7] Further, the scientific article is an example of persuasive writing, an argument to convince the reader that the results are both correct and important. It is not my purpose here to provide a detailed rhetorical analysis of the scientific article (this has been done by others from several perspectives[8]), but merely to indicate that the rhetorical perspective is useful in understanding science.

For example, Aristotle recognized three kinds of arguments: forensic, epideictic, and deliberative.[9] All three are found in scientific prose. The forensic argument is designed to establish facts; this is the primary purpose of the research article. The epideictic argument is concerned with value, judging various alternatives; this is the primary purpose of the review article. Deliberative arguments are designed to establish policy; this is one of the purposes of a proposal. In fact, all three kinds of arguments are usually used in any piece of scientific writing. The research article usually begins with a review of previous work—judgments are made concerning the relevance and value of the various antecedent papers. The body of the paper is usually one or more forensic arguments constructed to convince the reader that the data are complete and correct, and to establish their relevance to the questions being investigated. In some cases deliberative arguments are used to relate the specific research problem to larger public issues.

From another perspective the tasks of scientific research can be thought of as rhetorical tasks. What needs to be explained? What constitutes an explanation? How does an explanation constrain what counts as evidence? These are the questions that any writer or any scientist struggles with. Scientific discourse can be profitably thought of as an example of the rhetoric of assent, the art of discovering and sharing warrantable assertions.[10]

MODELS AND METAPHORS

While experimental science progresses through the development of instrumentation and technique, I think it can be demonstrated that conceptual science progresses primarily through the development of models and metaphors. In their classic study of metaphor, Lakoff and Johnson point out that "the essence of metaphor is understanding—experiencing one kind of thing in terms of another."[11] In science, a simple model or an incisive metaphor leads to what is often referred to as "physical intuition." For example, the driven, damped harmonic oscillator is a model, or metaphor, for many kinds of physical phenomena.[12]

Chemistry is a metaphor-rich science. The conceptual basis of chemistry is the understanding of matter in terms of atoms and molecules, so we need ways to think about entities that we cannot experience. For example, chemical bonding is the result of the interactions between nuclei and electrons, which are described by the formal theory of quantum mechanics. It is difficult, perhaps impossible, to think in terms of the solutions to partial differential equations, so chemists have developed a series of metaphors, or models, to describe the behavior of electrons. Perhaps the primary metaphor is the orbital, a metaphorical picture of the electron density of electrons in atoms and molecules. Chemists are adept at using this metaphor to understand and predict bonding in molecules. An even more primitive metaphor is the Lewis dot structure, which has developed into a powerful symbolic language, particularly in organic chemistry.

The most successful chemical metaphors seem to be simple, flexible, and visual. A chemical article is often full of drawings and graphs, especially molecular structures.[13] Model kits to build "ball and stick" models of molecules are important tools. The physicist writes a Hamiltonian description; the chemist draws a picture.

Our understanding of role of metaphor in science is in its infancy, but the power and pervasiveness of metaphor in human thought suggests that there is much to be learned about the practice of science, and perhaps about the philosophy of science, by examining the development and use of scientific metaphors.[14] This study will lead back to writing as thinking, because good writing uses metaphors creatively. "Writing up" is a poor metaphor for what scientists should do. They should write.

ACKNOWLEDGMENTS

These ideas have been developed over several years in conversations with Linda Bensel-Meyers, Brian P. Coppola, Roger Jones, and Donna W. Sherwood. I am grateful to the Camille and Henry Dreyfus Foundation for generous financial support of my work on writing in chemistry.

REFERENCES

1. STAFFORD, W. 1978. Writing the Australian Crawl. University of Michigan Press. Ann Arbor, MI. p. 17.
2. MERMAN, N.D. 2002. Writing physics. *In* Writing and Revising in the Disciplines. J. Monroe, Ed. Cornell University Press. Ithaca, NY. pp. 15–28.
3. MATTUCK, R.D. 1967. A Guide to Feynman Diagrams in the Many-Body Problem. McGraw-Hill. New York.
4. HOFFMANN, R. 1988. Under the surface of the chemical article. Angew. Chem. [International edition in English.] **27**: 1593–1602; HOFFMANN, R. 1995. The Same and Not the Same. Columbia University Press. New York.
5. POLANYI, M. 1964. Personal Knowledge. Harper Torchbooks. New York.
6. GRATZER, W. 2000. The Undergrowth of Science: Delusion, Self-Deception and Human Frailty. Oxford University Press. Oxford, UK.
7. MEDEWAR, P.B. 1964. Is the scientific paper fraudulent? Saturday Review. August **1**: 42–43.
8. LOCKE, D. 1992. Science as Writing. Yale University Press. New Haven, CT; GROSS, A.G. 1996. The Rhetoric of Science. Harvard University Press. Cambridge, MA; PENROSE, A.M. & S.B. KATZ. 1998. Writing in the Disciplines: Exploring the Conventions of Scientific Discourse. Bedford/St. Martins. Boston.
9. Rhetoric: I, 3 (1358^a36-1358^b7). *In* The Complete Works of Aristotle, Volume II. 1984. Jonathan Barnes, Ed. Princeton University Press. Princeton, NJ. Also discussed in GROSS, Rhetoric of Science.
10. BOOTH, W.C. 1974. Modern Dogma and the Rhetoric of Assent. Notre Dame University Press. South Bend, IN; see also ELBOW, P. 1973. The doubting game and the believing game: an analysis of the intellectual enterprise. *In* Writing Without Teachers. P. Elbow, Ed.: 147–191. Oxford University Press. New York; POLANYI, Personal Knowledge.
11. LAKOFF, G. & M. JOHNSON. 1980. Metaphors We Live By. University of Chicago Press. Chicago.
12. MARSHALL, A.G. 1978. Biophysical Chemistry: Principles, Techniques and Applications. J. Wiley & Sons. New York.
13. HOFFMANN, R. 2002. Writing (and drawing) chemistry. *In* Writing and Revising in the Disciplines J. Monroe, Ed.: 29–53. Cornell University Press. Ithaca, NY; HOFFMANN, R. & P. LASZLO. 1991. Representation in chemistry. Angewandte Chemie [International edition in English.] **30**: 1–16.
14. BHUSHAN, N. & S. ROSENFELD. 1995. Metaphorical models in chemistry. J. Chem. Edu. **72**: 578–582.

Statement Analysis in Chemistry

CLAUS JACOB

School of Chemistry, University of Exeter, Exeter, United Kingdom

ABSTRACT: Close philosophical examination of chemical research reports ("statement analysis") can lead to clarification of ambiguities and resolution of logical confusions, and also lead to other useful conclusions.

KEYWORDS: statement analysis; Exeter method; chemical symbolism; logic

Philosophers have long realized that chemical laboratory practice and chemical communication are interdependent. Whenever chemists make statements as part of their everyday activity, they generate potential objects of meta-chemical research programs that reflect on philosophical (for example, logical, epistemological, linguistic, ethical, aesthetic) questions. Although there are notable examples of meta-chemical reflections (for example, on Lavoisier's work), chemists generally write extensively about their *research*, but hardly at all about their research-related *writing*. Close analysis of chemical texts seems generally to be left to non-chemists, who are primarily focused on older texts. Such research, carried out with the benefit of hindsight, generally remains uncontroversial—and without influence on ongoing chemical research. This note briefly summarizes and reviews some of our recent studies in statement analysis (S-analysis) of *current* research publications in chemistry.[1–3] These investigations were primarily driven by a chemist's desire to be "philosophically correct" in chemical communications. Most areas of chemistry cannot rely on the kind of "mathematical truth" available in other sciences (e.g. physics). Scientific validity in chemistry is therefore frequently associated with a "convincing argument" that can be made for or against a particular idea. This holds particularly true for the field of biochemistry, where an additional problem for scientific reduction exists

Our studies of the *in vitro/in vivo* problem[1] have shown that confusion in biochemical research is often associated with the use of *equivocal terminology* at the chemistry/biology interface. Glucose tolerance factor (GTF) isolated from kidney is a chromium nicotinate complex.[1,4] Similar chromium picolinate complexes therefore might be expected to be of therapeutic use in glucose-related disorders (for example, in the treatment of diabetes). In clinical trials, however, chromium picolinate complexes have not shown any intrinsic activity. The GTF story is, of course, highly

Address for correspondence: Claus Jacob, School of Chemistry, University of Exeter, Stocker Road, Exeter EX4 4QD, United Kingdom. Voice: +44 1392 263462.
c.jacob@ex.ac.uk

complex.[4] Nevertheless, considering some of the underlying reasoning is highly instructive.[1] "GTF isolated from kidney" is not the same as "GTF in intact kidney." Purification methods used to extract the factor have been so severe that they have changed the original *in vivo* species to an "isolation procedure artifact." Secondly, "similarity" between chromium nicotinate and chromium picolinate complexes is chemical and not physiological. Similarities in chemical structure do not imply similar biological activity. The case for chromium picolinate complexes as boosters of glucose tolerance in humans is therefore not as strong as *in vitro* GTF research might suggest—although it cannot be denied outright, either. (S-analysis should not be understood as a tool that can be used to hinder speculative research.)

We have used similar examples to demand the clear and explicit distinction between chemical, biochemical, and biological expressions in biochemical research statements.[1] Once biochemists have a closer look at what is really said (and in which particular context terms such as "function" or "enzyme" are actually used) many misunderstandings can be avoided. Help of a philosopher is not necessarily required to define the meaning of a word such as "protein," but might help to pinpoint equivocal uses of such words and explain implications such misuse might have.

S-analysis is valuable for, but not limited to, defining the precise meaning of individual (bio)chemical expressions. The analysis of whole arguments is equally possible and interesting. Some research papers contain explicit lines of reasoning that can be described and analyzed as part of a meta-chemical investigation. We have recently discussed Henry Phillips's arguments on optical isomerism.[2,5] While trivalent sulfur compounds such as asymmetric sulfoxides occur as optically active isomers, asymmetric trivalent nitrogen compounds such as amines undergo rapid inversion and therefore can't be separated into isomers. Phillips's initial set of criteria for optical activity, that is, a central atom with three dissimilar groups attached and a lone electron pair at the central atom, could not explain this difference. He therefore added the criterion "positive charge at the central atom" (the notion of a "semipolar bond" in sulfoxides assigns a positive charge to sulfur and a negative charge to oxygen). Comparing optically inactive amines with active sulfoxides and tetravalent ammonium and carbon compounds, Phillips then dropped the lone electron pair criterion and associated optical activity with a central atom with three dissimilar groups attached and a positive charge at the central atom. This also implied optical activity of esters (semipolar carbonyl carbon–oxygen bond) that was impossible to confirm experimentally.[5]

A closer look at Phillips's arguments is instructive since it raises several epistemological questions.[2] For example, all of his sets of criteria are confronted with "experimental anomalies": while the first set can't explain the difference between certain sulfoxides and amines, his final set falsely predicts optical isomerism of esters. In the sulfur/nitrogen case, Phillips accepts the experimental findings and changes his set of criteria. In contrast, he eventually decides to keep his set of criteria in spite of experimental counter-evidence and blames the experimental separation methods for "ester isomers" as being insufficient.

In the argument for the removal of the lone electron pair criterion, Phillips also compares two distinctively different causes of optical isomerism: while tetravalent ammonium and carbon compounds have four atoms arranged around a central atom, trivalent sulfur compounds have three atoms and a fourth "ligand" in the form of a lone electron pair (which allows inversion). To conclusively decide whether the lone electron pair criterion is redundant, Phillips would have to compare optically active compounds with optically inactive compounds that differ from the optically active compounds only in this specific criterion. The comparison of trivalent sulfur compounds with tetravalent ammonium and carbon compounds is not sufficient to decide on the lone electron pair criterion for optical activity in trivalent compounds.

The discussion of Phillips's manuscript shows that S-analysis does not have to focus on historic landmark texts. As long as readers have sufficient chemical as well as philosophical knowledge, discussions of specific, specialized chemical texts are possible—and possibly beneficial. Few chemists or philosophers will have touched such articles in the past, and experience with analysis of individual research texts might help us to understand "how chemistry works" in a more general context, that is, how chemists explain, systematize, rationalize, and communicate their results.

Another fruitful area of S-analysis is provided by chemical symbolism. Chemists' use of symbols harbors many epistemological questions. Some of our studies have therefore focused on the rules governing the transformation of chemical symbols, formulas, and reaction equations.[3] Chemical symbolism can be described as model language exhibiting interdependent sets of syntactic and semantic rules. The former are particularly interesting since they determine which formulas and reaction equations are *formally* correct, almost regardless of which compounds are represented and which processes are described. This raises a number of epistemological questions central to chemical explanation.

We have explored a procedure ("Exeter method") that places chemists and philosophers in a multidisciplinary environment to jointly perform S-analysis.[1–3] Clearly defined roles and responsibilities of participants allow them to work together. Philosophers are in charge of logical analysis and determination of possible fallacies. Chemists are responsible for explaining the contents and contexts of statements, for revealing any implicit or tacit knowledge used in arguments, and for proposing alternative statements.

This evaluation of chemical statements pays tribute to the specific character of scientific validity in areas of chemistry such as analytical, synthetic, and biochemistry. Although a strict "mathematical" or logical validity of the scientific conclusions cannot be achieved, the argumentation itself has to follow basic logical rules. The following two examples briefly illustrate the kind of problems chemists face when logical consistency is overlooked.

The Jahn-Teller effect is used to identify energetically unstable molecular geometries and to explain why certain nonlinear molecules have a distorted geometry. In a widely used textbook of inorganic chemistry, this effect is described as follows[6]: "If the ground electronic configuration of a nonlinear molecule is degenerate, then the molecule will distort so as to remove the degeneracy and achieve a lower energy."

This statement is rather interesting for two reasons. First, it implies a distortion process that does not take place since the molecules in question are always found distorted. This is a rather important issue because students frequently misunderstand the definition given above (in a survey of 26 Exeter undergraduate chemistry students, all of them thought a measurable distortion process actually takes place).

Second, the reasoning behind this conditional statement logically reflects a theorem of the propositional calculus, that is, $Q \rightarrow (P \rightarrow Q)$ (with Q "the molecule will distort" and P "the ground electronic configuration of a nonlinear molecule is degenerate"). Interestingly, Q (i.e., the distortion of these molecules) is (empirically) true, therefore $P \to Q$ is true regardless of the truth value of P. It is therefore logically irrelevant if P is true or false, that is, whether a pre-distorted molecule with a degenerate ground electronic configuration exists or not. In contrast, this is a highly relevant issue from a chemical point of view. This description therefore provides an interesting example of a genuine, important entanglement of logical truth and empirical truth that might be significant in chemistry.

The second example supports this notion. The supposed discovery of cold fusion in 1989 contained, among other things, a line of arguments that is highly instructive from a logical point of view.[7] The scientists in question thought they had electrochemically triggered the nuclear fusion of two deuterium nuclei to heavier tritium and helium nuclei, but this did not agree with experimental evidence on enthalpy generation. They therefore concluded that reactions leading to tritium and helium "are only a small part of the overall reaction scheme and that other nuclear processes must be involved."[7]

The logical structure of this argument is rather interesting: Tritium and helium formation (reaction R_1) generates a certain amount of enthalpy H_1 ($R_1 \to H_1$). Experimental measurements of the enthalpy rule out R_1 (**not** H_1, therefore **not** R_1, Modus Tollens). This argument should therefore have been used to reject the idea that R_1 is taking place. Instead, the logical operation of addition is used: Postulating the occurrence of other nuclear processes (R_2) results in the disjunction $R_1 \vee R_2$ ($\vee$ is logical *or*) as an explanation for the unexpected amount of enthalpy generated. This leads to a logical disjunction that is true if either R_1 or R_2 (or both) are true. Since R_1 alone cannot account for the experimental findings and the processes behind R_2 are not defined, the truth of the disjunction and indeed R_2 cannot be inferred logically. In contrast, by assuming the truth of the disjunction and applying the rule of the Disjunctive Syllogism, the *impression* is created that **not** R_1 would support the involvement of R_2.

This is, of course, logically unsound—but provides a questionable (and occasionally used) strategy for coping with experimental counterevidence by postulating the involvement of "exciting, previously unknown" reactions and apparently "proving" their presence experimentally by ruling out the more conventional explanations.

Admittedly, explicit logical analysis of statements is cumbersome and time-consuming in practice, and it might be more advantageous to teach basic philosophical concepts to chemists to enable them to spot and correct "wrong" statements.

Steady increases in the number of undergraduate "Philosophy of Chemistry" courses at universities are encouraging. Our course in Philosophy of Chemistry for second- and third-year undergraduate chemists is highly popular among students (about forty students, that is, two-thirds of all chemistry students, sign up for this module). The course teaches basic epistemological concepts (for example, induction and deduction, context of discovery and context of testing, and verification and falsification); explains and practices S-analysis (analysis of logical structure of sample arguments); looks at chemical symbolism; and discusses ethical aspects of chemical research. Although there is no appropriate textbook for this course yet, recent articles from *Hyle* (especially the chemoethics texts published in the journal's seventh volume) and *Foundations of Chemistry* can easily be used as teaching support literature. Students' exceptional enthusiasm for the course, paired with a steep initial learning curve (at the beginning, students have virtually no idea about most of these issues) bodes well for the future of Philosophy of Chemistry in teaching and research.

REFERENCES

1. JACOB, C. 2002. Philosophy and biochemistry: research at the interface between chemistry and biology. Foundations of Chemistry **4:** 97–125.
2. HARLE, A.E.J. & C. JACOB. 2002. Historical and epistemological aspects of optical isomerism on trivalent sulfur and nitrogen atoms. Ambix **49:** 127–147.
3. JACOB, C. 2001. Analysis and synthesis, interdependent operations in chemical language and practice. Hyle **7:** 31–50.
4. VINCENT, J.B. 2000. Elucidating a biological role for chromium at a molecular level. Acc. Chem. Res. **33:** 503–507.
5. PHILLIPS, H. 1925. Investigations on the dependence of rotary power on chemical constitution. Part XXVII. The optical properties of n-alkyl p-toluene-sulphinates. J. Chem. Soc. **127:** 2552–2587.
6. SHRIVER, D.F., P.W. ATKINS & C.H. LANGFORD. 1994. Inorganic Chemistry. 2nd ed. Oxford University Press. Oxford, England.
7. FLEISCHMANN, M. & S. PONS. 1989. Electrochemically induced nuclear fusion of deuterium. J. Electroanal. Chem. 261: 301–308.

Beyond the Dimensionality of Visualization in Chemistry

ANDRZEJ BUREWICZ AND NIKODEM MIRANOWICZ

Department of Chemistry, Adam Mickiewicz University, Poznan, Poland

ABSTRACT: This article elaborates a key aspect of applied visualization in chemistry: dimensionality. On the one hand, the dimensionality of visualization is restricted by the medium of information transfer; on the other hand, several techniques make it possible to achieve visualizations of parameters of multidimensional objects of chemical research in ways that, in a sense, transcend the limits of the physical space–time continuum. Effectiveness of chemistry research, as well as of chemical education, requires proper perception of parameters of objects being investigated. The appropriate application of visualization techniques can be crucial.

KEYWORDS: visualization; transformation of data; molecular models; dynamic representation; interactivity

INTRODUCTION

In visualizing the structure of a scientific object, a researcher usually presents an image of the object in the four-dimensional physical continuum. However, in visualizing relationships, scientists often reach for additional parameters. Structural visualization, which mostly takes the form of structural formulas or images of models, often can use, quite satisfactorily, fewer dimensions than the original object occupies—but visualization of relations between parameters, generally presented as graphs or diagrams, often needs a larger number of dimensions.[1]

Contemporary tools of data analysis (as well as new methods of conceptual work and rapid development of experimental techniques) have led to considerable unification of the methodologies of the sciences. This can be seen in the modern understanding of interdisciplinarity.[2] Many properties of interest to chemists are not directly measurable, but are accessible only through multiple processing of physicochemical information. Chemistry is a multidimensional science, in more than just a trivial sense. The multidimensional treatment required for subjects of chemical research may well be decisive for the autonomy of chemistry.

Address for correspondence: Department of Chemistry, Adam Mickiewicz University, ul.Grunwaldzka 6, 60-780 Poznan, Poland. Voice: +48 (61) 8291375; fax: +48 (61) 8291375.
burewicz@amu.edu.pl, nmiran@amu.edu.pl

PERCEPTION OF RESEARCH OBJECTS

One of the most significant factors influencing effectiveness in chemistry research, as well as in chemisty education, is proper perception (by the end recipient, whether scientist, teacher, or student) of the parameters of the objects being investigated. The usual subject of research is not a chemical object itself, but rather an image of such an object, subjected to certain transformations. Chemists transform information at several stages of their research work. Final conclusions are often reached on the basis of multiply transformed images of the chemical object. Only adequate processing at intermediate stages—of data and the image itself—can provide an adequate basis for valid conclusions.

DIMENSIONALITY

The objects of chemical investigation have diverse dimensionality. This is not only the consequence of the space–time in which those objects exist, but also to variation in the interests of investigators: information needs to be appropriately presented to be useful to a given researcher. Chemistry deals with the composition, structure, properties, and transformations of substances. Typically, composition is finally represented by one or more molecular formulas: a linear sequence of symbols indicating components and relative amounts. Research information concerning composition usually may be treated as mono-dimensional or two-dimensional, even if these dimensions are sometimes multilayered or multiparametric. In some situations, semistructural formulas are needed to include information that is necessary to describe chemical composition appropriately.

Changes of composition of chemical compounds may be described by sequences of formulas: one-dimensional chemical equations or two-dimensional schemes of mechanisms of chemical reactions. When time, as an additional parameter, is treated linearly, the dimensionality of description increases. Sometimes results are expressed as factors dependent on chemical structure or elements of structure (for example, properties vary with substituent group). Geometric structure is normally represented in three-dimensional space (if dynamic change of the structure is also described, the representation becomes four-dimensional).

Experimental information on properties of a substance (one of the most important aspects of the chemical sciences) is often available in many different numbers of dimensions. Sometimes properties investigated experimentally are represented by single, isolated pieces of information (basic properties such as mass and temperature). In most situations, data consists of sets of values (often very large) that describe changes of properties with time, or with other linear or nonlinear parameters (such as temperature and pH) The number of dimensions involved in accumulating and processing such data can be very large, and can undergo changes at any one of many stages of research activity (including any aspect of scientific analysis, synthesis, or evaluation). The details of such transformations depend on many factors, including the level of inference that the problem requires.

VISUALIZATION

All chemical reactions, and other processes of interest to a chemist, are treated within four-dimensional space-time. Nonetheless, to perceive experimental information effectively and to make valid research conclusions, transformation of that information is required. Transformation of research parameters so as to facilitate generation of images is designated as "visualization." Adequate visualization is defined as dimensional transformation of data concerning a research subject into appropriate numbers, values, and images, that allow an investigator to reach valid conclusions.[3] This involves simplifying a complex situation, and mostly has to do with locating all information that is necessary for effective inference within a single image. For the chemist, visualization has an especially "ontological" aspect. When chemists succeed in visualizing an object, they often consider that things such as that object "really exist."[4]

Numbers are essential in creating a final image and are effective intermediaries between measurements made on chemical objects and their expression as "values" (values are often represented by numbers, but can be also coded into color, shape, mesh, distances, etc.).[5] Thus, numbers are a record of quantities that are transformed into visual information in a later stage of processing. Numbers can be coded, transformed, simplified, and subjected to initial processes of inference. Only small sets of numbers can be effectively (or easily) subjected to analysis by a researcher. It is decidedly more effective to transform numbers into charts, graphs, and so forth, that express relationships. (Imagine how inconvenient it would be to analyze spectral data as tables of numbers rather than in a spectrogram.) Symbols, drawings, structures, and models are the basic code of the "language of chemistry," much more basic than numbers.[6]

Frequently, visualization can be automated to a considerable degree, but a proper decision concerning suitable methods for data processing is essential. Typically, chemical researchers collect parameters of a reaction or process and apply appropriate processing to them, resulting in visualizations such as spectrograms, curves, structures, and models. This applies to many research areas, from examination of the properties of chemical compounds to research on the effectiveness of educational programs. In a subsequent stage, scientific work demands data validation and evaluation, as well as analysis in terms of hypotheses. A similar transition occurs when research data goes outside numerical parameter ranges that are comfortable and familiar (and within which measurement is known to be precise). At these higher levels of creative work, the chemist sometimes needs to move beyond established stereotypes of scientific practice; that is, automation and utilization of technical tools may become difficult or impossible. Still, even at these levels of research, visualization, if it still can be applied, may well be valuable in generating creative ideas. Spectacular examples of such illumination, manifested in original intellectual visualizations, are well known from the history of science.[7]

DEALING WITH DIMENSIONALITY

Another level at which the dimensionality of visualization may be described is the level of notation. Since the late medieval period, paper has been basic medium for recording and transferring scientific information. This medium, like others, sets its own conditions of application.[8] Paper is convenient, precise, durable, and versatile; nonetheless, it remains two-dimensional. Usually, multidimensional data must be transformed into two-dimensional form. There are many methods of reducing the dimensionality of information. Most simply, such reduction depends on simplifying and shortening information by cutting off selected parameters. Words are well suited to this function. An oral record is one-dimensional in physical form, but capable of describing various space–time contexts. Chemist have developed many conventional means of representing multidimensional objects. Chemical notation is quite highly developed

Visualization by reduction of dimensions is called "illustration." Illustration techniques are often used to deal with macroscopic reality (laboratory descriptions, procedures, etc.). Since reduction alone does not always yield satisfactory two-dimensional illustrations, many inventions (static interposition, constancy of size, shapes, lighting-shading effects, aerial perspective, graduated textures, and matting and gradation of color[9]) have been developed to help.

Elaboration of specific chemical notation and use of technical representation methods are generally quite sufficient for most applications, but such approaches have limitations. A major limitation concerns registration of the fourth dimension—time. Notation of mechanisms of chemical reaction, transformations of structures, vibratory states, and so forth, requires the use of additional graphic effects. The development of graphic techniques, partly through use of photographs but mainly by use of computers, allows representation of four-dimensional physico-chemical parameters on a two-dimensional medium. Use of dynamic two-dimensional media such as CRT and LCD screens (and even three-dimensional media such as holographic projectors) gives an even broader range of possibilities. Those methods are generally concerned with illustration and representation of images of real objects. When notation, registration, and illustration contain many more than four parameters, it is necessary to expand these techniques, or to develop a totally different approach.

Dynamic notation (a three-dimensional notation comprising two spatial dimensions and the dimension of time), provided primarily by means of computer technology, allows for the recording of data in a way that is enriched by one dimension. The dynamic aspects of presentation are mostly used in showing changes of structures in time, but also for the introduction of parameters that do not have a dynamic nature in their original form.[10] The first case is represented by the introduction of dynamic models of chemical reaction mechanisms, where the dimensionality of models is shown by graphic means, while the dynamism of the reaction is shown by means of a temporal framework. In another example, the parameters of processes (or chemical structures) undergoing change over time are presented in graphs. The dynamics of time-dependent parameters are reflected in the variation of the image of the graph with time. The dynamic change of the form of the graph may also refer to the variability of an additional reaction parameter, one that was originally not a temporal time parameter, such as temperature, pressure, or pH. FIGURE 1 shows examples of multidimensional description of information (aspects of the mechanism of the Hofmann rearrangement)[11] that originally had different dimensionality.

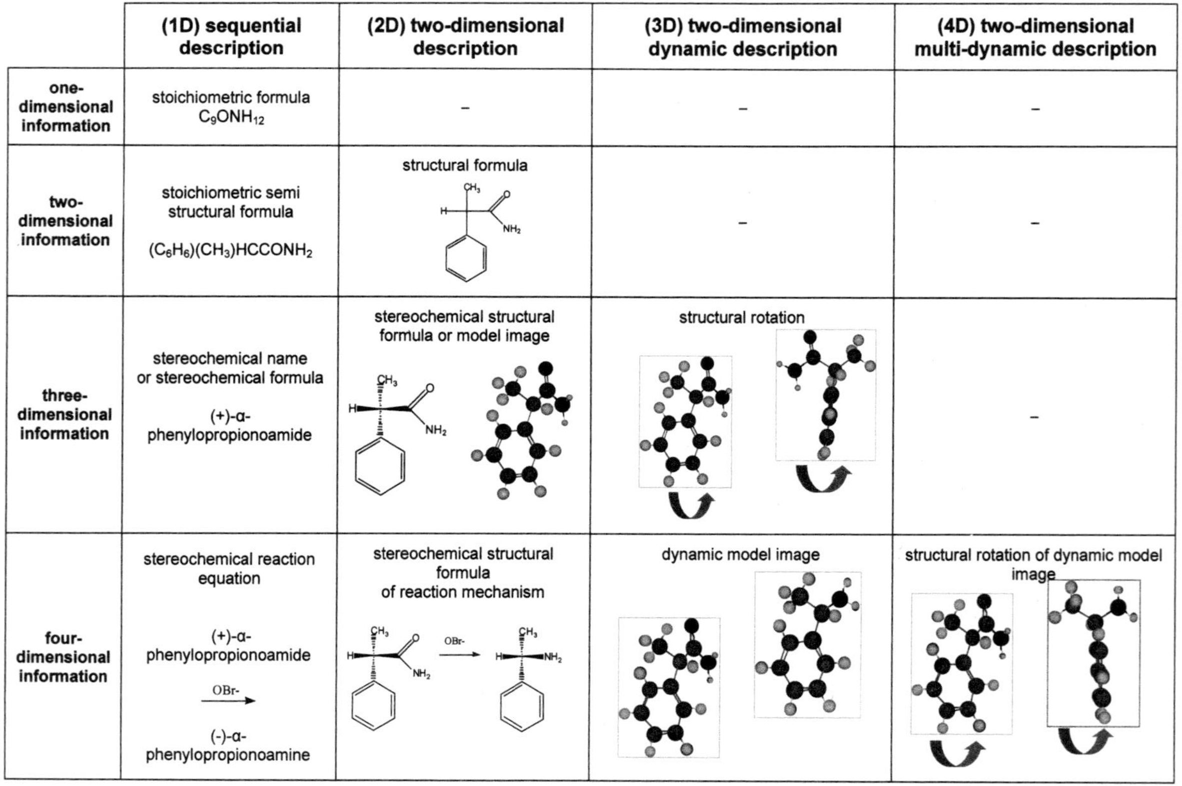

	(1D) sequential description	(2D) two-dimensional description	(3D) two-dimensional dynamic description	(4D) two-dimensional multi-dynamic description
one-dimensional information	stoichiometric formula C_9ONH_{12}	–	–	–
two-dimensional information	stoichiometric semi structural formula $(C_6H_6)(CH_3)HCCONH_2$	structural formula	–	–
three-dimensional information	stereochemical name or stereochemical formula (+)-α-phenylopropionoamide	stereochemical structural formula or model image	structural rotation	–
four-dimensional information	stereochemical reaction equation (+)-α-phenylopropionoamide $\xrightarrow{OBr^-}$ (−)-α-phenylopropionoamine	stereochemical structural formula of reaction mechanism	dynamic model image	structural rotation of dynamic model image

FIGURE 1

INTERACTIVE PRESENTATIONS

Use of dynamic two-dimensional information carriers (such as computer screens) disturbed the stability of a long-established notation system based on representations on paper, but it also provided researchers and educators excellent tools for multidimensional recording of physico-chemical data, and supported new types of multidimensional visualization that facilitate conceptual analyses and syntheses. For instance, stereoscopic visualization allows perception of properties of an object that are invisible in a two-dimensional visualization. An even more significant advance is that dynamic data carriers are frequently also interactive. In those cases, it is possible for the user to manipulate the temporal parameters of the visualization as the process unfolds. This feature allows researchers, teachers, and students to intervene directly and immediately in the representation of scientific objects, and facilitates discovery and understanding in new and profound ways.

CAVEAT PICTOR

Dynamic techniques of visualization still operate in no more than three dimensions (two spatial dimensions of the screen plus one temporal dimension). Paper continues to be a convenient carrier of scientific information. Researchers and educators are sometimes more confused than helped by the use (by themselves or others) of multidimensional techniques. Unfortunately, the distinct purposes of these (old and new) methods of information transfer are often overlooked, their functions treated interchangeably, and hence ineffectively.[12] In this context, the advice that scholars, educators, and scientists should employ technical means only in the narrow range of their best applicability appears well founded.

REFERENCES

1. BUREWICZ, A. & N. MIRANOWICZ. 2002. Categorization of visualization tools in aspects of chemical research and education. Int. J. Quantum Chem. **88:** 549–563.
2. WEINGART, P. & N. STEHR. 2000. Practicing Interdisciplinarity. University of Toronto Press. Toronto.
3. BLOOM, B. 1956. A Taxonomy of Educational Objectives. McKay. New York.
4. MIRANOWICZ, N. 2000. Categories of Chemical Visualization. Fourth Summer Symposium on the Philosophy of Chemistry and Biochemistry. August 7–10, 2000. Poznan, Poland.
5. DEHAENE, S. 1997. The Number Sense. Oxford University Press. New York.
6. JANICH, P. & N. PSARROS, Eds. 1995. Die Sprache der Chemie. Koenigshausen und Neumann. Würzburg.
7. Among well-known examples are Kekule's snake, Newton's apple, and Archimedes' bath.
8. MIRANOWICZ, N. & A. BUREWICZ. 1996. Stages in the construction of stereograms of molecular models. J. Mol. Graphics **14:** 73.
9. BRODLIE, K.W., et al., Eds. 1992. Scientific Visualization, Techniques and Applications. Springer. Berlin.
10. MCLUHAN, M. 1965. The Gutenberg Galaxy: The Making of Typographic Man. University of Toronto Press. Buffalo.
11. MORRISON, R. & R. BOYD. 1973. Organic Chemistry. Allyn & Bacon. Boston.
12. BOWIE, J. 1995. Data Visualization in Molecular Science: Tools for Insight and Innovation. Addison-Wesley. New York.

Justifying Instrumental Techniques of Analytical Chemistry

DANIEL ROTHBART[a] AND LADISLAV KOHOUT[b]

[a]*Department of Philosophy and Religious Studies, George Mason University, Fairfax, Virginia 22030, USA*

[b]*Department of Computer Science, Florida State University, Tallahassee, Florida 23206, USA*

> ABSTRACT: In this paper, we argue that the foundations of chemistry rely as much on the methods of measurement as they do on categories of chemical substance. To some degree, chemists perform the work of knowledge engineering: designing complex systems for the efficient retrieval of information. Indeed, in some cases, methods of instrumental detection move to the forefront of attention. For example, researchers are expected to deploy optimization methods designed to maximize desired signal and minimize the damaging effects of noise. But in his important contributions to the development of high-resolution NMR spectrometers, Hans Primas used stochastic methods to reveal *beneficial* effects of noise for characterizing physical systems, demonstrating the value of noisy signals for nonlinear physical systems in chemistry.
>
> KEYWORDS: instrumentation; noise; information; analytical chemistry

Chemistry is sometimes defined as the science of "chemical" reaction of one assembly of atoms to yield another. On the basis of this definition, foundations of chemistry are established by marking a portion of the world between the realms of biology and physics. We argue that the foundations of chemistry depend as much on measurement techniques as they do on subject matter. The design of an information system has profound implications for the ontology of chemical substance. Knowing that data are valid is inseparable from knowing how data-gathering instruments operate. Insight into the experimenter's relationship to the microscopic realm is gained by exploring the designs of such mediating technologies. The old distinction between pure science, as discovery of a given-reality, and technology, as deliberate production of artifacts, collapses for contemporary research. Science discovers because it produces artifacts.[1]

Address for correspondence: Dr. Daniuel Rothbart, Department of Philosophy and Religious Studies, MSN: 3F1, George Mason Unviersity, Fairfax, VA 22030. Voice: 703-993-1293; fax: 703-993-1297.

drothbar@gmu.edu

KNOWLEDGE ENGINEERING

According to Fritz Paneth, who was an eminent experimental chemist and co-founder of radiochemistry,[2] the foundations of this science are underpinned by commitments to basic substance at the level of atoms and molecules. A basic substance comprises the indestructible elements present in chemical compounds, devoid of sensible properties but endowed with capacities to produce an infinite number of possible attributes.[3] Absent from Paneth's conception of chemistry is any reference to experimental technique.

In a resurgence of interest in experimentation, many scholars today place instrumentation at the forefront of scientific practice.[4] During an experiment, nature is put on trial and forced to behave in a new way, so that its hidden properties may be discovered. Reliability of experimental findings is inseparable from the technology deployed to produce new states. The notion of $techn\bar{e}$, referring to the art and knowledge of making something through the skillful manipulation of material, must be given a central place in the philosophy of chemistry. Far from existing in opposition to ontology, $techn\bar{e}$ relies on ontological commitments concerning the portions of the world that are commissioned for production. Since instrumental techniques have invaded the relationship between researchers and the assembly of atoms, the pragmatics of technical efficacy must be given serious attention in the foundations of the sciences.

In analytical chemistry, the limits to research are continually being pushed back, enlarging the scientific world, challenging common sense concepts with seemingly "bizarre" discoveries. As instrumental techniques improve, the objects examined with instruments become increasingly distant from us: distant in space and in time; distant, ultimately, because of their strangeness to laws of the workaday, macroscopic world in which we live. Of course, the world of chemistry is becoming increasingly strange, challenging our common sense ideas through the production of bizarre substances. The pharmaceutical industry, for example, will soon be producing artificial enzymes, which retain many of the essential properties of self-assembly found in natural enzymes.

Chemists contribute significantly to standards of research through the design of intelligence systems. In certain respects, chemists perform the work of knowledge engineers—agents who develop, or improve upon, techniques for extracting information from the environment. The production of artifacts through the manipulation of a sample is essential to such techniques. As knowledge engineers, chemists determine what must be done to generate valid signals, what kinds of interference should be used to probe a sample's properties, or what kinds of results should be expected.

Foreknowledge of the strengths and weakness of such material is pivotal for designing an efficient information system, as knowledge engineers move back and forth from past results to future possibilities. When severe design problems threaten the advancement of instrumental techniques, research chemists often turn to an ostensibly "remote" subject matter, drawing upon knowledge that appears to be irrelevant in relation to the familiar disciplines under consideration. Analogies to familiar instruments can, in some cases, offer valuable insight, leading to substantial breakthroughs in design techniques. The design of NMR spectroscopy by Hans Primas is presented in the next section of this paper as a case in point.

> Signal generator ➜ Transmission channel ➜ Readout

FIGURE 1. Simple schematic of an information system.

Analytical chemists get no closer to the properties of a compound than a sequence of signals, produced from a dynamic unity of energy and the specimen's properties. We think of a desirable signal as a variable of information, constrained by features of an information system.

As Pierre Laszlo argues, these electrical ghosts (in the form of signals) stand for the real material samples; a molecule's properties are detected as the spectra of signals are read and understood.[5] The number and type of molecules in a sample, for example, are transformed via instrumentation into the signal's peaks and wavelength location. The operational principles associated with a particular instrumental technique determine the signal's validity. Other elements of the laboratory environment play an active role in the experimental process. A proposed solution path may be inadequate compared to the goal, necessitating new decisions and actions. In general, feedback procedures must be repeated until the information content has reached the level at which the optimum solution can be found (FIG. 1).[6]

But the damaging effects of noise are never completely absent from the scene. Accuracy of data is never determined in isolation from undesirable interference. Every measurement has certain ineliminable levels of noise associated with it.[7] Whenever analytical instruments are used, researchers are pressed to identify threats to the system's integrity. In practice, there is no possibility of detecting an analytical signal in an absolute and value-neutral sense. In order to separate desirable signal from a noisy one, we have to examine how noise will affect the detection process. In any experiment, some experimental phenomena carry information about a specimen's properties, but other phenomena can contaminate the results. A desirable signal is an information-carrying variable that has its source in a specimen's properties, based on a specimen's physical response to an instrument's probing manipulations. A noisy signal does not carry information about the specimen, but carries extraneous information based on some random influence.[8] A noisy signal, by definition, has a contaminating effect on the accuracy and precision of an analytic signal by negatively influencing the evolution of the signal in the instrument.[9]

The integrity of an analytical signal can be corrupted by the electrical wiring circulating the walls of the laboratory and building in which the research is performed. The environmental interference of radiation can be reduced by blocking, grounding, or decreasing the lengths of conductors used in experiments. One technique involves electrostatic shielding, in which some of the wires in the instrument are surrounded by a conducting material. Shielding is designed to minimize the effects of uninvited intrusion that arise from the instrument circuitry. Electromagnetic radiation is absorbed by the shield rather than by the enclosed conductors. The conductors are wrapped with aluminum or copper foil, with a drain wire in electrical contact with the foil, providing almost perfect protection. But other sources of noise cannot be eliminated by external protection. Thermal noise, for example, is interfering voltage having its origin in thermally induced motion of charge carriers.[10]

The physical sciences alone cannot account for the pivotal distinction between desirable signals and noisy ones. From the perspective of electronic circuitry, there is no difference between a desirable signal and noise; both are explained as variations of voltage with time; we might say that they are both signals. Under the best circumstances, the total noise that is identified can be minimized and the analytical signal maximized; one can never measure an analytical signal in an absolute sense. Furthermore, a signal's validity rests on factors that go beyond electronic circuitry. The production of a signal requires a solution to an optimization problem, centering on attempts to maximize desired signals and minimize the effects from noise. Such attempts, in turn, require hypothesis-testing to determine the causal source of the progression/interference, towards the goal of maximizing the desirable signal and minimizing the effects of noise.[11]

NOISE IN NMR SPECTROMETERS

For most instrumental techniques, researchers are expected to deploy optimization methods designed to maximize the desired signal and minimize the damaging effects of noise. But in his contributions to the development of high-resolution NMR spectrometers, Hans Primas demonstrated the value of noise to the system's efficiency, providing information about a specimen's dynamical properties.

NMR spectroscopy is based upon the measurement of absorption of electromagnetic radiation in the radio-frequency region of roughly 4 to 600 MHz. Isidor Rabi received a Nobel Prize in 1944 for his resonance method for recording the magnetic properties of atomic nuclei. In 1946, Felix Bloch at Stanford University and Edward M. Purcell at Harvard University independently observed that nuclei absorb electromagnetic radiation in a strong magnetic field as a consequence of the energy level splitting induced by the magnetic field. Yet at the time of Bloch's and Purcel''s first experiments, chemists found little use for NMR in their research studies.[12]

The situation gradually changed after 1950. Proctor and Yue[13] published a paper that claimed the "dependence of a nuclear magnetic resonance frequency upon chemical compound" and Dickinson found dependence of the F^{19} nuclear resonance position on chemical compound. These discoveries of chemical shifts motivated other chemists to investigate the efficacy of this technique for identifying properties of chemical compounds. The first attempts to commercialize NMR were made by Varian Associates, which sold the first commercial NMR spectrometers with a proton resonance frequency of 40 MHz using an electromagnet with field strengths of about 1 Tesla (10 kG).

Before 1953, Primas knew virtually nothing about nuclear magnetic resonance, and he had never before designed a complex electronic device. He not only learned "all the craft" necessary for his highly successful designs, but also enriched the field by creative transfer of knowledge from the Information Theory and System Science field to the field of NMR Spectroscopy.[15] In 1953 Hans Primas began working on the redesign of NMR spectrometers with Heinrich Günthard at the Laboratorium für Organische Chemie at ETH Zürich. They faced the difficult problem of producing a stable and homogeneous magnetic field. Primas and Günthard decided to use a permanent magnet. Homogeneity of the magnetic field of the permanent magnet played an important role for acquiring high resolution of the signal. But the inherent difficulties of achieving optimal homogeneity and minimal line width prompted Primas to study line shapes in inhomogeneous and time-dependent magnetic fields. Time dependency involves the dynamics of a physical system, and this, coupled with inhomogeneity, contributes to uncertainty in the system. In order to deal with these problems, Primas used the linear response theory coupled with stochastic signals perturbing the physical system, following the Wiener-Khintchine theorem. Wiener and Khintchine proved that the correlation function of a stationary stochastic process can be expressed in the form of integral

$$B(\tau) = \frac{1}{4\pi} \int_{-\infty}^{\infty} F(\omega)e^{i\omega\tau}d\omega = \frac{1}{2\pi} \int_{0}^{\infty} F(\omega)\cos\omega\tau d\omega$$

where $F(\omega)$ is bounded function $F(\omega) \geq 0$]. Furthermore, every function $B(\tau)$ which can be expressed by the above integral is a correlation function of some stochastic stationary process. $F(\omega)$ has a clear physical meaning: it is the spectral density of the mean power of the stochastic process.

Intuitively speaking, a correlation function relates the values of a signal at different times, so that the correlation describes a process in a time domain. But the function $F(\omega)$, which represents the spectrum of a signal, is the mathematical result of a Fourier transform of $B(\tau)$. $F(\omega)$ describes the signal in the frequency domain, and gives just the values of distribution of the energy of a signal or process on the frequencies of the elementary harmonic components.

Because the NMR signals were invariably weak and noisy, Primas was intrigued by the stochastic processes associated with electronic measurements and data transmission. In an important breakthrough in NMR spectroscopy, he discovered that the detection of noise could have beneficial effects on physical systems. Transferring certain techniques from the conventional domain of telecommunications and control engineering to an entirely different domain of chemistry, Primas established new applications for physical chemistry. He discovered the great value of random noise for identifying nonlinear physical characteristics of system in chemistry. Although this process of noise detection is used to separate signals from noise, the use of the Wiener-Khintchine theorem to describe an atomic system as a stochastic process was highly innovative. From the perspective of this theorem's application, not only the signal, but also the noise provides information about a specimen's properties. Indeed, Primas provided the first description of the response of a nuclear spin system to a stochastic perturbance based on the filtration of signals from noise.

The important contributions that Primas made to NMR spectroscopy can be summarized as follows[15]:

1. He discovered a new method for the direct calculation of the spectrum of the radiation absorbed or emitted by a quantum mechanical system.[16]

2. The second contribution extends the perturbation treatment used above by applying a systematic operator notation without every having to use a specific matrix representation.

3. For the first time, Primas utilized superoperators in a systematic way.[17]

4. Through the use of stochastic processes, noise could itself be used in a beneficial manner for characterizing the physical system.[18]

Primas was long intrigued by the work of Norbert Wiener, Claude Shannon, Andrei Kolmogorov, A.I. Khintchine, and others on stochastic processes in the context of electronic measurement and data transmission. The work of these major figures in stochastic system theory, information theory, and cybernetics greatly advanced the state of knowledge about ways in which signals can be filtered from noise. Primas relied heavily on optimized data processing and filtering theory to better understand the character of weak and noisy signals in NMR spectrometers. Random noise had been used to characterize technological artifacts. Primas' original idea was that the same apparatus can be used in testing physical quantum mechanical systems. Primas provided an account of the response of a nuclear spin system to a stochastic perturbation for the first time. Through the creative use of knowledge from separate domains, he relied on linear systems response theory, Fourier and Laplace transformation theory, and the associated frequency domain methods for determining system response and stability.[19] In 1963, he left the field of NMR spectroscopy to pursue theoretical research in quantum chemistry. Ernst, one of his students and collaborators, fittingly describes Primas' achievements in the field of NMR spectroscopy in the ten years between 1953 and 1963:

> [Hans Primas] achieved in ten years more than other successful scientists create during a life time. He has fertilized NMR instrumentation in an essential time of its development and he has laid some of the founding cornerstones of modern NMR methodology. Sometimes it is difficult for an outsider to understand why a scientist leaves a field of research in which he has been exceptionally successful. It is gratifying to know that one of the great theoreticians experienced such a wide range of aspects ... [such as] the practical chemist and ... the electronic instrument designer, inventor mastering engineering mathematics, ... finally leading to the activities for which Hans Primas is mostly known today.[20]

SETTING STANDARDS THROUGH INSTRUMENTAL DESIGN

If instrumental technology lies at the core of laboratory studies of the microworld, then we look to the designs of such devices for idealized standards for research. Design plans can be read as epistemic maps for technologically mediated research, revealing assumptions about the standards of inquiry. Each design plan conveys some understanding of an instrument's power, an experimenter's skills, and a specimen's capacities. In the interrelation between ontology and technology, a specimen itself functions as an experimental tool, known for its capacities to produce, create, and generate detectable states under various manipulations. Sometimes it functions as an agent for change; other times as a reagent under the influence of other forces. At all times, a specimen is endowed with capacities for change, as if it were an artifact of engineering. In an attempt to compensate for the widening "distance" between experimenter and specimen, physical chemists assume that the portion of the world under investigation is endowed with the same kinds of capacities that are attributed to their own creations. When analytical instruments are used in research, a specimen operates as a signal-generating agent, a source of information that is detected through technological inducements. A specimen is exploited for its signal-producing capacities, resulting in desirable signals and noisy ones. Noise is not always harmful to the system, as illustrated above from the innovative techniques discovered by Primas for increasing the stability and improving the homogeneity of the magnetic field, techniques that contributed to important advances in NMR spectroscopy.

ACKNOWLEDGMENTS

We would like to thank the two anonymous referees for their valuable observations and constructive comments on an earlier draft of this paper.

REFERENCES

1. LELAS, S. 1993. Science as technology. Br. J. Phil. Sci. **44:** 423–442.
2. PANETH, F.A. 1965. Chemical elements and primordial matter: Mendeleeff's view and the present position. *In* Chemistry and Beyond. H. Dingle & G.R. Martin, Eds.: 53–72. Wiley. New York. p. 66.
3. PANETH, F.A. 1962. The epistemological status of the chemical conception of element. Br. J. Phil. Sci. **XIII:** 1–14, 144–160. p. 151.
4. AGAZZI, E. 1999. From technique to technology: the role of modern science. Techné **4(2):** 1–6; BAIRD, D. 1991. Baird Associates' commercial three-meter grating spectrograph and the transformation of analytical chemistry. Rittenhouse **5:** 65–80;. BAIRD, D. 1998. Scientific instrument making, epistemology, and the conflict between gift and commodity economies. Techné **4(3–4):** 1–16; LELAS, S. 1993. "Science as technology"; QUERALTÓ, R. 1999. Technology as new condition of the possibility of scientific knowledge. Techné **4(2):** 11–18; SCHUMMER, J. 1997. Challenging standard distinctions between science and technology: the case of preparative chemistry. Hyle **3:** 81–94.
5. LASLO, P. 1998. Chemical analysis as dematerialization. Hyle **4:** 29–38
6. KOHOUT, L.J. 1990. Perspectives on Intelligent Systems: A Framework for Analysis and Design. Chapman & Hall. London.
7. MALMSTADT, H.W. & C.G. ENKE. 1963. Electronics for Scientists. W.A. Benjamin, Inc. New York. p. 193.
8. COOR, T. 1968. Signal to noise optimization in chemistry—part one. J. Chem. Ed. **45(7):** A533.
9. SKOOG, D.A. & J.J. LEARY. 1992. Principles of Instrumental Analysis. Harcourt Brace Jovanovich. Fort Worth, TX. p. 46.
10. MALMSTADT, H.W. & C.G. ENKE. 1963. Electronics for Scientists, p. 193.
11. ROTHBART, D. 2000. Substance and function in chemical research. *In* Of Minds and Molecules: New Philosophical Perspectives on Chemistry. N. Bhushan & S. Rosenfeld, Eds.: 75–89. Oxford University Press. Oxford, UK.
12. SKOOG, D.A. & J.J. LEARY. 1992. Principles of Instrumental Analysis, p. 310
13. PROCTOR, W.G. & F.C. YUE. 1950. The dependence of a nuclear magnetic resonance frequency upon chemical compound. Phys. Rev. **77:** 717.
14. ERNST. 1999. Hans Primas and nuclear magnetic resonance. *In* On Quanta, Mind, and Matter: Hans Primas in Context. H. Atmanspacher, A. Amann, and U. Müller-Herold, Eds.: 10–38. Kluwer Academic Publishers. Boston.
15. Ibid.
16. PRIMAS, H. & HS.H. GÜNTHARD. 1958. Eine Methode zur direkten Berechnung des Spektrums der von quantenmechanischen Systemen absorbierten bzw. Helv. Physc. Acta **31:** 413–434; PRIMAS, H. 1959. A new method for analyzing spectra in high resolution NMR spectroscopy. *In* Proceedings of the Conference of Molecular Spectroscopy. R. Thornton & H.W. Thompson, Eds.: 19–25. Pergamon Press. London.
17. PRIMAS, H. 1961. Eine verallgemeinerts Störungstheories für quantenmechanische Mehrteilchenprobleme. Helv. Physc. Acta 34: 331–351.
18. PRIMAS, H. 1961. Ueber quantenmechanische system mit einem stochastischen Hamiltonoperator. Helv. Physc. Acta **34:** 36–57.
19. ERNST, Hans Primas, p. 17.
20. ERNST, Hans Primas, p. 35.

Negotiated Identities of Chemical Instrumentation

The Case of Nuclear Magnetic Resonance Spectroscopy, 1956–1969

JODY A. ROBERTS

Science & Technology Studies, Virginia Polytechnic Institute and State University, Blacksburg, Virginia 24061, USA

ABSTRACT: What is an NMR spectrometer? Beginning with this seemingly simple question, I will explore the development of nuclear magnetic resonance spectroscopy between the years 1956 and 1969 from two vantage points: the organic chemists who used the new instrument, and Varian Associates—the makers of the first NMR spectrometers. Through an examination of the articles and advertisements published in the *Journal of Organic Chemistry*, I will draw two conclusions. First, organic chemists and Varian Associates (along with other actors) are co-responsible for the development of nuclear magnetic resonance spectroscopy (i.e., NMR spectroscopy was not created by a single actor). Second, by changing the way NMR spectrometers are used, organic chemists attempted to change to the identity of the instrument. Similarly, when Varian Associates advertised their NMR spectrometers in a different way, they, too, attempted to change the identity of the instrument.

KEYWORDS: nuclear magnetic resonance spectroscopy; chemical instrumentation; chemical technology; science advertisements; Varien Associates

INTRODUCTION

What is an NMR spectrometer? In this paper, I explore the development of an answer to this question between the years 1956 and 1969. Specifically, I will be looking at the introduction of NMR spectroscopy into the field of organic chemistry. I will examine the relationship established between organic chemists and Varian Associates—the first makers of NMR spectrometers—to understand how the development of NMR spectroscopy was a cooperative effort. I will then address questions concerning the identity of the instrument. I will make the point that changes in the use of the instrument by organic chemists and changes in the advertising scheme of Varian Associates changed the answer to the question, "What is an NMR spectrometer?"

I have focused my attention here on the abstracts and advertisements from the *Journal of Organic Chemistry* (*JOC*) that reference NMR spectroscopy between

---

Address for correspondence: Jody A. Roberts, Science & Technology Studies, Virginia Polytechnic Institute and State University, Blacksburg, VA 24061. Voice: 540-231-7879; fax: 540-231-6367.

jody@vt.edu

1956 and 1969. Within the abstracts, I have traced the uses of the instrument in each of these instances. By examining the abstracts from the *JOC*, I provide a glimpse of what knowledge NMR made available to organic chemists and how this changed during the course of this period. By looking at the advertisements placed in the journal by Varian Associates, I elucidate what the company wanted an NMR spectrometer to be—that is, how they wanted it to be used—and how they contributed to the transformation of the instrument.

ABSTRACTS FROM THE *JOC*: USING NMR SPECTROMETERS

The question "what is an NMR spectrometer?" invariably entangles itself with the question "what is an NMR spectrometer used for?" In this section, I will trace the relation between these two questions from the perspective of the organic chemists through their publications in the *Journal of Organic Chemistry* (*JOC*) between the years 1956 and 1969. I have divided the chemists' use of NMR spectrometers into five categories: structure analysis and determination; the expansion of NMR studies to elements other than H^1; making NMR a quantitative tool; the application of NMR spectroscopy to stereochemical studies; and the ability to systematically produce spectra.

Structure Analysis and Determination

In the early years, NMR spectrometers were used primarily for structural analysis and determination. The first study to utilize NMR spectroscopy and to be published in the *JOC*, entitled "9,10-Dihydro-9,10-methanoanthracene," appeared in 1956.[1] The researchers conducting this study—Wyman R. Vaughan and Masao Yoshimine from the University of Michigan—used NMR spectroscopy to confirm the structure of a newly synthesized compound. Other studies conducted that year included the use of an NMR spectrometer to confirm the structure of an organolithium compound[2] and Feist's acid[3] and an examination of the nuclear magnetic resonance spectrum of helvolic acid.[4] Understanding what organic chemists analyzed in their labs is as important as understanding how they analyzed it. As we see from the few examples mentioned above, the primary use of NMR spectroscopy in the early years was relatively straightforward. Early NMR spectroscopic studies focused on the analysis of newly synthesized organic compounds and the confirmation of structure for previously known compounds. But within one year, organic chemists expanded the use of NMR spectroscopy to analyze steroids,[5] deuterated compounds,[6] and natural products.[7] This expansion of use was important because it demonstrated the ability of NMR spectroscopy to be used on large, complex molecules. In addition to expanding the use of NMR spectroscopy to larger and more complicated molecules, organic chemists began applying NMR spectroscopy to the study of other atoms within organic compounds.

Beyond 1H NMR: Expanding NMR to Other Elements

For NMR spectroscopy to become a truly indispensable tool for organic chemistry, it needed to expand access to knowledge beyond routine proton analysis. As the desire to analyze a broader range of molecules increased, so did the capabilities of the NMR spectrometers.

In 1964, there were at least five articles in the *JOC* devoted to the application of NMR spectroscopy beyond traditional 1H analysis—utilizing both 19F and 31P resonance frequencies for structural investigations.[8]

The expanded capabilities of Varian's spectrometers—such as the addition of new radiofrequency controllers—allowed organic chemists to broaden the research application of NMR instrumentation. Previously "off-limit" structures were now open to investigation. This expansion of application had tremendous benefits for chemists working in fields related to medicinal chemistry, natural product synthesis, and other biologically active compounds. For instance, chemists involved in the synthesis and analysis of steroids used 19F investigations in addition to 1H studies to understand the conformation and three-dimensional structures of these highly complex molecules.[9] The development of N15 isotope "tracers" allowed chemists to follow conformational changes in complex molecules.[10] The willingness of chemists to apply NMR spectroscopy to new analytical problems prompted Varian to accommodate chemists with new instruments—with more powerful magnets, time-averaging computers, and new radiofrequency controllers—while keeping the old instruments from slipping into obsolescence.

NMR as a Quantitative Tool

As a quantitative instrument, NMR spectrometers offered chemists expanded access to information about sample mixtures and increased problem-solving capabilities through the ability to calculate the percent composition of a sample.

In 1960, Varian Associates ran a series of ads presenting chemists with the quantitative aspect of their NMR spectrometers. Varian ran three ads that year in the *JOC*, and all of them discussed this capability. The following year reflected the success of Varian's advertising campaign. In an article by Robert Filler and Saiyid M. Naqvi,[11] the authors described the use of a Varian NMR spectrometer to determine the quantity of a specific compound within a mixture. By demonstrating the instrument's ability to analyze for a specific compound within a mixture, the authors introduced a powerful tool to organic chemistry. Because the instrument could be manipulated to distinguish between multiple compounds, the chemist was freed (in some ways) from having to separate compounds prior to analysis. This was important since separation methods are often long and tedious—and in the worst case they destroyed the original sample. Removing this step allowed for faster and more diversified analyses.

In addition to content analysis, chemists also attempted to use NMR spectrometers for additional types of analyses, such as molecular weight determination.[12] However, the bulk of the chemists' quantitative uses for NMR centered on the analysis of chemical mixtures, and the determination of the chemical constituents within the mixture. This tool proved particularly useful when dealing with mixtures of stereoisomers and assisted in the qualitative structural analysis of these chemicals.

The Complex World of Stereochemistry

As early as 1961, organic chemists used NMR spectrometers to assign conformational structures and for the analysis of isomeric mixtures.[13] The ability to use NMR spectrometers to investigate the stereochemistry of a molecule facilitated the application of this technology to fields within (and bordering on) the bounds of organic chemistry.

Chemists have long sought naturally occurring chemicals with biological activity. Because the shape of a molecule could determine whether or not a chemical will produce certain effects (such as particular biological activities), chemists applied NMR spectrometers more frequently to fields such as natural product synthesis, biochemistry, and steroid synthesis. For this reason, the need for conformational—and not just molecular—information related to these products was imperative. There were no fewer than 90 articles devoted to the topics of alkaloid and steroid stereochemistry published in the *Journal of Organic Chemistry* between the years 1956 and 1969. As the capabilities of the instruments changed, so too did the analysis of these complex molecules. Chemists began to incorporate new features of the spectrometers into their studies. Increased use of 19F and 31P analyses allowed chemists to increase the scope of their investigations into a broader range of natural products and biologically active molecules. The expansive spectrum of chemicals to which NMR was now being applied demanded an instrument capable of producing spectra more efficiently.

From One Sample to One Hundred

As NMR spectrometers became more powerful, the speed of analysis also increased. The time for analysis decreased from nearly a day to only minutes. This increase in efficiency led chemists to broaden their investigations from single samples to entire families of compounds. Gone (for the most part) were the days of articles reporting on the NMR spectrum of a single compound. Instead, entire tables of data were presented. Rapoport and Bordner's 1964 article[14] lists NMR spectra for seven compounds. An article by Subramanian, Emerson, and LeBel in 1965[15] lists 16, and a 1967 article written by House, Latham, and Whitesides[16] lists spectral properties for 50 chemicals. What made these dramatic differences possible?

The introduction of Varian's new A-60 NMR spectrometer in 1962 was an attempt to address many of the chemists' concerns and desires for NMR analysis. The A-60's design embodied the key ingredients for any successful instrument—a low price and quick, reliable results. The A-60 did not represent the latest in high-powered NMR spectrometers. Instead, the A-60 (and later the A-60A) embodied Varian's attempts to bring NMR spectroscopy to the masses. While Varian continued to modify and expand its line of spectrometers, the A-60 remained the company's mainstay.[a] The year following the A-60's introduction, Varian experienced a 65% increase in profits and the number of NMR related publications in the *JOC* increased from 67 in 1961 to 115 in 1962 to 144 in 1963.[b]

The introduction of new, smaller, easier to operate instruments reinforced other transformations already under way in chemical laboratories. Researchers spent less time running samples and more time analyzing the results taken from instrumental analysis. Instrumental analysis became an everyday procedure usually undertaken

---

[a]In 1963, Varian Associates had spectrometer sales totaling $7,734,000. The A-60 accounted for $3,224,000 of this. See Lenoir and Lécuyer,[17] p. 322, Table 3.

[b]See Lenoir and Lécuyer,[17] p. 322

by technicians and not chemists themselves. The analysis of organic compounds became mechanized. Following the publication of the two-volume *High Resolution NMR Spectra Catalog* by Bhacca et al. in 1962,[18] chemists were able to quickly compare their results with the spectra of known compounds. The ability to contribute to this rapidly growing body of knowledge was facilitated both by changes in the practices of the chemists in their laboratories and also by the instruments now made available by Varian.

ADVERTISEMENTS FROM THE *JOC*: SELLING NMR SPECTROMETERS

Because Varian Associates held an exclusive patent on NMR spectrometers, James Shoolery, head of Varian's Applications Laboratory in Palo Alto, California, did not have to worry about competing with other NMR instrument makers. Instead, the marketing and research teams at Varian fought for a share of the broader instrument market. The advertisements of Varian between the years 1956 and 1969 reflect this position. Varian's advertisements did not focus on their instrument's abilities to outperform those of competitors. Instead, Varian emphasized the indispensability of the information that their instruments could provide. The primary function of the NMR advertisements, then, was to create a sense of necessity for this exclusive information. To this end, Varian Associates utilized three main forms of advertising: the "NMR at Work" series; the "Instrument Information Memo"; and what I call (for lack of a better term) traditional advertising.

"NMR at Work"

In the early 1950s, Varian began running a series of advertisements known as "NMR at Work." The purpose of the series was to introduce this new analytical tool and the information that it could provide to the world of chemistry (FIG. 1).

The "NMR at Work" series first entered the *Journal of Organic Chemistry* in June of 1958 (NMR at Work #48). The title of the advertisement sets the tone for the rest of the page: "Are you missing the information that NMR spectroscopy can furnish?"

In addition to providing an abstract for what NMR spectroscopy *can* do, the Varian ad offers a specific example of what can be done with their instrument. Indeed, the entire "NMR at Work" series involves the discussion and resolution of specific problems encountered by chemists in their analytical work. By depicting a specific problem that other analytical techniques could not solve, Varian demonstrated the importance (if not necessity) of using NMR as an analytical tool. The "interpretation" section of the advertisement discusses the potential problems in analyzing a sample such as this, and then shows how the unique capabilities allow NMR to overcome these obstacles in its analysis.

The effects of the pictorial representation of the sample should not be ignored. James Shoolery, head of Varian's Applications Laboratory, had the spectral output of the NMR instruments designed specifically to appear like the output from other instruments being used by organic chemists at this time (e.g., infrared [IR] spectrometers).[c] By showing organic chemists what to expect from an NMR spectrometer, Varian accomplished the first step towards establishing their instrument as a new tool for organic analysis—familiarity.

FIGURE 1A. The first "NMR at Work" to appear in the *Journal of Organic Chemistry*. Thanks to Varian Inc. for permission to reprint this image.

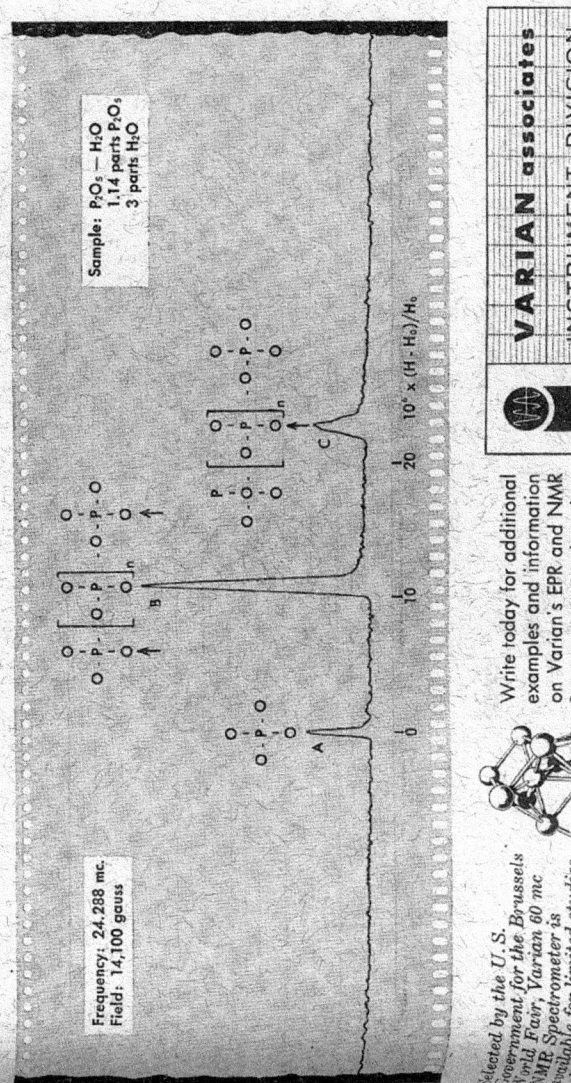

FIGURE 1B. "NMR at Work."

Over the course of the next several years, Varian Associates redesigned their "NMR at Work" series several times. Two points should be mentioned. First, while the overall theme of the series did not change, the examples presented did. Varian Associates slowly—but continuously—introduced chemists to new potential uses for NMR spectrometers. Second, the examples themselves were not always provided by Varian Associates. Chemists working in the field began to submit spectra and possible problems to be presented in the advertisements. While one cannot deny the importance of the issue of access represented by outside contributions, it is perhaps more important that Varian was able to convert the "NMR at Work" series into a forum for practicing chemists—a space for new examples to be presented and (eventually) for discussion of those examples.[d] This series provided a space where new information was communicated and examples furnished for later comparison. In this sense, the advertisement itself had become a journal article—something that organic chemists could look forward to reading so that they could stay up to date on the latest developments in the abilities and techniques for resolving specific analytical problems—problems that many of them encountered on a daily basis, themselves. In short, Varian created a journal inside of a journal.

Moving beyond the "NMR at Work" Series

In addition to the "NMR at Work" series, Varian Associates also introduced a new set of publications into the *JOC*—the "NMR/EPR Spectroscopy Instrument Information Memo" (*JOC* 1963 v28 n7). The memo format provided Varian with an opportunity to address in greater detail the specifics concerning their instruments, their use, why a chemist needs one, how to get one, and how to get access to one of Varian's if the chemist can't get one. Not surprisingly, the first piece in this particular memo compares 60-Mc operation with 100-Mc operation. The piece outlines the heightened abilities of the new HR-100 in structure elucidation. Varian also announced the opening of two additional Applications Laboratories away from the original location in Palo Alto: one in Pittsburgh, Pennsylvania and the other in Zug, Switzerland. Researchers with analysis problems are encouraged to schedule an appointment to meet with a Varian applications chemist at one of the three sites so that they can tackle the problem together using the latest in NMR technology. The piece also addresses advances in the ability for NMR spectrometers to "decouple" protons in samples using a radiofrequency oscillator. Of more interest, however, is the introduction of Varian's "living instruments."

In the last piece found in the first memo, Varian describes their instruments as "living." The concept of a living instrument is meant to imply the versatility and adaptability of the instrument to new tasks and new technologies. The Applications Lab is charged with maintaining this vitality; specifically their activity centers on:

> ... [t]he maintenance of a laboratory staffed by research and applications chemists who work in various areas of chemistry. Their responsibilities include independent research,

---

[c]See Lenoir and Lécuyer,[17] pp. 320–321.

[d]In "NMR at work" # 92, Varian Associates received an alternative explanation for their example presented in # 91. The issue is discussed and the alternative presented as a perhaps better explanation than what Varian had previously concluded. This "discussion" inside the advertisements provides a striking example what Varian Associates had hoped to establish in their "NMR at Work" series.

collaborative research on problems presented by scientists outside the company and the evaluation of newly designed instruments *from the viewpoint of a working chemist trying to solve real problems*. The design of new instruments and accessories *compatible with instruments presently installed* in customer laboratories throughout the world. (*JOC* 1963 v28 n7) [italics in original]

Put simply, the job of the Applications Lab is to remain active in the field and to design instruments accordingly. Additionally, current customers should not have to purchase a new instrument every time Varian develops a newer model. Instead, every model should hold the potential for upgrading. The piece specifically points out that an NMR spectrometer "operating at 30, 40, or 60 Mc may be converted to 100-Mc operation with the same performance specifications as our new HR-100" (*JOC* 1963 v28 n7). For more information, readers are pointed towards the "NMR at Work" series. This complementary piece to the "NMR at Work" series moved more towards traditional advertising while attempting to remain technical in style and content.

While the nearly constant presence of Varian Associates remained in the *Journal of Organic Chemistry*, many of the familiar features of Varian's advertising campaign had begun to fade away. In 1965, Varian ran only two ads from the "NMR at Work" series (#93 and #94). Both ads focused on further uses for the 100-Mc instruments. In place of the "NMR at Work" series and the "Instrument Information Memo," Varian resorted to a more conventional approach to advertising.

Getting Back to Basics

In the May 1965 issue of the *JOC*, Varian Associates introduced an updated version of the A-60, the A-60A. The advertisement for the new A-60A had much to say—textually and pictorially.

For the chemist who already used the A-60, external features of the design had not been changed, inducing a sense of comfort and familiarity. For those who had been either frustrated with certain drawbacks of the A-60, or reluctant to purchase the original, Varian described the improvements built into the new model. Sensitivity had been increased. The instrument could operate with smaller sample concentrations. And, perhaps most importantly, the new A-60A was unaffected by the magnetic and temperature environment of the lab. Varian then added, "Two more things you'll like" (*JOC* 1965 v30 n5). The first feature allows the instrument to operate for days without tuning. The second ensures an accurate scan that doesn't need someone to double-check it. These features were important because by removing the need for constant retuning and ensuring confidence in the first scan, Varian was making NMR accessible to a wider audience. A laboratory would not need to employ a specialist just to run and read NMR samples. Instead, vast amounts of data could be collected and processed by technicians who then passed along important and exceptional data to the laboratory chemist. The picture that accompanies the ad re-emphasizes this move towards routine NMR spectroscopy.

Additional features also stick out from the advertisement for the A-60A, including the size of the instrument with respect to the operator. Early NMR spectrometers were enormous instruments. Most laboratories simply could not accommodate one. The magnets were so large and so heavy that the physical structure of the laboratory often times had to be reinforced. Varian had made it one of its primary goals to overcome these physical restrictions, and the A-60A did just that.

In the January 1966 issue of the *JOC*, Varian again ran its ad for the new A-60A NMR spectrometer. While this was not the only ad Varian ran in the 1966 volume of the *JOC*, it was the only ad for their NMR spectrometers. Instead, Varian focused this year on marketing their new mass spectrometer to organic chemists. Three ads for their mass spectrometer appeared in the journal's pages, and while a new era was beginning for Varian, another was quietly fading away.

The End of an Era (?)

The last of the "NMR at Work" series to appear in the *JOC* during these years did so in the November issue of 1966. While brandishing a new design, #97 discussed further studies conducted at 100 MHz using 13C NMR spectroscopy. The number in the series, however, is, itself, noteworthy. The publications in 1965 concluded with #94 in the series. For four years, the *JOC* contained the entire set of the "NMR at Work" series. This marketing strategy on the part of Varian had now changed.

Varian's final advertisement for NMR spectrometers appeared in the *JOC*'s March issue in 1967. The ad depicts the A-60A with its "standard accessories." These "standard accessories" included the service engineer, spare parts and supplies supervisor, marketing engineer, research scientist, and applications chemist—demonstrating Varian's involvement in all aspects of their instrument's "life." In keeping with its original mission as both an instrument manufacturer and a research facility, Varian continued to sell itself (and its employees) along with the instruments. Varian was done advertising to the organic chemists—perhaps because it didn't have to any more.

CONCLUSION

In light of this investigation, I would like to discuss two points: (1) the partnership between the organic chemists and Varian Associates and (2) the changing identity of the NMR spectrometer.

Organic Chemists, Varian Associates, and the Co-development of NMR Spectrosopy

In the story outlined here the organic chemists and Varian Associates worked together to create what became known as the field of NMR spectroscopy. Both groups played integral roles as co-creators. It is for this reason that the story of NMR spectroscopy cannot be told completely from one perspective. In several instances, between the years 1956 and 1969, major developments were made possible by close cooperation between organic chemists and Varian involving the pursuit of common goals.

In the time between 1956 and 1969, these collaborative efforts transformed NMR spectrometers from instruments with tremendous potential capabilities, but a much narrower spectrum of practical applications, to dynamic instruments capable of a multitude of operations.

What Is an NMR Spectrometer?

Every time a chemist used an NMR spectrometer in a new way, or Varian advertised their instruments as capable of some new function, an attempt was made to alter the identity of the NMR spectrometer. Thus, answering the question "What is an NMR spectrometer?" becomes increasingly difficult outside of its historical and cultural context. The NMR spectrometer used by Norman Allinger in 1956[4] to investigate the structure of helvolic acid was definitely physically different from the NMR spectrometer used by Pletcher and Cordes in 1967[19] when they examined the basicity and coupling constants of three groups of compounds. But, the instrument did not undergo only physical changes.

An instrument is more than a collection of technological artifacts assembled to perform a designated function—although, it is also that. There is something more to the instrument that changes when the chemist (or physicist, or biologist, or doctor, or geologist...) attempts to use it in a novel way. The instrument changes conceptually for the researcher and— through the researcher's interaction within her/his community—the perceived identity of the instrument changes on a much broader scale. If we accept that the answer to the question "What is an NMR spectrometer?" is necessarily tied to the question "What is an NMR spectrometer used for?" then it becomes clear that this expansion of possible uses has the potential to change our response to the former question—and, thus, we could have a *different* instrument. Although the instrument remained technologically the same, conceptually the instrument was altered.

Changes in advertisements, too, had the potential to change the identity of the instrument. When Varian began marketing the NMR spectrometer, the company spoke of the capabilities of the instrument. They were guaranteeing the chemists access to a certain set of information if they chose to use their instrument for analytical work. The details of those advertisements changed as the instrument was re-tooled and re-interpreted. As I noted above, Varian not only set temporary guidelines for organic chemists to follow, they also responded to the demands of those same chemists. Researchers and designers at Varian were inclined to work with marketers and financial officers to ensure that the chemists received what they wanted. These various members of the company, then, were forced to confront the same question that organic chemists faced, "What is an NMR spectrometer?" Without an answer to this question, there would have been no advertisements, no new model developments, and no technological artifact that we call an NMR spectrometer.

These identity changes cannot be marked by single events. No, the identity of the instrument is something that is constantly being negotiated, and the factors involved in this negotiation were simply too numerous and too complicated to allow for this type of demarcation. The identity of the instrument, after all, involved much more than the design and use of the instrument. The identity of the instrument had to be linked to theoretical understandings involved with the design and use of the instrument, the experimental practices of those using the instrument, and—perhaps most importantly—the broader societal location of the instrument. While organic chemists and researchers at Varian could attempt to change what the NMR spectrometer was, they were confined by their own existence within a broader social fabric. Through these means, then, the identity of the instrument was negotiated.

REFERENCES

1. VAUGHAN, W. & M. YOSHIMINE. 1956. 9,10-dihydro-9,10-methanoanthracene. J. Org. Chem. **21:** 263–264.

2. DIXON, S. 1956. Elimination reaction of fluoro.ovrddot.olefins with organolithium compounds. J. Org. Chem. **21:** 400–403.
3. BOTTINI, A.T. & J.D. ROBERTS. 1956. Nuclear magnetic resonance spectrum of Feist's acid. J. Org. Chem. **21:** 1169–7110.
4. ALLINGER, N.L. 1956 Nuclear magnetic resonance spectrum of helvolic acid. J. Org. Chem. **21:** 1180–1182.
5. HERZOG, H.L., C.C. JOYNER, M.J. GENTLES, et al. 1957. 11-Oxygenated steroids. XVIII. Wagner-Meerwein rearrangement of 17-alpha-hydroxysteroids. J. Org. Chem. **22:** 1413–1417.
6. BISSELL, E.R. & R.E. SPENGLER. 1957. Organic deuterium compounds. I. 1,2-dimethoxy-D6-ethane-D4 and 2-methoxy-D3-ethanol-D4. J. Org. Chem. **22:** 1713–1714.
7. ZAVARIN, E. 1958. Extractive components from incense cedar heartwood (*Libocedrus decurrens*). VI. On the occurrence of 3-libocedroxythymoquinone. J. Org. Chem. **23:** 1198–1204.
8. SOLOMAN, W.C. & L.A. DEE. 1964. Steric course in cycloaddition of chlorotrifluoroethylene. J. Org. Chem. **29:** 2790–2791; NEWALLIS, P.E. & E.J. RUMANOWSKI. 1964. Fluoro ketones. II. Reaction of amides with fluoroacetones. J. Org. Chem. **29:** 3114–3116; LOGOTHETIS, A.L. 1964. Aziridines from diazomethane and fluorine-substituted imines. J. Org. Chem. **29:** 3049–3052; MARK, V. & J.R. VAN WAZER. 1964. Tri-tert-butyl phosphite and some of its reactions. J. Org. Chem. **29:** 1006–1008; BOSE, A.K., K.G. DAS & P.T. FUNKE. 1964. Fluoro compounds. II. Reactions and nuclear magnetic resonance studies of some fluorobromo esters. J. Org. Chem. **29:** 1202–1206.
9. BOSWELL, G.A., JR. 1966. Synthesis of 5-alpha,6,6-trifluoro-3-keto and 6,6-difluoro-delta-4-3-oxo steroids. J. Org. Chem. **31:** 991–1000.
10. LESTINA, G.J., G.P. HAPP, D.P. MAIER, et al. 1968. 15N isotope tracer study on 3-anilino-1-phenyl-2-pyrazolin-5-one. J. Org. Chem. **33:** 3336–3337.
11. FILLER, R. & S.M. NAQVI. 1961. Enol Content of γ-fluoro-β-keto esters by proton magnetic resonance. J. Org. Chem. **26:** 2571–2573.
12. BARCZA, S. 1963. Molecular weight determination by nuclear magnetic resonance (N.M.R.) spectroscopy. J. Org. Chem. **28:** 1914–1915.
13. HUITRIC, A.C. & J.B. CARR. 1961. Proton magnetic resonance spectra and stereochemistry of 2-(O-tolyl)cyclohexanol. J. Org. Chem. **26:** 2648–2651.
14. RAPOPORT, H. & J. BORDNER. 1964. Synthesis of substituted 2,2′-bipyrroles. J. Org. Chem. **29:** 2727–2731.
15. SUBRAMANIAN, P.M., M.T. EMERSON & N.A. LEBEL. 1965. Nuclear magnetic resonance studies of 5,6-dihalo-2-norbornenes and 2,3-dihalonorbornanes. J. Org. Chem. **30:** 2624–2634.
16. HOUSE, H.O., R.A. LATHAM & G.M. WHITESIDES. 1967. The chemistry of carbanions. XIII. The nuclear magnetic resonance spectra of various methylmagnesium derivatives and related substances. J. Org. Chem. **32:** 2481–2496.
17. LENOIR, T. & C. LÉCUYER. 1995. Instrument makers and discipline builders: the case of nuclear magnetic resonance. Perspect. Sci. **3:** 276–345.
18. BHACCA, N.S., L.F. JOHNSON & J.N. SHOOLERY. 1962. High Resolution NMR Spectra Catalog. Vol. 1. Varian Associates. Palo Alto, CA; BHACCA, N.S., D.P. HOLLIS & L.F. JOHNSON, et al. 1962. High Resolution NMR Spectra Catalog. Vol. 2. Varian Associates. Palo Alto, CA.
19. PLETCHER, T.C. & E.H. CORDES. 1967. Basicity and coupling constants of certain Acetals, ketals, and ortho esters. J. Org. Chem. **32:** 2249–2247.

Chemical versus Biological Explanations

Interdisciplinarity and Reductionism in the 19th Century Life Sciences

JOACHIM SCHUMMER

Department of Philosophy, University of South Carolina, Columbia, South Carolina 29208, USA

ABSTRACT: This paper analyzes four controversies in the 19th century life sciences: the nature of fermentation, the nature of infectious diseases, the generation of life from inanimate matter, and vitalism. All these controversies appear to concern chemical versus biological explanations, suggesting that reduction of biology to chemistry was the common underlying issue. My analysis rejects such interpretations, including the labels for explanation, and instead points out sophisticated forms of interdisciplinarity between chemistry, medicine, and biology in the first three debates. I argue that the philosophically favored perspective on reductionism, historically induced by a few physicians in the fourth debate, leads us astray from understanding interdisciplinary research.

KEYWORDS: interdisciplinarity; reductionism; 19th century life sciences; fermentation; infectious diseases; generation of life; vitalism

INTRODUCTION

Nowadays, the disciplinary structure of the sciences established in the 19th century is increasingly undermined by goal-oriented interdisciplinary research. From biomedical research through materials science to nanoscale science and technology, chemical concepts, theories, and methods play a pivotal role in most of these research fields, even if the term "chemistry" is usually avoided. Historians, and especially philosophers of science, are challenged to leave their disciplinary focus behind if they wish to contribute to an understanding of recent research. Moreover, philosophers of science—with their focus on the conceptual and epistemological structures of each of the sciences—may encounter difficulties in understanding interdisciplinary relations other than asymmetric logical relations, such as the various forms of reduction or supervenience. What we need is a broader perspective.

Address for correspondence: Joachim Schummer, Department of Philosophy, University of South Carolina, Columbia, SC 29208. Voice: 803-777-3731; fax: 803-777-9178.
js@hyle.org

In this paper I analyze interdisciplinarity in the 19th century life science. Then, discipline formation was still in progress—scholars from various young disciplines worked together, or struggled with each other on similar issues. The advantage of considering this period is that the landscape of the life sciences was far less complex than it is nowadays, while the major disciplines then involved, that is, chemistry, medicine, and biology, were roughly the same as they are today. I use the term "scientific discipline" in the sense of a social institution with effective means for research, communication, and teaching, including self-reproduction—and with individuals involved who share a common area of problems, methods, concepts, and knowledge. This usage so combines sociological and cognitive aspects. Taking this broad perspective, I aim at a better understanding of forms of interdisciplinarity, which, as I will argue, is disguised by the perspective on reductionism favored by philosophers.

Let us first take a brief look at the general historical context. Starting around 1810 in the German states, universities changed from secondary teaching institutions to tertiary teaching *and* research institutions. Moreover, their faculties of philosophy upgraded from merely providing preliminary teaching service to having equal rank with the higher faculties of theology, law, and medicine. Philosophy faculties offered an increasing range of scientific studies when chemistry and natural history moved in from the medical faculties. Although the philosophical faculties became the home of the formation of most modern disciplines, the medical faculties continued to be important to the life sciences.[1,2] Here too, radical changes took place, including a new research orientation in large part modeled on experimental research in chemistry. Apart from experimental pathology, hygiene, diagnostics, therapy (pharmacology), and so on, it was particularly the new physiology that shaped the scientific-experimental image of medicine. However, many fields that we today assign to biology remained under the umbrella of medicine, such as embryology, cytology, and bacteriology. Although chemistry could incorporate important parts of the life sciences (i.e., organic chemistry, chemistry of plants and animals, and then biochemistry), subjects like physiological chemistry remained with medicine, notwithstanding considerable overlap.

By mid-19th century, there was a freshly marked out multidisciplinary field of the life sciences, consisting of (organic) chemistry, biology, and medicine, still full of disciplinary dynamics and mutual overlap. At that time, researchers of the several new disciplines engaged in four controversies, all being about the same general issue —whether a certain field of phenomena could (or should) be explained by reference to inanimate or animate matter. Historians of science frequently present these stories as competitions between "chemical" and "biological" explanations, which suggests that the common underlying issue was the reduction of biology to chemistry.

In the following, I first sketch the four controversies very concisely (interested readers may find more details in the references[3–18]), with just enough detail to point out some analogies. Although these analogies might suggest the adequacy of the labels for explanations and reductionism as the underlying issue, I raise serious doubts about that in the DISCUSSION section. My emphasis is rather on interdisciplinarity— in this respect, three of the four cases show interesting common features. I argue that focus on the issue of reductionism, although favored by recent philosophy and suggested by the fourth controversy, leads us astray from understanding much more intricate relationships between the sciences.

TABLE 1. Opponents in the debate on fermentation

	Substance theory ("chemical explanation")	Microbe theory ("biological explanation")
Chemists	Joseph L. Gay-Lussac Justus Liebig Marcelin Berthelot Eduard Buchner	Louis J. Thenard Christian F.P. Erxleben? Friedrich Kützing (pharmacist) Pierre J.A. Béchamp Louis Pasteur
Physiologists, biologists	Wilhelm Kühne	Theodor Schwann
Engineers		Charles Cagniard de Latour

FOUR CONTROVERSIES IN THE 19TH CENTURY LIFE SCIENCES

Fermentation

The first controversy I deal with involved the nature of fermentation and ferments (TABLE 1). Fermentation is a transformation of substances that requires the presence of a component, the ferment, which is not transformed in the process. A typical example, which was also in the focus of the controversy, is the transformation of sugar to alcohol and carbon dioxide in the presence of yeast. A kind of substance transformation, fermentation had long been a subject matter of chemistry, but also of physiology because most steps of digestion had been taken as fermentation. Up to the early 19th century, there was broad agreement that ferments like yeast were more or less homogeneous inanimate materials.

In the 1830s, mainly supported by microscopic studies based on the newly developed achromatic lens systems, that classical view began to be doubted because yeast showed a globular structure, and seemed to grow like a living being in the course of fermentation. According to the new theory, the life processes of microorganisms, the ferments proper, caused fermentation. Opposed to that, Justus von Liebig (1803–1873) developed a complex theory of (auto-) catalysis, claiming that yeast, as inanimate matter, was both a waste product and a catalyst of fermentation. At first, this so-called "chemical explanation" had many followers, particularly among chemists. By the 1840s, further studies, particularly microscopy and chemical investigations of metabolism, increasingly convinced researchers (and finally Liebig himself) that yeast was a living organism. Thus, the "biological explanation" appeared to be the winner of the controversy—but that was only the prelude.

The main controversy developed over the question of whether the cause of fermentation, the "ferment proper," was the entire living organism itself ("biological explanation") or an inanimate substance that might be isolated from the yeast organism ("chemical explanation"). The main opponents were two chemists, Justus von Liebig and Louis Pasteur (1822–1895). Since the particular issue of brewer's yeast evaded experimental clarification, the controversy was soon put on a general level to include all kinds of fermentation. Pasteur investigated various species of yeast and argued that each had its own particular fermentation effect—dependent on its particular metabolism—which he tried to clarify by chemical analysis. Liebig referred to apparently inanimate ferments, such as pepsin, which the physiologist Theodor Schwann (1810–1882) had isolated from gastric juice shortly before, and argued that finding the ferment proper was only a matter of skilful experimental isolation. Pasteur responded by excluding such cases from "fermentation proper," which, according to his definition, necessarily involved a living organism.

Although the methodological agreement on the validity of experimental proofs (here, chemical isolation) favored Liebig's view to some degree, the debate on fermentation reached a state that evaded a general decision on experimental grounds. Provided an open set of yet to be discovered fermentations, it is logically impossible to prove or disprove by experimental means that there exists at least one kind of fermentation that resists isolation of a specific substance as the ferment proper. The particular issue of brewer's yeast turned out to be extremely difficult. Indeed, the successful isolation of the ferment proper (then called enzyme, from Greek *zyme*, "yeast") was not achieved before 1896. Eduard Buchner (1860–1917) received the Nobel Prize for chemistry in 1906 for this achievement.

Since fermentation is a kind of chemical substance transformation, we may properly call it a chemical issue. However, the controversy, which almost ran through the whole 19th century, also involved physicians, physiologists, biologists, engineers, and pharmacists. They applied a broad scope of experimental and observational technologies, each from their own disciplinary background—such as chemical analysis, chemical investigation of metabolism, microscopy, the breeding and cultivating of microorganisms, taxonomic distinctions of species, and so on. Regardless of their disciplinary background or position in the controversy, researchers accepted and used these various methods and their conceptual and theoretical bases as valid tools of argumentation. Moreover, there is no correlation between disciplinary background and view defended in the controversy (TABLE 1). Therefore, unlike the opposition "chemical versus biological explanation" suggests, the controversy was truly interdisciplinary—both with respect to the researchers involved and to the methods and conceptual bases accepted.

Infectious Diseases

My second instance of controversies in the 19th century life sciences was about a typical issue of medicine or pathology, the nature of infectious diseases, followed-up by debates on the mechanisms of vaccination and the immune system (TABLE 2). According to the dominating view since antiquity, those diseases were transmitted and caused by poisonous substances (in Latin also called *virus*, "toxic slime"). Opinions differed only as to whether these poisons came from the environment (the miasmatic theory, with malaria as its paradigm) or from other infected persons (the contagion theory, with syphilis as its paradigm). A serious alternative came up only in the mid-19th century, again supported by microscopy. According to this view, the proper transmitters of diseases were microorganisms or microbes. Despite the different set of issues, there were many obvious analogies to the controversy on fermentation; also, some persons were involved in both debates.

Growing empirical evidence (e.g., from microscopy, cultivation of microbes, controlled infection of test animals) supported the *microbe theory*, so that defenders of

TABLE 2. Opponents in the debate on infectious diseases aand vaccines

	Poison theory/Serum theory ("chemical explanation")	Microbe theory/Cell theory ("biological explanation")
Chemists	Justus Liebig	Louis Pasteur
Physiologists, physicians, biologists	Max Pettenkofer Jean-Joseph Henri Toussaint Emile Roux Alexandre Yersin Robert Koch Emil Behring Shibasaburo Kitasato	Jacob Henle Robert Koch (prelude only) Ilya Metchnikoff

the *poison theory* increasingly accepted the involvement of microbes (e.g., cholera bacteria). Again, that was only the prelude, after which the camps divided anew. Proponents of the revised poison theory argued that, while microbes were undoubtedly involved, the disease-causing agent proper was a poison that might be isolated from microbes. The main proponents of that view, which historians of medicine are accustomed to call "the chemical explanation," were the physician Robert Koch (1843–1910) and his school. The leader of the camp of the so-called "biological explanation," according to which entire living microbes are necessarily involved, was again Louis Pasteur, a chemist. Contrary to Pasteur's original assumption, researchers successfully isolated a toxin from diphtheria bacteria, which, upon injection into test animals, caused the typical symptoms of diphtheria. That made adherents to the poison theory claim that every infectious disease could finally be reduced to toxins, if the analyses were only performed with sufficient rigor. Defenders of the microbe theory rejected such instances as diphtheria and claimed that infectious diseases proper involved (by definition) a living organism. (Still, nowadays, malaria might be a case in point.) Once more, the controversy reached a level of generality where definite decisions on experimental grounds became impossible.

The medical controversy on infectious diseases initiated at least two other related controversies, for each of which historians of medicine have assigned the labels of "chemical" and "biological" explanations. Followers of the microbe theory believed that all vaccines for infectious diseases must contain living or "moderated" cells, and they explained the mechanism of the immune system by the effects of corporeal cells ("cytophages" or later "leukocytes"). Proponents of the poison theory picked up the old idea of antidotes. They tried to isolate vaccines from the cell-free blood serum of infected animals, and explained the immune system by the effects of natural antidotes. Both parties collected sufficient evidence for their views, such that we still have two kinds of vaccines nowadays—and a historically rooted divide of the immune system into a cellular and a humoral branch.

Despite its being a distinct issue of medicine, the controversy on infectious diseases once more shows the characteristics of interdisciplinarity mentioned previously. Besides a majority of physicians, also chemists and biologists were involved, all of whom brought in their own disciplinary concepts and methods, each accepted or even used by others, regardless of disciplinary background or position in the debate.

TABLE 3. Opponents in the debate on the generation of life

	Abiogenesis ("chemical explanation")	Preformation theory ("biological explanation")
Chemists		Joseph Priestley
		Justus Liebig
		Louis Pasteur
		John Buchanan
		John Tyndall
		Svante Arrhenius
Physiologists, physicians	"Naturphilosophen" (e.g. Oken, R. Treviranus, Burdach)	Eduard Pflüger
	Max Schultze	Rudolf Virchow
	Henry Charles Bastian	
Biologists, naturalists	George-Louis de Buffon (18th c.)	Charles Bonnet (18th c.)
	Erasmus Darwin	
	Jean-Baptiste de Lamarck	George Cuvier
	Felix Pouchet	Ferdinand Cohn
	Ernst Haeckel	
	Thomas H. Huxley	
Biologists, naturalists	George-Louis de Buffon (18th c.)	Charles Bonnet (18th c.)
	Erasmus Darwin	
	Jean-Baptiste de Lamarck	George Cuvier
	Felix Pouchet	Ferdinand Cohn
	Ernst Haeckel	
	Thomas H. Huxley	

Generation of Life

The third controversy that I briefly deal with is about a core issue of biology, already discussed at book-length by Aristotle, that is, the generation of living beings. Although some critical views were advanced already in the 17th and 18th centuries, the prevailing view up to mid-19th century was that some primitive living beings could more or less spontaneously arise from inanimate matter by abiogenesis (or self-organization, as chemists would probably call it nowadays). People found support of abiogenesis not only in the Bible and Aristotle, but also in ample everyday life evidence—for instance, when worms suddenly crept out of feces or putrefying meat, or when swarms of small animals appeared in rotting water. Indeed, spontaneous generation of life, which incidentally also implied the possibility of artificial creation of life, was a triviality, both for learned and uneducated people—and it was in accordance with, if not part of, most philosophies of nature.

By the mid-19th century, however, the counter-position gained extraordinary strength (TABLE 3). According to that view, which was defended already by Leibniz, all life arose from life—or is at least preformed in germs, eggs, semen, or spores (the preformation theory). Again, historians of science have used the labels "biological" and "chemical" for the opposing views, such that the two main opponents in mid-19th-century France seemed to act in changed roles: the biologists Felix Pouchet defended the "chemical" view, whereas the chemist Pasteur held to the "biological" view.

As compared to the former controversies, the debate over abiogenesis had strong theological and metaphysical overtones, and it was closely connected to many other biological issues of the time, such as systematics, evolution theory, and embryology. Nonetheless, opponents on both sides used (and accepted) arguments based on an experimentally sophisticated level. For instance, proponents of abiogenesis tried to prove by infusion experiments that primitive life forms arose from inanimate matter, such as dried grass, water, and oxygen. They heated some grass at 300 °C in oxygen-free atmosphere and infused it under mercury into a disinfected flask that contained nothing but oxygen and freshly synthesized water. Opponents of abiogenesis, like Pasteur, sought indirect experimental evidence by showing, for instance, that disinfection was imperfect or that spores, much too small to be observed through microscopes then, could enter a flask despite all care. Notwithstanding the sophisticated experimentation, that controversy was also impossible to settle on experimental grounds because of the generalized level of the claims on both sides.

By the end of the 19th century, the debate reached a new dimension when biologists like Ernst Haeckel (1834–1919) and Thomas H. Huxley (1825–1895) rejected the cell theory of Theodor Schwann, according to which the cell was the smallest unit of life. Some biologists believed they had found the "essence of life" in colloid cell juices (protoplasm) or so-called "life molecules"—which were assumed to emerge spontaneously from inanimate matter. Such views increased the gap between biologists and biochemists for some time well into the 20th century. Biochemists, on the one hand, dived into the complexities of molecular processes within the cell, which they tried to grasp by dynamic systems approaches, always facing the limits of chemical understanding. Biologists and physiologists, on the other hand, sought for a simple molecular "essence of life." (Today's pop-science view—according to which the essence of life is the DNA molecule—is probably a late echo of that simplistic essentialism of life.) Their belief in abiogenesis gave biologists much more trust in chemical explanation than chemists who, faced with chemical complexities, mostly denied the possibility of abiogenesis. Strangely enough, the controversy on spontaneous generation of life reveals once more that biologists tended to favor the "chemical" view, whereas chemists were inclined to the "biological" view.

Nonetheless, we have a debate on a biological issue here that, save for an intermediary period, shows again the characteristics of interdisciplinarity with respect to researchers, concepts, and methods involved, and which greatly worked for the benefit of biology.

Vitalism

The fourth controversy I briefly discuss concerned vitalism (TABLE 4). Unlike the former issues, vitalism evades the simple assignment to one or more scientific disciplines and belongs rather to metaphysics or speculative philosophy of nature. The general issue was, whether there was a substantial difference between animate and inanimate matter, such that the explanation of life phenomena required reference to particular factors not required for explanations of nonlife phenomena. Anti-vitalists rejected that, whereas vitalists answered in the affirmative (strong vitalism) or at least allowed for the possibility (weak vitalism). Historians of science and philoso-

TABLE 4. Opponents in the debate on vitalism

	Anti-vitalism ("chemical-mechanical explanation")	Vitalism (weak and strong) ("biological explanation")
Chemists	Marcelin Berthelot Garrit Jan Mulder	Jöns J. Berzelius Leopold Gmelin William Prout Justus Liebig Franz Döbereiner Louis Pasteur
Physiologists, physicians, biologists	Emil du Bois-Reymond Ernst von Brücke Hermann von Helmholtz Carl Ludwig Karl Vogt Jacob Moleschott Ludwig Büchner	Johannes Müller Jacob Henle Rudolf Virchow Xavier Bichat (François Magendi) Claude Bernard

phy frequently narrate the controversy in terms of an opposition between physical-chemical explanation and biological explanation, or as a debate over the issue of whether biology can be reduced to chemistry (and ultimately to mechanics).

Apart from Cartesianism and some radical forms of materialism in the French Enlightenment, virtually all natural philosophies with a significant impact on medicine had included vitalism in one or the other form (e.g., Aristotle, Galen, Avicenna, Paracelsus, van Helmont, Stahl, Haller, Schelling, and so on). They differed, however, as to whether the particular factors were (i) particular (ponderable or imponderable) substances; (ii) particular forces attached to certain substances; or (iii) matter-independent (transcendent) mental or psychological principles. In the first half of the 19th century, almost all organic chemists as well as the founders of experimental physiology in the German states (Müller, Henle, Virchow) and France (Bichat, Magendie, Bernard) were vitalists at least in the weak sense. It is true that they all tried to explain life phenomena as far as possible without reference to particular factors. However, they met with considerable barriers to understanding and explaining life phenomena, including barriers to synthesizing organic substances. Thus, they did not exclude *a priori* the possibility of particular life substances, or forces attached to certain substances, such as magnetic forces appearing only at certain substances. In so doing, they were also in strong opposition to the *Naturphilosophie* of Schelling and his followers, which was extremely influential on German medicine then and which came under the label of "vitalism" too. According to that view, a transcendent "principle of life" governed "nature" overall, with various manifestations on different levels such as mechanical force, life, and mind. For most experimental scientists, both the metaphysical existence claim of a transcendent principle and the *a priori* negation of life forces or substances were beyond the standards of scientific argumentation. When they referred to particular life factors, they did that frequently in a provisional manner to mark the limits of knowledge of the time. And when they explicitly confessed to strong vitalism, by claiming the existence of life forces or substances, they did that mostly outside the scientific discourse proper—in popular lectures or books.

TABLE 5. Biographical dates of the major anti-vitalists

Name	Dates	Doctorate	Full professor in physiology (first app.)	Teacher in physiology	Education by chemists
Emil du Bois-Reymond	1818–96	Dr. med. 1843 (Berlin)	1858, Berlin	Joh. Müller	Mitscherlich, G. Magnus
Ernst von Brücke	1819–92	Dr. med. 1842 (Berlin)	1849, Wien	Joh. Müller	Mitscherlich, G. Magnus
Hermann von Helmholtz	1821–94	Dr. med. 1843 (Berlin)	1849, Königsberg	Joh. Müller	Mitscherlich, (G. Magnus)
Carl Ludwig	1816–95	Dr. med. 1840 (Marburg)	1849, Zürich	Ludwig Fick	Assistant to Bunsen in Marburg
Karl Vogt	1817–95	Dr. Med. 1839 (Bern)	1847, Giessen	K. Vogt, senior	First study under Liebig in Giessen
Ludwig Büchner	1824–99	Dr. med 1848 (Tübingen)	(1854, PD in Tübingen)	?	First study under Liebig in Giessen
Jacob Moleschott	1822–93	Dr. med. 1845 (Heidelberg)	1856, Zürich	Jacob Henle	Assistant to Mulder in Utrecht

By mid-19th century, the debate on vitalism in the German states grew to unparalleled dimensions when seven young physicians, all in their early twenties and with many obvious biographical parallels (TABLE 5), entered the stage to proclaim anti-vitalism or mechanical materialism and to sharply denounce any talk of life forces, substances, or principles. They were very active in giving public speeches, writing manifestos (including the foundation manifesto of the first German "physical society"), and popular philosophy texts, such as Ludwig Büchner's *Kraft und Stoff* ('Force and Matter') which probably topped any other philosophical text of the time regarding circulation and number of editions. Although these physicians indirectly criticized also the weak vitalism of their influential physiology teachers (e.g., Müller and Henle), their target was rather the *Naturphilosophie* tradition in medicine. For, with the help of their teachers, all of them made a rapid academic rise and obtained most of the new chairs of physiology in the German-speaking countries. Of the powerful people in physiology then, only Liebig, a strong vitalist, called the anti-vitalists "*dilletanti* in the sciences" because of their naive claims to the omnipotence of chemical and mechanical explanation.[19] (Note that they proclaimed mechanical explanation of physiological phenomena when chemical structure theory was not even developed.) Not surprisingly, when they became leading physiologists, they were unable to translate their metaphysical ideas into successful research programs, as they failed to provide any mechanical explanation of a life phenomenon. However, they were extremely successful in applying and developing further experimental approaches to physiology and in ousting the influence of *Naturphilosophie*.

Overall, the debate on vitalism had little to no impact on experimental research that was comparable to the other controversies discussed previously. Rather, it was primarily as a struggle within the medical faculties, articulated in philosophical terms: metaphysically as materialism versus idealism, and methodologically as experimental research versus speculative *Naturphilosophie*. In addition, like the late 18th century French physicians La Mettrie and Cabanis, it was related to atheism and republicanism, such that three of the seven physicians were so much involved in the political quarrels in the German states in 1848 that they emigrated to Switzerland afterwards.

DISCUSSION

Interdisciplinarity on Disciplinary Issues

Behind all of the four controversies, there is the common general issue as to whether some field of phenomena can (or should) be explained with reference to animate or inanimate matter. The first three controversies show further common features, all of which are missing in the debate on vitalism.

(1) All opponents, no matter of which disciplinary background, presupposed that there was a substantial difference between animate and inanimate matter; otherwise there would have been nothing to debate. (For an anti-vitalist who, as defined above, rejected the distinction between animate and inanimate matter, there was, strictly speaking, no controversy.)

(2) All of the three controversies were interdisciplinary in the sense that researchers of different disciplines were involved who, regardless of their disciplinary background, used and accepted the experimental methods of different disciplinary origin as valid elements in scientific discourse. Thus, in each controversy, an interdisciplinary consensus on scientific methods developed, or, as Hans-Jörg Rheinberger has called it, a common "experimental culture."[20]

(3) All of the three controversies went through several steps of stating views more precisely, refining experimental approaches, and generalizing theses up to a level that they evaded decision on experimental grounds, which nonetheless fostered further experimental research.

(4) All of the three controversies were related to each other in such a way that ideas in one area could be used to support views in others. For instance, in the fermentation controversy, defenders of the "chemical" explanation explained the growth of yeast as a secondary effect (not the cause) of fermentation by spontaneous generation ("biological" explanation).

(5) In spite of structural similarities and their interdisciplinary characteristics, each of the three controversies was about a certain issue that is clearly related to a single discipline, which also benefited most from the debates: (*a*) a kind of substance transformation, fermentation belongs to the classical issues of chemistry; (*b*) a kind of disease, infectious diseases belong to the classical issues of medicine; (*c*) a kind of generation of life, abiogenesis belongs to the classical issues of biology.

In sum, the controversies prove that interdisciplinarity is possible even if the issues under debate are clearly related to a single discipline. They further prove that researchers from different disciplines can struggle with each other on a common methodological basis that defines how arguments are to be generated and applied properly. In addition, they illustrate that the disciplinary background does not imply a bias with respect to the views under debate, as the labels "chemical" and "biological" explanations suggest.

The Special Role of the Debate on Vitalism, Reductionism, and the Labels for Explanations

The debate on vitalism plays a special role among the controversies discussed in this paper because it shows none of the characteristics of interdisciplinarity, nor does the issue clearly belong to any of the scientific disciplines. Further, the debate stood rather outside of the 19th century scientific discourse because it did not keep the conceptual, methodological, and argumentation standards established in the experimental sciences then. Both the negation and the affirmation of the existence of some life forces, substances, or principles, without compelling experimental evidence, were metaphysical statements of the sort which the 19th century experimental sciences had successfully overcome. Similarly, bold claims regarding the omnipotence of certain explanatory approaches, without providing ample evidence by instances of successful explanations, were relicts of 17th century speculations about nature which no longer had a place in the 19th century experimental sciences. Weak vitalism, leaving it open to future research whether particular factors are required for the explanation of life phenomena or not, was the only rational view then, if any was.

The particular attention that 20th century philosophers have paid to the debate on vitalism suggests that the issue belongs to philosophy. Let us consider the arguments in favor of that.

First, one might say that, because 20th century philosophers have dealt with vitalism, albeit mostly from a historical point of view, it is a genuine topic of philosophy. Back to the 19th century, however, it was a handful of physicians who picked the quarrel as an attempt, I assume, to reform medicine. Professional philosophers, such as Albert Lange (1828–1875)[21]—if they were not philosopher-turned-physician, as was Hermann Lotze (1817–1881)—entered the scene only later. However, they had much more general concerns, carefully avoided making bold claims, and approached the topic from an epistemological point of view. If philosophy were about making bold claims that modern sciences had banned from their discourses for methodological reasons, it would simply be an anachronistic way of doing "science."

Second, one might argue that, because the issue of vitalism was connected to metaphysical and theological issues and could not be decided by scientific means, it qualifies as a genuine philosophical topic. Yet, the same is also true of the issues of the other three controversies which we can nonetheless clearly assign to scientific disciplines. For instance, the issue of abiogenesis was closely related to theological issues of creation; Lamarck favored abiogenesis because it was in accordance with his teleological view of evolution, whereas Cuvier rejected abiogenesis because it did not fit his teleologically based systematics; and so on. Furthermore, as I have argued above, all three controversies were pushed to a level of generality that evaded decision by experimental means. Therefore, the relation to metaphysics and its experimental indecisiveness does not as such make an issue a genuine philosophical one.

Third, strong vitalism and strong anti-vitalism, in the sense of stating and negating the existence of an extra-factor to be considered in the explanation of life phenomena, could be translated into methodological norms, for instance: "Seek/avoid reference to animate matter in scientific explanations of life phenomena!"

From the point of view of methodology, the debate on vitalism then appears as a methodological meta-discourse on science that governed all the three other controversies and urged participants in the debates to take either side. In each case, methodological anti-vitalism would recommend what has been called the "chemical" explanation, whereas methodological vitalism favored the "biological" explanation. Further, as strong anti-vitalism implies ontological reduction of life phenomena to phenomena of inanimate matter, methodological anti-vitalism recommends reductionist approaches to life phenomena, and vice versa. Now, if one takes biology as the science of life phenomena and chemistry as the science of inanimate matter, anti-vitalism corresponds to the reductionism of biology to chemistry. I assume that this is why philosophers have paid particular attention to the debate on vitalism and why historians have chosen the labels of "chemical" and "biological" explanations.

Let us finally consider why the idea of a methodological meta-discourse is a misleading fiction, both from historical and philosophical points of view.

(1) Historically, the scientists involved in various debates did not match the simplistic polarization. For instance, Liebig advanced good reasons to take fermentation and infection as processes of inanimate matter ("chemical explanations"), but he was skeptical about abiogenesis and a defender of strong vitalism ("biological explanations"), as were most of his colleagues in organic chemistry then. If we adopt the attitude of normative methodology suggested above, we would have to call most scientists irrational, although their scientific arguments are perfectly understandable.

(2) The opposition between biology/physiology as the sciences of life phenomena and chemistry as the science of inanimate matter is an anti-modern fiction that lacks any historical evidence. If it were correct, there would never have been any of the interdisciplinary controversies I have discussed. Interdisciplinarity requires overlap where different disciplines deal with the same subject matter and use the same concepts although their methods and perspectives might differ to some degree. In fact, concepts of inanimate matter have belonged to biology and physiology as much as concepts of animate or organized matter to chemistry.

(3) Consequently, labeling explanations as "chemical" or "biological" based on whether references are made to inanimate matter or not would lack any historical and philosophical justification and lead us to absurdities. Note that these labels refer neither to disciplines (as sociological entities) nor to issues, methods, particular concepts, theories, and subject matter of certain research fields. If, for instance, a trained biologist suggests an explanation for a biological issue based on applying biological methods, we would have to call it a "chemical" explanation whenever the explanation refers only to inanimate matter. Furthermore, the controversies I have discussed above would urge us to conclude that biologists and physiologists favored "chemical" explanation, whereas chemists favored "biological" explanation. If one needs disciplinary labels for explanations, I would suggest referring to disciplinary *issues*, such that every explanation related to a biological issue is by definition a biological explanation.

(4) If philosophers adopt the idea of a methodological meta-discourse about chemical versus biological explanation, anti-vitalism versus vitalism, or reductionism versus anti-reductionism, they would not only be at odds with the history of science, but also use arbitrary definitions for the sciences and invent strange labels for explanations. Such simplistic dichotomies also prevent us from understanding the complex relationships between the actual sciences. Interdisciplinarity evades dichot-

omies, such as the reductionism issue, which makes us blind for analyzing the intricate forms of interdisciplinarity. The three 19th century cases I have briefly discussed in this paper are only simple examples without much reference to theories, as compared to the complex forms of interdisciplinary research nowadays. Here is a field of growing importance for philosophers of science.

REFERENCES

1. KOHLER, R.E. 1982. From Medical Chemistry to Biochemistry: The Making of a Biomedical Discipline. Cambridge University Press. Cambridge, England.
2. BROMAN, T.H. 1996. The Transformation of German Academic Medicine, 1750–1820. Cambridge University Press. Cambridge, England.
3. LIPPMANN, E.O. v. 1933. Urzeugung und Lebenskraft. Zur Geschichte dieser Probleme von den ältesten Zeiten an bis zu den Anfängen des 20. Jahrhunderts. J. Springer. Berlin.
4. LIEBEN, F. 1935. Geschichte der physiologischen Chemie. Deuticke. Leipzig-Wien [reprinted 1970: Hildesheim.New York].
5. PARTINGTON, J.R. 1964. A History of Chemistry, Vol. IV. Macmillan. London.
6. HALL, T.S. 1969. Ideas of Life and Matter [2 vols.] Chicago University Press. Chicago.
7. FRUTON, J.S. 1972. Molecules and Life: Historical Essays on the Interplay of Chemistry and Biology. Wiley. New York.
8. FLORKIN, M. 1972. A History of Biochemistry, Parts I & II. Elsevier. Amsterdam [Vol. 30 of Comprehensive Biochemistry].
9. GARDNER, E.J. 1972. History of Biology, 3rd ed. Burgress. Minneapolis.
10. FARLEY, J. 1974. The Spontaneous Generation Controversy from Descartes to Oparin. John Hopkins Univerity Press. Baltimore, MD.
11. LEICESTER, H.M. 1974. Development of Biochemical Concepts from Ancient to Modern Times. Harvard University Press. Cambridge, MA.
12. JAHN, I. 1990. Grundzüge der Biologiegeschichte. Fischer. Jena.
13. TEICH, M. & D.M. NEEDHAM. 1992. A Documentary History of Biochemistry: 1770–1940. Fairleigh Dickinson University Press. Rutherford, NJ.
14. GEISON, G.L. 1995. The Private Science of Louis Pasteur. Princeton University Press. Princeton, NJ.
15. PORTER, R. 1997. The Greatest Benefit to Mankind: A Medical History of Humanity, from Antiquity to the Present. Harper Collins. London.
16. BOTSCH, W. 1997. Die Bedeutung des Begriffs Lebenskraft für die Chemie zwischen 1750 und 1850. Ph.D. Thesis. University of Stuttgart, Germany.
17. JAHN, I., Ed. 1998. Geschichte der Biologie: Theorien, Methoden, Institutionen, Kurzbiographien, 3rd ed. G. Fischer. Jena, Germany.
18. LEIBER, T. 2000. Vom mechanistischen Weltbild zur Selbstorganisation des Lebens. Alber. Freiburg & München.
19. LIEBIG, J.v. 1878. Chemische Briefe, 6th ed. Winter. Leipzig & Heidelberg.
20. RHEINBERGER, H.-J. 1997. Toward a History of Epistemic Things. Synthesizing Proteins in the Test Tube. Stanford University Press. Stanford, CA.
21. LANGE, F.A. 1902. Geschichte des Materialismus und Kritik seiner Bedeutung in der Gegenwart, 7th ed. Baedeker. Leipzig [1st edn 1865; 2nd ed. 1873].

Richard Rufus's Theory of Mixture
A Medieval Explanation of Chemical Combination

MICHAEL WEISBERG AND REGA WOOD

Department of Philosophy, Stanford University, Stanford, California 94305-2155, USA

ABSTRACT: Richard Rufus of Cornwall offered a novel solution to the problem of mixture raised by Aristotle. The puzzle is that mixts or mixed bodies (blood, flesh, wood, etc.) seem to be unexplainable through logic, even though the world is full of them. Rufus's contribution to this long-standing theoretical debate is the development of a modal interpretation of certain Averroistic doctrines. Rufus's account, which posits that the elemental forms in a mixt are in accidental potential, avoids many of the problems that plagued non-atomistic medieval theories of mixture. This paper is an initial examination of Rufus' account.

KEYWORDS: Richard Rufus; mixture; mixt; combination; accidental potential; potential; elements; medieval chemistry; Aristotelian science; Averroes; medieval chemistry; Aristotelian science

Richard Rufus of Cornwall[1] (d. 1259?) plays an important and as yet unappreciated role in medieval discussions of natural science. This paper focuses on Rufus's novel attempt to make Aristotle's hylomorphic theory consistent with the possibility of mixture—which is analogous to what modern chemists call chemical combination. After explaining Rufus's treatment of the problem of mixture and showing that it is an elaboration and refinement of Averroistic doctrines, we will assess a claim made by the eminent medieval scholar Anneliese Maier,[2] who suggested that no non-atomistic medieval theory could explain how elements can be combined in uniform bodies or mixts. She further claimed that such theories merely state the fact that mixture is possible, but do not give any explanation of how a mixt reflects the state of the elements. Richard Rufus, as we will see, gave a plausible account, but let us first turn to Aristotle's statement of the problem.

ARISTOTLE'S PUZZLE

In *On Coming to Be and Passing Away*, Aristotle raises a particularly vexing problem for his theory of the elements. The problem concerns the possibility of homogenous mixed bodies. Specifically, he asks what happens to the elemental com-

---

Address for correspondence: Michael Weisberg, Department of Philosophy, Stanford University, Stanford, CA 94305-2155. Voice: 650-723-1157. After July 2003: Department of Philosophy, 433 Logan Hall, University of Pennsylvania, Philadelphia, PA 19104-6304.

mweisberg@csli.stanford.edu

ponents in a mixed body. Before turning to this problem, we want to note how we will use the terms "mixture" and "mixt" and how these are related to the philosophical and chemical literature about these issues. We will always use the term "mixture" as a process term. Rather than using "mixture" to refer to the product of the process of mixing, we follow Needham[3] and use the archaic English word "mixt." A mixt is the product of the process of mixture. Using "mixt" as the product term also helps us to distinguish mixts from the sorts of things chemists call mixtures. Chemists' notion of mixture includes both homogenous and heterogeneous samples. The sand on a beach, for example, is a chemical mixture in the modern sense because it is composed of several different kinds of grains. When a chemist uses the term "mixture" without further qualification, she is usually referring to a heterogeneous mixture like sand. An Aristotelian would describe sand as an aggregate and the process which produced it as juxtaposition, not mixture.[4] For Aristotelians, mixts are completely homogenous. With these issues clarified, let us consider Aristotle's argument.

Aristotle sets his argument up in the form of a trilemma. The full discussion can be found in *On Coming to Be and Passing Away*, but we will summarize it briefly.[5] Imagine combining two elements together to form a mixt. There are three possibilities: they both continue to exist after they have been combined to form a mixt, neither continues to exist after they have been combined, or one of the elements continues to exist while the other is destroyed. Let us consider each of these possibilities.

Horn A: If both elements still exist once they have been combined, then they must have been unaltered in the process of mixture. But if they remained unaltered, then they couldn't really have been combined because a mixt requires complete homogeneity.[6] Therefore, mixture is impossible if the elements still exist once they have been combined.

Horn B: If neither element exists once they have been combined, then the elements must have been destroyed in the process of mixture. But if they were destroyed during the process of mixture, then the elements cannot take part in the mixt. If the elements cannot take part in the mixt, then mixture could not have taken place, for there is nothing to combine. Therefore, mixture is impossible if the elements no longer exist once they have been combined.

Horn C: If one element dominates or overwhelms the other, then only one of the elements remains and the other is destroyed. Aristotle gives the example of a single drop of wine being overwhelmed by a large body of water.[7] If one element dominates the other, only one element remains and a mixt cannot be generated. Therefore, mixture is impossible when one element overwhelms another.

In each case, mixture is impossible. Since we have exhausted the possibilities, we must conclude that mixture is impossible.

Aristotle, of course, thinks that there is a solution to this puzzle, a possibility which we overlooked when setting up the trilemma. He writes that "each of the things which were, before they were mixed, still is, but potentially, and has not been destroyed."[8] We were wrong to think that the elemental components must either exist or not exist. We overlooked the possibility that the elements may be only potentially, not actually present. After discussing the trilemma, Aristotle tells us that it is a potential, not actual element that continues to exist in the mixt.[9] So it is neither true that the elemental components continue to exist actually nor that they cease to exist completely. Rather, in a mixt, elemental forms exist as potentialities.

Aristotle correctly emphasizes the distinctness of mixture from generation, corruption, and domination.[10] Unfortunately, his solution seems only to push the problem back a few steps. Even if we accept the idea that the elements are potentially, not actually, present in a mixt, we are still very far from understanding how mixture is possible. Specifically, we are still left with at least three related puzzles concerning mixture:

1. In the processes of mixture, how can different mixts have less or more of the same element, given that the elemental ingredients exist as potentials, not as actual components of the mixt.

2. How can potential elements affect the properties of actual mixts?

3. Why aren't all mixts the same given that they all contain the four elements as potentialities?

These puzzles and associated worries about the metaphysics of mixts set the stage for thousands of years of discussion.

Still today students of ancient philosophy disagree about how to understand Aristotle. Kit Fine recently offered an interpretation of Aristotle on mixture,[11] which suggests that the mixed form is a ratio of the elements in the mixt. Responding, Alan Code argued that Fine's solution does not escape Horn B, since it cannot distinguish between mixture and generation and corruption. It cannot, he argues, "rule out the possibility that the original matter was destroyed," and a "new compound popped into existence."[12]

More seldom considered today are medieval contributions to the discussion. The recent collection of articles on the commentary tradition for Aristotle's *On Coming to Be and Passing Away,* edited by J. Thijssen and H. Braakhuis, laments the fact that this topic is seriously under-studied. Their introductory collection includes only one article on the problem of the mixture, a piece by de Hass which focuses on Philoponus, Alexander of Aphrodisias, Proclus, and Simplicius.[13]

Interesting as it is, this piece by de Hass offers little to the student of the thirteenth century commentary tradition, since, when Rufus was active, between 1230 and 1255, virtually all these works were unavailable. By contrast, Anneliese Maier surveyed the views actually discussed in the thirteenth century. Moreover, the twenty-five medieval authors whose solutions to the problem of the mixture she described include some of Rufus's contemporaries.[14] For that reason, the best way to evaluate Rufus's contribution to the debate is by considering his views in the context provided by Maier.

Maier began her study by describing the alternative solutions inherited by medieval authors from the Islamic world of Avicenna and Averroes. We will follow Maier's lead and begin our discussion with Averroes because Rufus's theory of mixture is a development of certain Averroistic doctrines.

AVERROES: ELEMENTS AS QUASI-SUBSTANCES

It is easier to understand Averroes' views about the status of the elements in a mixt if we begin by considering his theory of elemental change. Let us consider the transition from air to fire. The primary qualities of air are hot and moist whereas the primary qualities of fire are hot and dry. Thus in order to be transformed into fire, air must lose its quality of moistness.

Aristotle and the medievals were aware that such transitions, invoked in discussions of both generation and mixture, took place by degrees.[15] When you burn wood, it doesn't instantaneously become fire. It proceeds through stages of getting hotter and hotter until finally it bursts into flame. The observation is commonplace, but giving a theoretical account of this turned out to be very difficult. The hylomorphic theory requires a transition from one *form* to another, because matter is postulated as the unchanging substratum. To explain elemental change, therefore, one has to be able to explain how the transition between forms can admit of degrees.

The metaphysics of hylomorphism suggests no obvious way to allow for degrees of change, but as we have said, observations of nature require it. This appears to be a serious problem. However, we already know what form the solution will have to take. There will have to be some part of our theory of the elements that can change in degrees. This will give us the necessary theoretical tool with which we can offer a solution to the problem.

Averroes' solution to the problem of gradual change is known as *formae fractae*, which means "broken forms" or "fractured forms." The view is that the elemental forms are not of the same character as ordinary substantial forms. They are *diminished* in some way. Averroes elaborates this point as follows:

> We say that the substantial forms of these elements are diminished in respect of perfect substantial forms; they are, as it were, an intermediate between forms and accidents. And therefore it is not impossible that their elemental substantial forms should be mixed, in such a way that another form should arise from their commingling, as many intermediate colors are made from the mixture of white and black.[16]

For Averroes, elemental change is possible because elemental forms can be enhanced or diminished. For normal substantial forms, this is impossible. Having the elemental forms be halfway between a substance and an accident, however, is supposed to make room for the possibility of greater and lesser degrees of the form. In order to see how this is possible, we need to know that in typical Aristotelian metaphysics, accidents admit of degrees, while substances do not.

First let us consider full-blown substances like the human. One woman cannot be more human than another. All women are of the same species, human. Similarly, for all substantial forms, an object either instantiates the form or it does not. There is no room for degrees.

The situation changes when we consider accidents. Consider an example discussed by Henry of Ghent[17]: the accident of being white. Clearly this accident comes in degrees. There are many degrees between pure white and pure black, indicating that the accident of whiteness comes in degrees. In addition, there are a variety of instantiations of whiteness within different individuals. Henry writes that "whiteness [varies from the] whiteness existing in this rock to the whiteness existing in that wood."[18]

We can now piece together what Averroes had in mind with his doctrine of *formae fractae*. Elements seem to have properties of both substances and accidents. Like substances, they can exist independently; like accidents, they come in degrees and can be combined. Averroes thus posits that while elemental forms are a special case of substantial forms, they have some characteristics of accidents. Therefore, they are "halfway between substance and accident."[19] This seems perfectly reasonable, but also deeply unsatisfying theoretically. Accepting the Averroistic doctrine at face value gives us little insight into the nature of this intermediate position between substance and accident. Without this knowledge, it will be difficult to determine whether this is a useful development of Aristotle's ideas. As it turns out, Richard Rufus elaborated on this doctrine in a far more sophisticated way than Henry, using it to provide a novel and important theory of mixture.

RICHARD RUFUS

In her essay on material substance, Maier discusses various thirteenth century refinements to the basic Averroristic doctrine. She credits Roger Bacon with giving the doctrine of *formae fractae* a modal interpretation.[20] This claim is misleading, since Richard Rufus also provides a modal interpretation to Averroes, and his early lectures influenced Bacon.[21] We will begin with a general discussion of modal interpretations of Averroes based on Rufus's discussion. We will then turn to some specific theoretical concepts that Rufus employs.

Rufus's modal interpretation of Averroes suggests that elemental forms can be more or less present by being more or less actual. This potentiality allows elemental forms to admit of degrees like accidents, but to remain substantial forms potentially. The degree to which an elemental form is potential or actual is the much needed dimension which admits of degrees that will allow us to account for the degrees of elemental change.

Consider again the transition from air to fire. According to this interpretation of Averroes, what happens as we heat the air is that the elemental form of fire becomes closer to actuality while the elemental form of earth becomes further from actuality. Both forms are present during the change, but they differ in how close to actuality they are. Rufus explains this as follows:

> But we should say that in some sense there is remission and intension there — namely, as when we refer to the intension of the form itself as it actually exists, but to the remission of a form itself as it potentially exists, as has been mentioned. Intense heat is caused by the actually existing form of fire; remiss heat is caused by the same form existing in potential, partly moved toward act. And yet the substantial form itself, in its essence as form, does not undergo intension and remission.[22]

Although it seems that the Rufus/Averroes view is a pretty good solution to the problem of elemental change, it is less straightforward how the theory works in the case of mixture. In order to assess and understand the view, we will need to begin by examining one of Rufus's key theoretical distinctions.

Rufus distinguishes between *essential potential* and *accidental potential*. This distinction has to do with how close to being actual a particular possibility is. Forms that are in accidental potential are very nearly actualized. Forms that are in essential potential, on the other hand, are far from being actual: they are "remote possibilities."

The distinction can be made clearer with an everyday example. Imagine three standard spring-loaded mousetraps. Mousetrap *A* is unloaded, meaning the swingbar has not been retracted. Trap *B* is "set," but has not gone off. Mousetrap *C* has been triggered and the swing-bar is swinging down on its target with tremendous force. Although we are talking about high-velocity swinging motion, not forms, we can use the example to make the relevant distinctions.

The swing-bar on A is in essential potential. Although it could be made to swing at high velocity if the trap were set, at the moment the high-velocity motion of the trap is far from becoming actual. B is in accidental potential. A hook, which we should think of for the purpose of this example as something external to the swing-bar, prevents the spring from swinging the bar with tremendous force. One might say that the natural state of the bar attached to this spring would be to swing. However, this state is prevented from becoming actual by the hook, which holds the bar in place. Finally, in the high-velocity motion of C, the potential has been actualized. The important lesson to be extracted from this case is that not all potentials are of the same character. Some are far more likely to be actualized than others.

Since we will be most concerned with understanding the nature of accidental potential, let us consider the second case in more detail. It is important to note that in trap B, there is only one *external* factor *preventing* the motion of the swing-bar from being actual. It needs no further *positive* action by an external agent to move; it only requires the removal of an external obstacle in order for the motion to be actualized. Rufus tells us that accidental potential is the state in which a form "could emerge in act by itself, yet ... is prevented by another." [23]

We need one further concept in order to interpret Rufus's theory of mixture, the concept of *necessity*. For Rufus, the term "necessity" is a term of art. A fairly clear indication of the term's meaning can be found at *DMet* 9.4, where he writes:

> But we should know that ultimate matter is a natural thing, which is a necessity to which no addition is possible. For such matter lacks nothing except only actuality, and that actuality adds no other essence.[24]

In other words, when something is a necessity, it has all the properties that it would have if it were actual. The only thing it lacks is actuality.

It seems that there should be a connection between accidental potential and necessity. Without even looking at the text, it seems pretty obvious that all necessities are also accidental potentials. This is confirmed in the text where Rufus says:

> When there is a necessity, then it is in accidental potential, and it will emerge in actuality by itself unless prevented. And it seems to me that we should suppose that this is the state of component forms in the mixt.[25]

The next natural question to ask, then, is whether necessity and accidental potential are coextensive terms or whether necessity is a proper subset of accidental potential.

Rufus commits himself in at least one passage to accidental potential and necessity being coextensive. He writes "And before there is this necessity, the form is always in essential potential and needs an agent [to be actualized]."[26] Because Rufus believes that among potentials, the distinction between essential and accidental potential is exhaustive, we can conclude that when something is no longer in essential potential, it is in accidental potential.[27] Hence, at the same time the matter becomes a necessity, it also is in accidental potential, and the two terms are therefore coextensive.

RUFUS'S THEORY OF MIXTURE

We now turn to Rufus's theory of mixture. In this section we will present Rufus's theory and discuss the extent to which it solves Aristotle's dilemma. Some of these details form the basis for answering Maier's challenge, which we will discuss in the concluding section of the paper. We begin by examining an early key text (*MMet* 7.16), where Rufus is quite explicit about the nature of mixture:

> We can say that components are in the mixt with an incomplete and diminished actuality, and so the conclusion is evident. But if we are to understand what we are asking about in this section, you should keep in mind that the components are neither actual nor in essential potential in the mixt. Rather they are in accidental potential or incomplete act. This does not, however, occur violently, [but rather] on account of the confusion of its forms in the third nature which is the form of the mixt.[28]

In this passage, Rufus very directly tells us that the elements are neither in essential potential nor are they actual; rather, they are in accidental potential. The elements are not actual substances; they are special kinds of potential forms that are being prevented from becoming actual by something external.

The next step in understanding Rufus's theory requires us to ask how the elements in a mixt can be in accidental potential and what is preventing them from becoming actual. Rufus writes (*In De Gen* 1.6.3):

> And it seems to me that we should suppose that this is the state of component forms in the mixt. Thus each could emerge in act by itself, yet each is prevent by another. Therefore the actual form does not exist, and yet what is there is not matter alone, but a potential moved toward form, and because it is a formal nature, force is consequent on this potential. So, too, when fire is generated from earth, heat is consequent on such a potential moved toward the form of fire in the same matter, [even] when it is still subsumed in the form of earth.[29]

This passage suggests that each elemental form could emerge in act if it was not prevented from doing so *by another elemental form*. In other words, the elements prevent each other from becoming actual. The elements are potentials moved toward (actual) forms, prevented from actuality by each other.

In the earlier passage we quoted from *MMet*, Rufus claimed that all four elements are present in mixts as necessities, or accidental potentials. This re-creates part of the problem that we claimed a modal interpretation of Averroes solves—namely, finding a dimension along which there can be degrees or concentrations of the elemental forms. Since according to the *MMet* text, the elements are all in accidental potential and since we know that accidental potential is the same as necessity, there seems to be no room for intension and remission. The upshot of this worry is that all mixts will be the same. All mixts will consist of the four elements in accidental potential. There will be no difference between different mixts.

Rufus specifically addresses this worry in a latter passage of *In De Gen* (2.4.3). He introduces the question that we have been worrying about as follows:

> Next we can pose the following doubt: If, as has been said, components are necessities in so far as they are present in a mixt, then all mixts will be similar, since short of being actual, a necessity is as complete as it can be.[30]

In this passage, Rufus is expressing exactly the sort of worry we discussed above. He suggests that all mixts will be similar if all the ingredients of the mixt are necessities.

Rufus continues by giving us his first response:

> We should say that this does not follow, since just as fire, light, flame and coal differ as more and less, in some sense, so too necessities differ from the [form] itself. Thus it can happen that some potential, which is a necessity, can be more and less.[31]

Rufus is here asking us to imagine the substances that contain only fire. Light, flame, and coal, he argues, are all species of fire, with different degrees of heat. Yet, he argues, something remains in accidental potential. Given that this can happen when only a single element is present, it seems likely that it could also happen when we have multiple elements.

Rufus is here making a legitimate comparison between mixture and elemental generation. Since we already know from experience that elemental forms can come in degrees, it is only a bit more complicated to see this in the case of mixts. The problem with this answer is that it doesn't really explain anything. What is at issue is not the fact that mixts differ from one another; but rather, *how* mixts could differ from one another. His continued response, however, sheds more light on the problem.

In the second part of the passage, Rufus writes:

> Again we can say that some mixt is such that one contrary is a necessity, but another is not, as happens in some mixt when there is corruption. And on this basis, we can verify what the Commentator [Averroes] says—namely, that the forms of elements are intermediate between substances and accidents or [rather] substantial forms and accidents. Let us understand this in so far as they are in the mixt. For they are there as potentials and not in ultimate actuality. In that sense they are deficient in respect of a substantial form absolutely [speaking]. But in so far as when joined together they can perfect matter, they are more than accidents.[32]

In this passage, Rufus gives a substantial explanation for the similarity problem, but one that is quite confusing. He tells us that in mixts, not all of the elements need be necessities. Perhaps this means that only one element needs to be a necessity. The others are all essential potentials, but can be closer to or further from actuality. These differences in actuality allow us to explain how different mixts have different properties and, thus, how they are distinct.

That only one element need be in accidental potential makes a good deal of sense. The problem is that this seems to conflict with the earlier passage from *MMet* where Rufus tells us that all of the elements in a mixt are in accidental potential. We know that the *MMet* text is earlier than the *In De Gen* text. Further, this passage is consistent with the earlier passage quoted from the same book that we have been discussing. An easy way out of the difficulty would be to claim that Rufus changed his mind in the later texts. If we could find an even later passage with the same view, we would be quite justified in going this interpretive route.

Another possibility, which seems more promising, is that Rufus resolves this issue by implicitly relying on a traditional explanation which allows mixts to be composed of the elements in different proportions. This militates against common prejudices about non-atomistic Medieval theories of mixture which are alleged to only have room for qualitative, not quantitative dimensions. But Rufus clearly discusses different proportions of the elements both in *In De Gen* 2.4.3 and in his Oxford theology lectures. So this seems likely to be his solution to the problem, but the subject still requires further investigation.

CONCLUSION

Maier claimed that all non-atomistic medieval theories were unable to solve the problem of mixture. They could assert that elements combined to form mixts, but they couldn't explain how mixts reflected the states of the elements. Rufus, as we have seen, has a theory about how the elements can remain in the mixt and how the mixt reflects the states of the elements. Like Averroes, he thinks elemental forms are in a certain kind of potential. However, he rejects Averroes' claim that elemental forms are quasi-accidental forms, and he is able to give a much more detailed account of the status of the elemental forms and prime matter in the mixt. There are still problems with Rufus's account, both in his actual theory and as we have interpreted it. It is clear, however, that had Maier been acquainted with Rufus's account of mixture, she would have had to modify her claim. Though unpolished, Rufus has a fundamentally satisfactory account of the potential existence of elements in a mixt.

ACKNOWLEDGMENTS

We would like to thank Paul Bogaard, Joseph E. Earley, Sr., D.S. Neil van Leeuwen, Neil Lewis, Paul Needham, and Allen Wood for discussing issues about mixture with us. We also thank Deena Skolnick for many helpful comments on earlier drafts of this paper. This research was partially supported by an NSF Graduate Research Fellowship (M.W.) and NEH Grant RZ-20539 (R.W.).

NOTES AND REFERENCES

1. For a general discussion of Richard Rufus's life and work, the reader may wish to consult Rega Wood's article on Richard Rufus of Cornwall *in* Blackwell Companion to Philosophy in the Middle Ages. T.B. Noone and J. J.Gracia, Eds. Blackwell Publishers. Malden, MA. [in press]. Information about the Richard Rufus of Cornwall Edition can also be found at http://www.stanford.edu/~regawood/rufus/
2. MAIER, A. 1952. Die Struktur der materiellen Substanz. *In* An der Grenze von Scholastik und Naturwissenschaft, 2nd ed., pp. 88, 137–139. Rome.
3. NEEDHAM, P. Duhem's theory of mixture in the light of the Stoic challenge to the Aristotelian conception. *In* Studies in History and Philosophy of Science. In press.
4. ARISTOTLE. Metaphysics VIII.2.1043a13 and On Sense and Sensibilia III.440b1-3. [*Editor's note*: References to Aristotle's works are given according to Bekker's 1831 system, showing page number, column letter, and line number. This information is generally included as marginal notes to current published translations.]
5. ARISTOTLE. On Coming to Be and Passing Away I.10.327a33-327b10. [*Editor's note*: This paper follows the practice of much current philosophical literature, and of the Forster translation, in using the English word "mixture" to translate the Greek $\mu\iota\xi\iota\varsigma$ (*mixis*). The Foster translation also uses "mixture" to render the Greek $\mu\iota\kappa\tau o\varsigma$ (*mixtos*)—for which this paper uses the antique English word "mixt." In contrast, the revised Oxford translation by H. H. Joachim (1984. The Complete Works of Aristotle. J. Barnes, Ed. Princeton University Press. Princeton, NJ) construes *mixis* (in On Coming to Be) by the English word "combination," and uses forms of the verb "to combine," rather than forms of "to mix." Most people familiar with the language used in contemporary chemical discourse would find the Joachim translation much more congenial than the Foster version. Michael Weisberg has pointed out that, in translating a passage in an earlier work of Aristotle (Topics 4.2, 122b30, Barnes, ed., page 206), Joachim uses the English "mixture" for *mixis* ($\mu\varepsilon\iota\xi\iota\varsigma$). (For the Greek text, see: Ross, D. 1974. Aristotles Topica et Sophistici Elenchi. Clarendon Press. Oxford.) In this passage, Aristotle contrasts *mixis* ($\mu\varepsilon\iota\xi\iota\varsigma$) with *krasis* ($\kappa\rho\alpha\varsigma\iota\varsigma$), "blending"—pointing out that merely putting two components together (mixing) does not necessarily mean that those components will blend (combine). It seems that Aristotle's use of *mixis* ($\mu\varepsilon\iota\xi\iota\varsigma$) in Topics is similar to the modern chemists' use of the English word "mixture" and quite different from his more technical use, in On Generation, of *mixis* ($\mu\iota\xi\iota\varsigma$) to mean a homogeneous ("blended") combination. (In a preface to his translation of On Coming to Be, Foster comments on the difficulty of translating technical terms, and also has high praise for Joachim's work.)

6. ARISTOTLE, On Coming to Be, I.1.314a20 and I.10.328a10
7. ARISTOTLE, On Coming to Be, 1.10.328a25-30
8. ARISTOTLE, On Coming to Be, I.10.327b2-27
9. ARISTOTLE, On Coming to Be, 1.10.327b23-27
10. ARISTOTLE, On Coming to Be, I.10.327b8
11. FINE, K. 1995. The problem of mixture. Pac. Philos. Q. **76**: 266–369.
12. CODE, A. 1995 Potentiality in Aristotle's science and metaphysics. Pacific Philosophical Quarterly **76**: 405–418; the forthcoming Symposium Aristotelicum for 1999, devoted to book 1 of On Coming to Be and Passing Away can be expected to continue the controversy. In press; ARISTOTLE, De generatione et corruptione I: Proceedings of the Symposium Aristotelicum, Deurne, 1999. Frans de Haas and Jaap Mansfeld, Eds. Oxford University Press. Oxford, UK.
13. DE HASS, F. 1999. Mixture in Philoponus: an encounter with a third kind of potentiality. In The Commentary Tradition on Aristotle's De generatione et corruptione. J. Thijssen and H. Braakhuis, Eds. Turnhout.
14. MAIER, Die Struktur, 3–140.
15. ARISTOTLE, On Coming to Be, I.10.328a33 and II.4.331a24
16. AVERROES. 1550. Aristotelis opera cum Averrois Cordubiensis commentariis 5, In De Caelo 3.67.ed. apud Junctas. Venice.
17. MAIER, Die Struktur, 38.
18. MAIER, Die Struktur, 38–39.
19. AVERROES,In De Caelo, 3.67.
20. MAIER, Die Struktur, 46.
21. Maier, of course, cannot be faulted for this mistake since the manuscript of Rufus's commentary on Aristotle's On Coming to Be was not known in her day. For more on the relationship between Rufus and Roger Bacon, see WOOD, R. 1997. Roger Bacon: Richard Rufus's successor as a Parisian physics professor. Vivarium **35**: 222–250.
22. RICHARDUS RUFUS CORNUBIENSIS. In De generatione et corruptione Aristot. 2.4.2: "Sed dicendum quod aliquo modo est ibi remissio et intensio—ita, scilicet, ut dicamus remissionem in ipsa forma actu exsistente; remissionem autem in ipsa secundum potentiam exsistente, ut prius tactum est. Et a forma ignis actu exsistente causatur calidum intensum, et ab eadem potentia exsistente, mota tamen in parte ad actum, causatur calidum remissum. Sed tamen ipsa forma [lectio dubia E] substantialis secundum quod est sub ratione formae non suscipit intensionem et remissionem" (Q312.18ra). Quotations are based on a provisional edition of Rufus' works edited by R. Wood and N. Lewis. Since this edition is unpublished, citations are to the manuscripts at Erfurt University, Amploniana Quarto 312.
23. RUFUS, In De Gen 1.6.3: "Et in tali statu, ut mihi videtur, debemus ponere formas miscibilium in mixto, ita ut cum quaelibet possit de se exire in actum, quaelibet tamen per aliam prohibetur" (Q312.16vb-17ra).
24. RUFUS. Dissertatio in Metaph. Aristot. 9.4: "Sciendum est autem quod materia ultima est res naturalis, quae est necessitas cui impossibile est additio. Nihil enim [tamen E] tali materiae deficit nisi solum actualitas, et illa [add. vero E] actualitas non addit essentiam aliam" (Q290.27vb).
25. RUFUS, In De Gen 1.6.3: "et cum est necessitas, tunc est in potentia accidentali et per se ipsam exiens in actum si non sit prohibita. Et in tali statu, ut mihi videtur, debemus ponere formas miscibilium in mixto" (Q312.16vb).
26. RUFUS, In De Gen 1.6.3: "Et antequam est necessitas, semper est forma in potentia essentiali et indiget agente" (Q312.16vb).
27. Of course unlike accidental potential, which has a very specific state attached to it, there are many ways for a form to be in essential potential. This is where the terms "near" and "remote potential" become useful. "Accidental potential" refers only to the state of lacking nothing but actuality. Forms that are in essential potential, however, can be near or far from becoming actual.
28. RUFUS. Memoriale in Metaph. Aristot. 7.16: "Dici potest quod miscibilia sunt in mixto actu incompleto et diminuto, et ita patet conclusio. Sed ut quaesita in parte ista pateant intellige quod miscibilia sunt in mixto non potentia essentiali nec in actu, sed in potentia accidentali sive secundum actum incompletum, non tamen violente, propter

confusionem formarum suarum in naturam tertiam quae est forma mixti" (Q290.49va).
29. RUFUS, In De Gen 1.6.3: "Et in tali statu, ut mihi videtur, debemus ponere formas miscibilium in mixto, ita ut cum quaelibet possit de se exire in actum, quaelibet tamen per aliam prohibetur. Sic ergo non est ibi actu forma, nec tamen sola materia, sed potentia mota ad formam—ad quam potentiam, quia ipsa est natura formalis, consequitur virtus. Et sic etiam ad talem potentiam motam ad formam ignis, cum ex terra generatur ignis, consequitur caliditas in eadem materia adhuc exsistente sub forma terrae" (Q312.16vb-17ra).
30. RUFUS. In De Gen 2.4.3: "Consequenter potest dubitari sic: Si, sicut dictum est, miscibilia prout sunt in mixto sunt necessitates, ergo omnes mixtiones erunt similes, quia quod necessitas est completum est quantum potest citra actum" (Q312.18ra).
31. RUFUS. In De Gen 2.4.3: "Dicendum quod non sequitur hoc, quia sicut ignis, lux et flamma et carbo differunt aliquo modo secundum maius et minus, sic et necessitates ad ipsa. Et sic contingit potentiam aliquam, quae est necessitas, esse et secundum plus et secundum minus." (Q312.18ra).
32. RUFUS. In De Gen 2.4.3: "Possumus iterum dicere quod aliqua mixtio est sic, in qua unum contrarium est sic in necessitate, alterum autem non, ut est in aliquo mixto quando est in corruptione. Et iuxta hoc possumus verificare hoc quod dicit Commentator, quod formae elementorum sunt media inter substantias et accidentia sive formas substantiales et accidentia,ut intelligamus hoc secundum quod sunt in mixtione. Sunt enim ibi ut potentiae et non in suis actibus ultimis, et in hoc deficiunt a forma substantiali simpliciter; eo autem quod simul coniunctae possunt perficere materiam, plus habent quam accidentia" (Q312.18ra).

The Lavoisier–Kirwan Debate and Approaches to the Evaluation of Theories

MICHAEL AKEROYD

*Division of Computing Technology, Bradford College,
Bradford, BD7 1AY, United Kingdom*

ABSTRACT: This paper applies concepts proposed by Larry Laudan for evaluation of scientific theories to problems that arose for Antoine Lavoisier's oxygen theory during the period from 1780 to 1790. Lavoisier's original theory (1779) had an *empirical* problem (why does dilute sulfuric acid produce hydrogen on contact with metals, whereas concentrated acid, even when heated, does not?). The solution to this puzzle led to a more refined *conceptual* problem, regarding the overall consistency of the theory. Conceptual arguments used by the French school of chemists to defend Antoine Lavoisier's oxygen theory against the phlogistic arguments (1784) of Richard Kirwan touch on the problem of criteria for satisfactory explanation.

KEYWORDS: explanation; empirical problems; conceptual problems; phlogiston

Cassandra Pinnick and George Gale[1] recalled Carl Hempel's[2] famous 1942 injunction that historians should reject "narrative explanations" and adopt a "covering law model" of explanation.

Hempel hoped to encourage the seeking of generalizations that might connect with the norms of philosophy, methodology, and maybe even with the practice of science. Present-day historians are unimpressed: for instance, Clayton Roberts states:

> Historical knowledge concerns the concrete, the particular, the unique.... Historians study the battle of the Marne, not battles in general; they seek the causes of the Enlightenment, not of enlightenments in general; they study the rise of Hitler, not of dictators in general.[3]

LAUDAN'S STRATEGY OF THEORY CHOICE

In a somewhat similar vein, Larry Laudan[4] holds that scientific theories generally develop incrementally—scientists find particular solutions for specific problems they encounter. A corollary of this concept is that theories should be evaluated on their ability to meet and resolve difficulties. Whitt[5,6] and also Kukla[7] have criticized Laudan's problem-solving approach to evaluation of rival theories. Both these writ-

Address for correspondence: Michael Akeroyd, Division of Computing Technology, Bradford College, Great Horton Road, Bradford, BD7 1AY, UK. Voice: 1274-433417; fax: 0044-1274-736175.
 m.akeroyd@bilk.ac.uk

ers propose technical devices for measuring a theory's conceptual growth or improvement, in preference to Laudan's strategy of assessing a theory's abilities in problem reduction. With Lipton,[8] I hold that it is usually easier to obtain reliable negative knowledge than dependable positive knowledge, so that Laudan's approach should be preferred.

Laudan proposed a taxonomy of types of problem that may confront a theory that is being evaluated:

> To begin with, I suggest that we separate *empirical* from *conceptual* problems.... At the empirical level, I distinguish between potential problems, solved problems and anomalous problems....[9]

"Anomalous problems" are actual problems that rival theories solve, but are not solved by the theory in question. Unsolved or potential problems need not be anomalous ones. In addition to empirical problems, theories may be confronted with *conceptual* problems. Such problems arise for theory T in any of the following circumstances:

(1) when T is internally inconsistent, or the theoretical mechanisms it postulates are ambiguous;
(2) when T makes assumptions about the world that run counter to other theories or to prevailing metaphysical assumptions, or when T makes claims about the world that cannot be warranted by prevailing epistemic and methodological doctrines;
(3) when T violates principles of the research tradition of which it is a part; or
(4) when T fails to utilize concepts from other, more general theories to which it should be logically subordinate.[9]

Conceptual problems, like anomalous empirical problems, indicate liabilities in theories (i.e., partial failure on their part to serve all the functions for which they were designed).

Four years previously, Laudan had elaborated other differences between conceptual problems and empirical problems: Conceptual problems are "characteristics of theories and have no existence independent of the theories which exhibit them, not even that limited autonomy which empirical problems sometimes possess."[10]

Empirical problems are "first order" questions about "substantive entities" in some domain. In contrast, conceptual problems are "higher order" questions about the well-foundedness of the conceptual structures (e.g., theories) which have been designed to answer first-order questions. One vivid example of a conceptual problem is the discovery that a theory is logically inconsistent, and thus self-contradictory.[11]

BACKGROUND OF THE PROBLEM

This paper deals with a brief period in the history of the Chemical Revolution—the period from the announcement in 1784 of an "updated" version of the phlogiston theory[12] by the Anglo-Irish chemist Richard Kirwan (1733–1812) to Kirwan's *abandonment* of that theory in 1791.

In this discussion of that historical period, modern names for chemicals will be inserted in parentheses, as an aid to the reader. In a well-known history of chemistry, John Partington[13] devotes eleven pages to acknowledging Kirwan's role in the discovery of stoichiometry (that branch of chemistry concerned with the exact proportions required for the combining weights of reagents in chemical reactions), while he describes Kirwan's revised version of phlogiston theory as "interesting" on only one page. Arthur Donovan's[14] collection of historical essays on the same period gives Kirwan only cursory treatment. Philip Kitcher[15] devotes ten pages to a detailed analysis of the conceptual issues involved with Kirwan's theory. He analyzes only metal/metal oxide transitions and concludes[16] that by 1784 Kirwan's arguments had started "to run into inconsistency." In his extensive survey of the Chemical Revolution, John McEvoy[17] often mentions Joseph Priestley as a defender of the phlogiston theory, but does not mention Kirwan. McEvoy—who classifies Larry Laudan, Thomas Kuhn, Imre Lakatos and Stephen Toulmin as "postpositivist' philosophers"[18]— concludes that:

> [t]his [multidimensional] model of the Chemical Revolution challenges the ingrained empiricist assumption, shared by positivists and post-positivists alike, that the *evaluation* of scientific theories should be governed by empirical considerations alone and supports the view that strictly nonempirical, or "conceptual," parameters play an important role in the articulation, development, and *evaluation* of scientific theories[19] [emphasis in the original].

This paper concerns problems that arose for the oxygen theory of acidity of Antoine Lavoisier (1743–1794) because of the observation that *cold, dilute* sulfuric acid reacts with metals with concomitant evolution of an inflammable gas (hydrogen)— whereas *hot, concentrated*, sulfuric acid *does not react* with metals to yield this gaseous product.

LAVOISIER'S THEORY OF ACIDS

The element oxygen held a preeminent place in Lavoisier's system, and also in his attacks on the Phlogiston Theory developed by Georg Stahl (1659–1734). In Lavoisier's original system, published in 1779, oxygen was the principle of combustion (thereby replacing Stahl's "phlogiston") and also the principle of acidity—oxygen was taken to be an essential component of all acidic substances. One of the early problems encountered by Lavoisier's theory was to explain why there was evolution of hydrogen when a metal (which did not contain hydrogen) reacted with cold dilute sulfuric acid (which, according to the original theory, did not contain hydrogen) to dissolve and form a salt. The phlogiston theory had an explanation of these observations: metals were considered to be compounds of "a calx plus phlogiston"—the principle of fire. When metals dissolved in cold dilute acid, phlogiston was released in the (flammable) gas that was produced.

In 1783, after experiments with iron and steam had suggested that water was a compound made up of hydrogen and oxygen, Lavoisier's coworker Pierre Laplace[20] (1749–1827) proposed that the hydrogen evolved in the reaction of dilute acid with metals came from the water of solution (as was possible according to the theory). Lavoisier adopted this *ad hoc* suggestion[20] and by so doing solved an empirical problem, but also created a conceptual problem (in Laudan's terminology).

KIRWAN'S DEFENSE OF PHLOGISTON THEORY

Around 1780, Richard Kirwan had accepted that Stahl's version of the phlogiston theory had been refuted by the arguments and experimental data of the Lavoisier

School, but he was still convinced that metals did in fact lose important properties on calcination[21] (conversion to a calx, what we now call oxidation). He therefore proposed a new hybrid theory incorporating (as he saw it) the best of the old and the new views. Kirwan accepted that, on combustion, sulfur absorbed "pure air" (oxygen in the Lavoisier system) to form vitriolic acid (sulfuric acid),[22] that water was a compound of inflammable air (hydrogen) and dephlogisticated air (oxygen),[23] and that ordinary atmospheric air was a mixture of three simple substances —dephlogisticated air (oxygen), phlogisticated air (nitrogen), and fixed air (carbon dioxide).[24]

In 1784 Richard Kirwan published a book[12] on a revised phlogiston theory that the Lavoisier School took seriously enough to arrange for a French translation to be published—along with appended rebuttals by French chemists to Kirwan's arguments. This French edition, including the rebuttals, was retranslated into English in 1789.[21] Little new experimental material had been discovered in the period from 1785 to 1787; it is not surprising that most of the argumentation and counter-argumentation in this volume was of a type that Laudan's taxonomy classifies as *conceptual*.

The key postulates in Kirwan's revised phlogistic system were:

(a) "inflammable air" (hydrogen) is pure phlogiston, and is contained in metals;
(b) inflammable air (hydrogen) and "vital air" (oxygen) could unite to form "fixed air" (carbon dioxide) *as well as* water;
(c) charcoal is *not* an element (carbon), but a compound of phlogiston and fixed air. (That is, charcoal was considered to be higher hydride of oxygen than water, in modern notation. H_nO (n>2) as compared with H_2O.[25])

In 1784[13] and again in 1789[25] Kirwan mocked Laplace's proposal—that the gas produced when metals contacted dilute acid came from solvent water—as *inconsistent*. It was known that gaseous hydrogen reduced cold, concentrated, sulfuric acid to sulfur dioxide, whereas metals required the assistance of heat to reduce sulfuric acid to sulfur dioxide. Kirwan asked, "If hydrogen has a greater affinity for oxygen than sulphur, how then can the metal reduce the cold water associated with the dilute acid more easily than the pure hot concentrated acid?"[26]

That is to say, since hydrogen can attract oxygen away from sulfur in the cold while a metal can perform that function only with the assistance of heat, we should expect that a metal reacting with dilute acid should require the assistance of heat to extract oxygen away from water, and thus liberate hydrogen gas. Both the modified argument of the Lavoisier School (that the hydrogen evolved derives from water) and the objection of Kirwan are well formed and logically sound, but those two arguments are inconsistent with each other.

Lavoisier's advocate Antoine Fourcroy (1755–1809) rebutted Kirwan's arguments as follows:

1. Most of his [Kirwan's] objections can be removed by a more intimate knowledge of our [Lavoisier's] theory.
2. Those that cannot be so removed are explained by no other theory.[27]

This seems to be a classic example of understanding the difference between a "potential problem" and an "anomalous problem." Fourcroy admits that the antiphlogistic system is not the "final word," a "perfected system," but is a theoretical system with potential for improvement.

It might be objected[28] that the problem cited by Kirwan is empirical, since his criticism relies on the empirical fact that gaseous hydrogen reduces concentrated sulfuric acid in the cold, whereas conceptual problems concern the *meaning* of theoretical terms. But what Kirwan pointed out was that Lavoisier's theory runs into *internal* contradiction in attempting to provide coherent simultaneous explanation of two empirical observations. Clearly, this was a *conceptual* problem for the Lavoisier theory.

Kirwan's system was not merely an *ad hoc* adjustment of Stahl's discredited phlogiston theory. It gave better accounts in two important areas of known phenomena than did the Lavoisier Theory:

1. Kirwan regarded "marine acid" (hydrochloric acid, also known today as muriatic acid) as a compound of phlogiston (hydrogen) and "dephlogisticated marine acid" (chlorine). This is consistent with the modern view. In contrast, the Lavoisier school held that marine acid was the oxide of an unknown element ("murium") and that the green gas it evolved on oxidation (chlorine) was a higher oxide of murium. This viewpoint was not disproved until Davy's electrochemical work of 1810.

2. Kirwan's theory had an explanation for the production of an inflammable gas (which was not hydrogen) during the heating of charcoal with either metal oxides or steam. This phenomenon was not satisfactorily accounted for by the Lavoisier School until the characterization of carbon monoxide in 1799.

Joseph Priestley (1733–1804), *circa* 1783, had observed that when steam was passed over incandescent charcoal, a "pure" inflammable air with a density half that of atmospheric air (now recognized as a mixture of carbon monoxide and hydrogen) was sometimes obtained. Often he obtained an "impure" inflammable air which contained fixed air (carbon dioxide). In order to rationalize Priestley's claim that "inflammable air" was formed, mixed with "fixed air," on passing atmospheric air over red-hot charcoal, the Lavoisier School maintained that Priestley had not sufficiently dried his air and that hydrogen liberated from water vapor present as an impurity had reacted with charcoal to form methane[29]:

$$H_2O + C \rightarrow H_2 + CO$$

$$2H_2 + C \rightarrow CH_4$$

In Kirwan's system, charcoal (carbon) was considered to be a compound of fixed air (water plus phlogiston) and pure phlogiston. This auxiliary hypothesis explained the fact that when charcoal was heated in a sealed metal screw-topped tube with saltpeter (potassium nitrate), a volume of fixed air obtained that weighed (by density measurements) more than three times the original weight of the charcoal. In modern terms this reaction is:

$$4KNO_3 + 5C \rightarrow 2K_2CO_3 + 2N_2 + 3CO_2$$

According to Kirwan, the charcoal released "its own" fixed air and phlogiston, which then reacted with the dephlogisticated air (oxygen) present in the saltpeter to form more fixed air—*not* water as one might expect. Although this ingenious hypothesis could not be refuted experimentally, it laid itself open to conceptual criticism from the Lavoisier School:

> One of the experiments upon which Mr Kirwan grounds his opinion... is that of Mr. Priestley, in which that philosopher, by passing air over charcoal in a red hot earthen

tube, obtained much inflammable air together with fixed air; and by burning this inflammable air with an equal volume of dephlogisticated air, he obtained a greater quantity of fixed air than that of the inflammable air made use of; but this result is not favourable to him, for how, according to his theory, can air change charcoal into inflammable air? It is much better proved that the inflammable gas, and the carbonic acid obtained, arise from the decomposition of the water dissolved in the air, and that this inflammable gas, holding much carbone [sic] in solution, afforded a large quantity of carbonic acid, and a small quantity of water in its combustion with vital air. Besides, we must not forget,

1. That no one has changed all the charcoal into inflammable gas, which is necessary, according to Mr. Kirwan's theory.

2. That the pure inflammable gas, or hydrogene [sic], affords only pure water on combination with oxigene [sic].

3. That it is absolutely necessary that charcoal should be present in the formation of carbonic acid or fixed air.

4. That when hydrogenous gas, which holds much carbone in solution, is burned, such as that disengaged in the experiment of Dr. Priestley, a much greater quantity of vital air is necessary, than to burn an equal volume of hydrogenous gas, and by this combustion a mixture of water and carbonic acid is obtained proportional to the quantities of hydrogene and carbone which form it. These positive facts absolutely destroy the theory of Mr. Kirwan, and agree perfectly with our own doctrine.[29]

Earlier in the book, Lavoisier used a conceptual argument against Kirwan's auxiliary hypothesis that fixed air was not an oxide of carbon but a higher hydride of oxygen. He stated forcefully:

We will show, in the discussions of the experiments of Mr. Kirwan, that universally whenever he obtained fixed air, or carbonic acid, charcoal was present: and that whenever charcoal was not present, the carbonic acid was not formed: lastly, that it is a gratuitous and unfounded supposition, that inflammable gas or hydrogene is one of the elements of charcoal.[30]

The force of this argument can be appreciated when, two years later, Kirwan announced his conversion to the ideas of the Lavoisier school.

I know of no single decisive experiment by which one can establish that fixed air is composed of oxygen and phlogiston, and without this proof it seems to me impossible to prove the presence of phlogiston in metals, sulphur or Saltpeterluft (nitrogen).[31]

In some ways Kirwan's 1791 concession to the Lavoisier School was unfortunate. In the domain of metal calcinations, Kirwan's approach was closer to modern ideas than Lavoisier's theory was. According to the "electron gas" hypothesis, metals are made up of a positive metal ions ("calx") and the "electron gas" ("phlogiston") responsible for metallic malleability, ductility, and luster. Metal oxides are regarded as aggregates of metal ions ("calx") and negative oxide ions (phlogisticated oxygen atoms). Kirwan[32] believed that:

Metallic substances by calcination, lose their phlogiston, which is nothing but pure inflammable air in a concrete state, and at the same time unite most commonly to fixed air, formed during the operation;

Since he believed that fixed air was phlogisticated oxygen (H_3O) it is obvious in this case that Kirwan's ideas were closer to the current viewpoint that the formula of zinc oxide should be written as $Zn^{++}O^=$ rather than ZnO. Kirwan confused his "fixed air" (phlogisticated oxygen) with the gas or "air" invisibly "fixed" in limestone, and liberated by the action of dilute acid (carbon dioxide).[33]

The modern "band theory,"[34] which explains electrical conductivity in metals, organic crystals, and some ceramics is also closer to Kirwan's viewpoint than that of Lavoisier and his school. The solid is treated as a giant molecule and a "band" represents the sum of all the individual atomic/molecular orbitals of the solid. A "full band" represents the sum of localized electron densities. Burdett states that "[a] metal is a solid with a partially filled energy band."[35] Shriver, Atkins, and Langford state that "[a] solid is an insulator if enough electrons are present to fill a band completely and there is a considerable energy gap before an empty orbital becomes available."[36]

Since, on this definition, metals are defined by the possession of full, partially filled and empty energy bands, and insulators (often metallic oxides) by the possession of only full or empty bands, it is clear that the phlogistic intuition that traditional metals were more complex than metal oxides (calxes) was reasonable. This statement is further reinforced by Burdett, who states: "The 'typically ionic' compound TiO_2 becomes metallic after treatment with a small amount of tertiary-butyl lithium."[37] This suggests that something added to a calx may make it metallic. Kirwan's intuitions and his subsequent explanations of a wide range of phenomena were not unreasonable. Calxes (metallic oxides in modern terminology) are indeed *simpler* than metals, as Kirwan held.

A PROBLEM RESOLVED

The question Kirwan posed to Lavoisier's original theory was, paraphrased:

> If metals are elements, if water is either an element or an inert solvent, and acids are compounds of non-metals and oxygen, why is something that you claim to be the element hydrogen evolved on mixing metals with acids?

This seems to be a first-order question about substantive entities. The experimental observation that dilute acids dissolve iron with evolution of inflammable gas has autonomy outside the domain of Lavoisier's theory: It is a problem for technology (e.g., the development of metal containers for storing food) as well as for theory. In contrast, Kirwan's criticism of Laplace's resolution of this problem (in this case water is not an inert solvent) highlights a new internal conceptual problem for Lavoisier's theory. If metals have sufficient affinity for oxygen to release hydrogen from dilute aqueous solutions of acids, why does not the same reaction happen for concentrated acid? A logical argument, satisfactory in one part of the theory's domain, fails to explain related facts in another part of that domain. This second problem has no autonomy outside the domain of the theory: it is a conceptual (rather than merely an empirical) problem.

Laudan admits that there is a spectrum ranging from straightforward empirical problems to straightforward conceptual problems with no sharp discontinuity between the two classes. He cites—as a "vivid" example of an intra-scientific conceptual difficulty—the observation that geological and biological evidence in the 19th century suggested that the world was very old, partially fluid under the surface, and losing heat only slowly, which conflicted with the thermodynamics and physics of Lord Kelvin.[38] There is no ambiguity over theoretical terms here. The conceptual problem arises because calculations based on a new, highly mathematical, theory contradict old data, previously classified as uncontroversial. If Laudan counts this example as "conceptual," then Kirwan's second criticism should also count as a conceptual problem for Lavoisier's theory.

A partial resolution of this conceptual problem was given by Lavoisier's colleague, A. de Fourcroy.[39] It is true that the reaction between iron and water is extremely slow in the cold and normally requires a temperature of more than 100 degrees Celsius, but, according to de Fourcroy, this is because a coating of iron oxide is formed during the early stages of the reaction and acts as a protective coating, hindering the further reaction with water. In the presence of *dilute* sulfuric acid, the acid dissolves the freshly formed iron oxide coating—*via* a well-known acid–base reaction—to form the soluble salt iron sulfate, thus exposing the metal surface to the additional attack by water. According to the most fully developed account of these questions given by the Lavoisier School, a metal could not react directly with an acid to form a salt: the metal had first to unite with oxygen. Oxygen could be obtained in three ways: by extraction from the acid present (as happened in reaction with hot concentrated acid), by removing oxygen from the solvent water (as happened in reactions of cold, dilute acids), or from oxygen *dissolved* in the solvent water (e.g., unreactive metals such as copper react slowly with weak acids such as acetic acid). Similar observations require diverse explanations, suggesting that the theory might have conceptual difficulties.

The particular conceptual problem of why dilute sulfuric acid had a different mode of action as compared with concentrated sulfuric acid was not satisfactorily answered by the Liebig Hydrogen Theory of Acids of 1839, the conceptual scheme that succeeded Lavoisier's original theory. A satisfactory resolution of this question was not obtained until 1887, when Svante Arrhenius (1859–1927) proposed that dilute sulfuric acid is dissociated in solution into positive hydroxonium ions and the negative counter-ions, whereas concentrated sulfuric acid reacted as the covalent molecule H_2SO_4, a vigorous oxidizing agent.

ACKNOWLEDGMENTS

I thank an anonymous referee and Joe Earley for their comments on a previous draft of this paper.

NOTES AND REFERENCES

1. PINNICK, C. & G. GALE. 2000. Philosophy of science and history of science: a troubling interaction. J. Gen. Phil. Sci. **31:** 109–125.
2. HEMPEl, C.G. 1942. The function of general laws in history. J. Phil. **39:** 35–48.
3. ROBERTS, C. 1996. The Logic of Historical Explanation. Pennsylvania State University Press. University Park, PA. p. 8.
4. LAUDAN, L. 1981. A problem solving approach to scientific progress. *In* Scientific Revolutions. I. Hacking, Ed.: 144–155. Oxford University Press. Oxford.
5. WHITT, L.A. 1989. Conceptual dimensions of theory appraisal. Stud. Hist. Phil. Sci. **19:** 517–529.
6. WHITT, L.A. 1992. Indices of Theory Promise. Phil. Sci. **59:** 612–634.
7. KUKLA, A. 1990. Ten types of scientific progress. Proc. Philos. Sci. USA 1990 **I:** 457–466.
8. LIPTON, P. 1995. Popper and reliabilism. *In* Karl Popper: Philosophy and Problems. A. O'Hear, Ed.: 31–43, see p. 39. Cambridge University Press. Cambridge, UK.
9. LAUDAN, "A problem solving approach," p. 146.
10. LAUDAN, L. 1977. Progress and Its Problems. Routledge and Kegan Paul. London. pp. 57–64.

11. Ibid., p. 57.
12. KIRWAN, R. 1784. An Essay on Phlogiston and the Constitution of Acids. J. Johnson. London.
13. PARTINGTON, J.R. 1962. A History of Chemistry, Vol. III. MacMillan, London. pp. 660–672, [see esp. p. 664].
14. DONOVAN, A. 1988. The Chemical Revolution: Essays in Reinterpretation. Osiris [new series] **4:** 15–231.
15. KITCHER, P. 1993. The Advancement of Science. Oxford University Press. Oxford, UK.
16. Ibid., p. 288.
17. MCEVOY, J.G. 2000. In search of the chemical revolution. Found. Chem. **2:** 47–73.
18. Ibid., p. 55.
19. Ibid., p. 64.
20. PARTINGTON, History of Chemistry, p. 444.
21. KIRWAN, R. 1789. An Essay on Phlogiston and the Constitution of Acids. A new edition to which are added, notes, exhibiting and defending the antiphlogistic theory; and annexed to the French edition of this work; by Messrs de Morveau, Lavoisier, de la Place, Monge, Berthollet, and de Fourcroy, translated into English, with additional remarks and replies by the author, pp. 166–168. J. Johnson. London [also published in facsimile edition in 1968 by F. Cass, London].
22. Ibid., p. 63.
23. Ibid., p. 43.
24. Ibid., p. 30.
25. Ibid., p. 54, p. 68, and see esp. p. 231.
26. Ibid., p. 231.
27. FOURCROY, in KIRWAN (1789), Essay on Phlogiston, p. 237.
28. This argument has been produced twice, once by a referee of a previous version and once by a philosopher when an earlier version of this paper was read.
29. FOURCROY, in KIRWAN (1789), Essay on Phlogiston, p. 226
30. Lavoisier, in KIRWAN (1789), Essay on Phlogiston, p. 57.
31. PARTINGTON, History of Chemistry, p. 664.
32. KIRWAN (1789), Essay on Phlogiston, p. 168.
33. A referee of a previous version made the comment that, as the loss of phlogiston in Stahl's original theory was associated with the loss of heat on combustion, Kirwan's proposal of a return of phlogiston in the form of "fixed air" was self-evidently incoherent. However, neither Lavoisier or Fourcroy picked up on any alleged incoherence and Kirwan never proposed that all the departed phlogiston returned in the form of fixed air.
34. BURDETT, J.K. 1997. Chemical Bonding in Solids. Oxford University Press. Oxford.
35. BURDETT, Chemical Bonding, p. viii.
36. SHRIVER, D.F., P.W. ATKINS & C.H. LANGFORD. 1994. Inorganic Chemistry, 2nd ed. xford University Press. Oxford. p. 97.
37. BURDETT, Chemical Bonding, p. 89.
38. LAUDAN, Progress and Its Problems, p. 57.
39. FOURCROY, in KIRWAN (1789), Essay on Phlogiston, pp. 242–243.

Social Background of the Discovery and the Reception of the Periodic Law of the Elements

Recognizing the Contributions of Dmitri Ivanovich Mendeleev and Julius Lothar Meyer

MASANORI KAJI

Tokyo Institute of Technology, Graduate School of Decision Science and Technology, History of Science and Technology Group, Tokyo, Japan

ABSTRACT: The favorable and relatively rapid reception of Mendeleev's periodic table of the elements can be attributed, in part at least, to his social connections. These connections were evident in the recently organized Russian Chemical Society. In addition, Mendeleev enjoyed the support of the editorial board of the journal of the German Chemical Society.

KEYWORDS: periodic table; Dmitri Ivanovich Mendeleev; Julius Lothar Meyer; Russian Chemical Society; social context of science

In November 1882, Dmitri Ivanovich Mendeleev (1834–1907) and Julius Lothar Meyer (1830–1895) both received the Davy Gold Medal from the Royal Society in London for "research on the classification of the elements."[1] They are both recognized as the discoverers, independently, of the periodic law of the elements.[2] Although their research paths often crossed, Mendeleev, who was Russian, and Meyer, who was German, possessed social backgrounds that were far apart.

Both Mendeleev and Meyer attended the First International Congress of Chemists at Karlsruhe in 1860. During that time, Mendeleev was participating in a two-year program of study in Heidelberg, and Meyer was a privatdocent at Breslau University. At the Congress, both of them received Cannizzaro's famous paper on the atomic weight system. They were both much impressed by the paper, and each wrote a chemistry textbook based on Cannizzaro's system after returning home.[3]

Meyer wrote a textbook on theoretical chemistry in 1864.[4] At the end of this book, Meyer arranged 50 elements in three separate tables. The arrangement was based on the valences and the differences in atomic weights of analogous elements. In 1866, Meyer left Breslau and became professor of chemistry in the School of Forestry at Neustadt-Eberswalde. In 1868, when he was preparing a new edition of his textbook, he managed to combine the three tables. He arranged 52 elements in 15 columns. Because of a delay in the publication of the new edition, this table was not published during his lifetime.[5]

Address for correspondence: Masanori Kaji, Tokyo Institute of Technology, Graduate School of Decision Science and Technology, History of Science and Technology Group, 2-12-1 Ookayama, Meguro-ku, Tokyo, 152-8552 Japan. Voice: +81-3-5734-2270; fax: +81-3-5734-2844.
mkaji@aqu.bekkoame.ne.jp

Mendeleev, after his return to Russia in 1861, wrote an organic chemistry textbook. In 1864, he became professor of chemistry at St. Petersburg Technological Institute. In 1865, he became professor of technical chemistry at St. Petersburg University. After changing his chair to general chemistry in 1867, he started to write a textbook on general chemistry, entitled *The Principles of Chemistry*. Writing this textbook, his major work, forced him to seek an appropriate system of arranging the chemical elements.

There was immense change and reform in Russia during the period from the middle of the 1850s until the end of 1860s—after the Crimean War (1853–1856) and including the emancipation of the serfs in 1861. This period is known as the "Era of the Great Reforms." During this time, the education system, especially at the higher levels, was reorganized.

In the late 1850s and the early 1860s, a group of young Russians, which included Mendeleev, gathered in St. Petersburg to promote chemistry research.[6] These young, active chemists established the Russian Chemical Society in St. Petersburg University on October 26, 1868.[7] Only a few weeks later, on November 6, the Society's inaugural monthly meeting was held in a chemistry lecture room of the University.

In early February 1869, while writing *The Principles of Chemistry*, Mendeleev compiled his first table of the elements based on atomic weights. It was also the first table to show periodicity of the properties of the elements. Soon afterwards, he wrote his first paper on the periodic law of the elements that he had discovered. In March 1869, Menshutkin, the secretary of the newly established Russian Chemical Society, read the paper in the monthly meeting of the Society for Mendeleev, who was absent. The paper was published in the second/third combined issue of the first volume of the new journal of the Society in May 1869.

Also in 1869, Mendeleev asked Beilstein, then professor of chemistry at St. Petersburg Technological Institute, to translate the summary of his first paper on the periodic law for publication in a German journal. One of Beilstein's students translated the paper into German. It must have been sent in March or April of 1869 to *Zeitschrift für Chemie*, one of the editors of which was Beilstein himself. According to the student who translated it, the summary was sent to the journal through none other than Lothar Meyer.[8]

At the end of the same year (1869), Lothar Meyer submitted his famous paper, "The Nature of the Chemical Elements as a Function of their Atomic Weights," which was published early in 1870.[9] Although Meyer admitted in the paper that his table was essentially the same as Mendeleev's,[10] his table of elements was more refined than Mendeleev's first table, especially in showing the so-called transition metals clearly. Meyer also had the correct atomic weight of indium—incorrect in Mendeleev's first table. Meyer succeeded in vividly conveying a periodic dependency of properties of the elements on their atomic weights by plotting the solid-state atomic volumes of those elements that could be isolated as elementary substances against their atomic weights. However, the conclusion of Meyer's paper was tentative and cautious.

It took some time before Mendeleev's discovery was recognized, even in the Russian chemistry community. For example, in the fall of 1869, Zinin, the first president of the Russian Chemical Society, advised Mendeleev to do "[real] work," meaning something experimental, preferably on organic chemistry, which was the mainstream research discipline at that time.[11]

The majority of Russian chemists recognized Mendeleev's work on the periodic law after two papers on the "natural system" of elements appeared in November and December 1870.[12] Mendeleev predicted properties of some undiscovered elements in detail. These papers, which succeeded his initial attempt to draw up a table of elements, can be regarded as the conclusion of his two-year research project. Even Zinin, who was once critical of Mendeleev's work, wrote a laudatory letter, dated February 18, 1871.[13]

The first article on the periodic law by a chemist other than Mendeleev was written by Savchenkov (1832–1903) and published in May 1871.[14] The article was based on Mendeleev's first paper on the periodic law and on his paper "The Natural System of Elements." One should note that Mendeleev not only predicted properties of undiscovered elements, but for a while, he also tried to find them. He collected various minerals both in Russia and in Europe and often asked his friends and colleagues in the University and some governmental institutions to help him collect samples.[15]

Mendeleev's work on the periodic law was conveyed through German-speaking subjects of Imperial Russia, including translation of papers or summaries. For example, Viktor von Richter, a native speaker of German who was also fluent in Russian, became the first correspondent-chemist of the German Chemical Society and sent many articles to that group about activities of the Russian Chemical Society, including Mendeleev's research. In 1874, von Richter published the first inorganic chemistry textbook in Russian based on the periodic law.[16]

After the appearance of Mendeleev's long paper on the periodic system in German in *Annalen der Chemie und Pharmacie* in 1871,[17] Mendeleev started to be recognized in Western Europe. Thanks to this paper, Meyer became assured of the correctness of the periodic law and tried to apply it fully to systematize inorganic chemistry. Meyer's paper "For Systemization of Inorganic Chemistry" in 1873 was one of its results.[18]

In 1880, Meyer[19] and Mendeleev[20] had a dispute over the priority and contributions towards the discovery and the development of the periodic law in the journal of the German Chemical Society. Here, Meyer suggested that there was an unfavorable atmosphere towards theoretical work in the German chemical community for "a paper without any new data."[21] However, Mendeleev was bold enough to send a very long paper "without any new data" to the same journal. For one thing, as Brauner (1855–1935), a Czech chemist and Mendeleev's friend,[22] wrote later, the younger members of the editorial board of the journal[23] strongly supported the publication of Mendeleev's paper. Together with powerful arguments in the paper itself (such as a detailed prediction of undiscovered elements that had persuaded Russian chemists to take Mendeleev's side), these favorable conditions in the German editorial board helped to promulgate Mendeleev's discovery. These "German connections," as well as the support and encouragement of the Russian chemical community, helped Mendeleev to concentrate his studies on the periodic law during this important period.

NOTES AND REFERENCES

1. SEUBERT, K. 1896. Lothar Meyer. Berichite der Deutschen Chemischen Gesellschaft **28:** 1119.
2. IHDE, A.J. 1964. The Development of Modern Chemistry. Harper & Row. New York. p. 243–251.

3. MEYER, L. 1864. Die modernen Theorien der Chemie und ihre Bedeutung für die chemische Statik. Verlag von Nrushke & Berendt. Breslau; Mendeleev, D. 1861. Organicheskaya khimiya [Organic Chemistry]. Izd-vo "Obshestvennaya pol'za." St. Petersburg.
4. This small book developed into Meyer's major work by being constantly revised throughout his life: 2nd ed., 1872; 3rd ed., 1876; 4th ed., 1883; 5th ed., 1884; 6th ed., 1896 (only the first volume was published).
5. This table was first published in 1895. SEUBERT, K., Ed. 1895. Das natürliche System der chemischen Elemente. Abhandlungen von Lothar Meyer und D. Mendelejeff. Ostwald's Klassiker der exakten Wissenschaften Nr. 68. Verlag von Welhelm Engelmann. Leipzig.
6. For more detailed analysis, see BROOKS, N.M. 1989. The Formation of a Community of Chemists in Russia, 1700–1870. Ph.D. dissertation. Columbia University. New York; BROOKS, N.M. 1998. The evolution of chemistry in Russia during the eighteenth and nineteenth centuries. In The Making of the Chemist. D. Knight & H. Kragh, Eds. Cambridge University Press. Cambridge. pp. 163–176. For an overview of the history of Russian chemistry, see SOLOV'EV, Y.I. 1985. Istoriya Khimii v Rossii [The History of Chemistry in Russia]. Nauka. Moscow. Also see KAJI, M. 1997. Mendeleev's Discovery of the Periodic Law of the Chemical Elements—The Scientific and Social Context of His Discovery (Japanese). Hokkaido University Press. Sapporo. pp. 30–56.
7. For the history of the Russian Chemical Society, see KOZLOV, V.V. 1961. Ocherki Istorii Khimicheskikh Obshchestv SSSR [The History of Chemical Society of the Soviet Union]. Izd-vo AN SSSR. Moscow. This date is in the Julian calendar, used in Russia until January 1918. It lags twelve days behind the Gregorian calendar in the nineteenth century. In this paper I use the Julian calendar for events in Russia and Gregorian dates for those outside of Russia.
8. KROTIKOV, V.A. 1969. Dve oshibki v pervykh publikatsiyakh o periodicheskom zakone D. I. Mendeleeva [Two mistakes in early publications on D.I. Mendeleev's Periodic Law]. Voprosy istorii estestvoznaniya i tekhniki, no. 29. p. 129. I would like to thank Dr. Michael D. Gordin for bringing this paper to my attention.
9. MEYER, L. 1870. Die Natur der chemischen Elemente als Function ihrer Atomgewichte. Annalen der Chemie und Pharmarcie Supplemente band. **7**: 354–364.
10. Meyer quoted the abstract in Zeitschrift für Chemie. MENDELEJEFF, D. 1869. Über die Beziehungen der Eigenshaften zu den Atomgewichten der Elemente. Z. Chem. **12**: 405–406).
11. There remains Mendeleev's undelivered letter to Zinin, dated December 24, 1869, which defended his position. KEDROV, B.M. 1959. Filosofiskii analiz pervuikh trudov D.I. Mendeleeva o preiodicheskom zakone (1869-1871) [Philosophical analysis of early works of D.I. Mendeleev on periodic law (1869–1871)]. Izd-vo Nauka. Moscow. pp. 243–244).
12. MENDELEEV, D.I. 1958. Periodicheskii Zakon [Periodic Law]. B.M. Kedrov, Ed. Izd-vo AN SSSR. Moscow. pp. 59–67, 69–101.
13. KEDROV, Philosophical analysis of early works, p. 246.
14. MENDELEEV, D.I. 1953. Nauchnui arkhiv, T. 1, Periodicheskii zakon. Estestvennaya sistema elementov. Rukopisi i tablitsui. 1869–1871 [Scientific Archive, Vol. 1, Periodic Law, Natural System of Elements, Manuscripts and Tables: 1869–1871]. Izd-vo AN SSSR. Moscow. pp. 749–761.
15. Ibid., pp. 187, 649. There remain some private and official letters for collecting mineral samples.
16. von Richter was a Baltic German who was born in Dobele, not far from Riga, an old Hanseatic town in the Baltic, and he graduated from Dorpat University, a German-speaking university in the Russian Empire. During the 1860s, he worked in St. Petersburg Technological Institute as a colleague of Mendeleev. VON RICHTER, V. 1874. Uchebnik neorganicheskoi khimii po noveichim vozzreniyam [The Textbook of Inorganic Chemistry in Newest Point of View]. Tip. Ivanaa Yavorskogo. Warsaw.
17. MENDELEJEFF, D. 1871. Die periodische Gesetzmässigkeit der chemischen Elements. Ann. Chem. Pharm. Suppl. **8**: 133–229.

18. MEYER, L. 1873. Zur Systematik der anorganischen Chemie. Ber. Dtsch. Chem. Ges. **6:** 101–106.
19. MEYER, L. 1880a. Zur Geschichte des periodischen Atomistik. Berichite der Deutschen Chemischen Gesellschaft **13:** 259–265; Meyer, L. 1880b. Zur Geschichte des periodischen Atomistik. Ber. Dtsch. Chem. Ges. **13:** 2043–2044.
20. MENDELEJEFF, D. 1880. Zur Gechichte des periodischen Gesetzes. Ber. Dtsch. Chem. Ges. **13:** 1796–1804.
21. MEYER, Zur Geschichte des periodischen Atomistik, p. 263.
22. Brauner wrote the section on the rare earths for the seventh and eighth editions of Mendeleev's *The Principles of Chemistry.*
23. They must be Emil Erlenmeyer (1825–1909) and Jacob Volhard (1834–1910) (MENDELEEV. Scientific Archive, Vol. 1. p. 710). After the retirement of Kopp from the editorship in March 1871, Erlenmeyer, Volhard, and Liebig were the editors of Annalen. ROCKE, A.J. 1993. The Quiet Revolution. University of California Press. Berkeley. Los Angeles, CA. p. 288. Erlenmeyer was very much a Russophile.

How G.N. Lewis Reset the Terms of the Dialogue between Chemistry and Physics

PAUL A. BOGAARD

Department of Philosophy, Mount Allison University, New Brunswick, Canada

> ABSTRACT: While quantum physics provided a mathematical framework for understanding the valency of chemical elements, to make use of these mathematical equations required approximations based on chemical experience. To better understand how this establishes limits to the presumption that chemistry can be derived from physics, this paper starts with the warnings in Coulson's widely read text, *Valence*, as to why there are pitfalls in the interpretation of what "valence" accomplishes. He directs some of these concerns back to Pauling's earlier text, *The Nature of the Chemical Bond*. Both Coulson and Pauling are aware that quantum theory's ability to approximate the energetic stability of chemical systems is of use to chemistry only if it can be correlated with the molecular structure that results from chemical bonding. While this may now be routinely expected, it was a struggle to see that this was the challenge chemistry posed to physics—and this achievement they both attribute to G.N. Lewis. Until the more mature forms of quantum mechanics arose in the 1920s, Lewis was skeptical that the dynamic models of the atom posed by Bohr could meet this challenge. How Lewis formulates this challenge initially in 1916 and then in his longer work, *Valence and the Structure of Atoms and Molecules*, will be examined.
>
> KEYWORDS: G.N. Lewis; C. Coulson; L. Pauling; quantum chemistry; chemical bond; valence; reduction

In his widely read text *Valence* (1952), Charles Coulson repeatedly warns against the misuse of certain theoretical constructs in our understanding of chemical valence. He questions whether real phenomena are being singled out by mathematical techniques that are suggestive of "hybridization" between electron orbitals, or "resonance" between ionic and covalent forms of bonding, or of the notion that an "exchange force" is generated by electrons trading places between bonded atoms.[1] The latter idea is suggested by the mathematical work of Heitler and London, who in 1927 proposed a way of writing the wave function for the hydrogen molecule as a linear combination of the two plausible ways to construe the two electrons—as if their places were exchanged.[2] As Coulson explained: "The identical character of the electrons compels us to choose [this formulation] and allow greater freedom to the

---

Address for correspondence: Paul A. Bogaard, Department of Philosophy, Mount Allison University, New Brunswick, Canada. Voice: 506-364-2339.

pbogaard@mta.ca

Ann. N.Y. Acad. Sci. 988: 307–312 (2003). © 2003 New York Academy of Sciences.

electrons. Such greater freedom is generally followed by a lowering of energy."[3] Thus we have, in brief, the familiar characterization of chemical bonding—not by some classical means of holding or forcing two atoms to remain together, but by the energetic advantage found in this quantum delocalization effect.

We also have a reason why the challenges of theoretical chemistry are often said to have been resolved by the work of mathematical physicists in the quantum mechanics of the 1920s and thereafter. But Coulson's other two warnings shed some light on this issue, especially because they seem to be directed at Linus Pauling, for whom both resonance and hybrid orbitals played crucial roles in his *The Nature of the Chemical Bond*. In a series of publications beginning soon after the work of Heitler and London and leading up to his influential text of 1939,[4] Pauling had insisted on the notion of resonance to emphasize the dynamic characterization of electronic interaction required by quantum chemistry. Pauling was persistent in applying his theoretical insight that the dynamism of chemical quantum systems tends towards a combination resonating between alternatives—whether it be between the restrictions of electrons to specific nuclei in a molecule (much like the notion of "exchange," mentioned above), between stereo-chemical alternatives in larger molecules, or between single and double bonds.

This is a characterization with which Coulson was comfortable, presumably, so long as chemists would remain wary of concluding that resonance itself describes what is really happening. Coulson could also appreciate Pauling's use of the hybridization concept: "Hybridization is the most effective way of preserving the concept of a localized bond with perfect pairing of orbitals on the two atoms of the bond."[5] The puzzle is that while delocalization always seems to be associated with lowering energy of the system, it also leads to the loss of all "pictorial sense" of a molecule's structure.[6] The latter may not be a loss of any consequence for simple molecules such as diatomic hydrogen (where the spatial orientation of one nucleus relative to the other makes little difference); however, in slightly more complex molecules like H_2O, and quite dramatically in the vast array of organic molecules, the structural arrangement of atomic nuclei, carrying their collective electrons with them, is indispensable to an understanding of molecular chemical behavior. Even Coulson, for all his efforts to be cautious about where theoretical constructs can lead, insists that "the chief advantage in a hybridized a.o. [atomic orbital] is its highly directional character, which [also] ... should yield a stronger bond."[7] Molecular structure—and the search for directionality of bonding that could produce it—is motivated neither by the mathematical formalisms of quantum theory nor by the physics which underlies quantum mechanics, but by experience with chemical substances.

That is the contention of this paper, and it will lead us back to the source that both Coulson and Pauling cite as the first to make this point clear. Coulson, in the 1950s, stated that "in practically the whole of theoretical chemistry, the form in which the mathematics is cast is suggested, almost inevitably, by experimental results."[8] That is, by chemical experience. There would be nothing contentious about this—theory and experimental work are invariably playing off one another—except that the success of quantum chemistry has often been taken as confirmation of chemistry's inherent dependence on physics. After all, the buildup of electrons in atoms of increasing atomic mass seems to match up precisely with the chemists' periodic table of the elements—periodicity that was discovered by comparing the valency of elements of increasing weight. And it was quantum physics that first offered a mathematical framework in which to model valence. It is the contention of this paper, nevertheless, that there are limits to this dependence, and these limits can best be understood in the terms Pauling uses in his *The Nature of the Chemical Bond*—key concepts both he and Coulson attribute to G.N. Lewis.

Pauling's major text of 1939, which was subtitled *An Introduction to Structural Chemistry*, was dedicated to Lewis, and presumably with the major achievements of the preceding two decades very much in mind, it set the agenda for quantum chemistry. The "fundamental principle," he declares, is this:

> [T]he energy value calculated by the equations of quantum mechanics with use of the correct wave function for the normal state of the system is less than that calculated with any other wave function ... the actual *structure* of the normal state of the system is that one, of all conceivable structures, that gives the system the maximum *stability*.[9]

The challenge for theoretical chemistry, as summed up here, is to provide the connection between stability and structure. The proper equation (with an appropriately constructed wave function) can point us towards the lowest energy value, hence that system's stability. But to carry chemical significance (for anything more complex than elemental atoms and the simplest molecules), it must correlate that energy value with the structure that such molecules maintain over time, and through various forms of interaction. In some sense, accounting for the stability of chemical substances had been the objective for decades, but as successful as chemistry had been in experimentally determining various molecular structures, accounting for their stability had proven to be elusive. This required accounting not only for valence, but also its structural results.

Both Coulson and Pauling point back to the theoretical work of such chemists as N.V. Sidgwick, I. Langmuir, and (earlier still) G.N. Lewis. Lewis himself, however, had seen that his interest as a chemist in understanding "valence" and the nature of the chemical bond was coming into conflict with theoretical physics as he saw it unfolding. Physics had opened up the new possibility that characteristics of the chemical elements are due to each atom's internal constituents, and in particular that bonding is the work of electrons internal to each atom. But Lewis was concerned that efforts to characterize the role of the electron within the atom, in his day, had not sufficiently taken into account the role of electrons in chemically bonding atoms together, or the structural consequences of such bonding, so crucial to chemistry.

A major gathering of both chemists and physicists was held in December 1916. On that occasion, the sections of Physics and of Chemistry of the American Association for the Advancement of Science, the American Physical Society, and the American Chemical Society, all met jointly. Addressing that assembly, Lewis openly complained that recent developments in physical theory were not taking sufficient account of what chemical experience had already demonstrated. He had been invited, apparently, to speak on his recent proposal that the chemical bond be understood as a pair of electrons shared by two atoms (which he had first published earlier that year), but he warned his audience that he was shifting his topic to "The Static Atom."[10] Bohr's dynamic atom was built on a planetary model with orbiting electrons, at the energetic levels required to account for spectroscopic evidence. Lewis acknowledged that Rutherford's evidence of a concentrated nucleus, and Bohr's model of rapidly revolving electrons "were the first to present any sort of acceptable picture of the mechanism by which spectral series are produced."[11] But, he warned, the chemical evidence required something altogether different. Molecular structure, especially in the case of organic compounds, was already well grounded in empirical work—and these structures could persist through long times and a variety of interactions.

> Now assuming that the electron plays some kind of essential role in the linking together of the atoms within the molecule, and ... no one conversant with the main facts of chemistry would deny the validity of this assumption.... It seems inconceivable that electrons which have any part in determining the structure of such a molecule could possess proper motion.[12]

Lewis's argument to this mixed audience of physicists and chemists was that a "static atom"—an atom in which the location of electrons was taken to be rigidly fixed—could also be made to account for the evidence from physics (for which he had several suggestions), and his was the only model that could account for the evidence from chemistry.

> [It] is these very electrons held in rigid positions in the outer shell of the atom which may, in case of chemical combination, become the joint property of the two atoms, thus linking together the mutually repellant positive atomic kernels and themselves constituting the bond which has proved so serviceable in the interpretation of chemical phenomena.[13]

Unfortunately, this controversial challenge to Bohr's dynamic atom seems to have drawn attention away from Lewis's idea of shared pairs of electrons, at least during the war years.

After the war, I. Langmuir launched a more effective promotion among chemists of the idea of the shared-pair bond—he coined the term "covalent" bond. A few years later, the "Pauli exclusion principle" helped to validate the idea of electron pairs for physicists. When Lewis returned to this debate, publishing *Valence and the Structure of Atoms and Molecules* in 1923, he was prepared to be more accommodating than he had been seven years earlier. In *Valence* (the title echoed by Coulson thirty years later), he resolutely promoted the notion of chemical bonding by shared pairs of electrons, insisting that the electrostatic attraction between ions be construed as a subordinate case of his notion of electron interaction in pairs. But he acknowledged that when Bohr moved away from the idea of electrons revolving in rings to electrons in shells—to specific states or orbitals—reconciliation between these two views became possible. If one considers the electron orbital as a whole,[14] he argues, that might finally provide chemistry with the structural significance required. As earlier, Lewis remained adamant about Coulombic forces:

> While electrostatic forces evidently play an important part in processes of ionization, and very likely also in numerous reactions which verge upon the ionic type, such forces are responsible neither for the fundamental arrangements of electrons within the molecule nor for the bonds which hold the atoms together.[15]

Rather, it might well be some account of the electromagnetism of such orbitals, he supposed, which could account for "a condition of maximum stability, when it possesses in its outer shell four pairs of electrons situated at the corners of a regular tetrahedron."[16] Perhaps the atom need not be conceived as completely static, he conceded, but it remains paramount that maximum stability carry directional implications, to provide for stable structures. Evidence from both physics and chemistry suggests that the "tendency to form pairs is not a property of free electrons, but rather that it is a property of electrons within the atom."[17]

Moreover, as these pairs fill Bohr's shells in elemental atoms (accounting for both the spectroscopic evidence from physics and the organization of chemistry's periodic table), they also provide the possibility of completing pairs in outer orbitals by sharing electrons between two atoms in a molecule. It is these shared orbitals that Lewis expects, when each orbital is taken as a whole, will provide the directionality needed to account for structure. In this way (as Lewis in 1923 could only speculate), the discontinuity that quantum theory introduces[18] might well account for the correlation between maximum stability with specific molecular structures.

Lewis could not have seen, as yet, the role of the Pauli exclusion principle or the success of the Schrödinger equation in capturing these "discontinuities." Nor would his image have been borne out, as yet, by the approximation techniques proposed by Heitler and London to capture the energetic stability of the hydrogen molecule. For Heitler and London's treatment, as for Bohr's earlier treatment of individual atoms, it was enough to account for the stability of a system in which no relative directions are privileged. Lewis's demands for stable molecular structure—the gauntlet tossed before his physics colleagues in 1916 on behalf of chemists—only began to be realized in the more restricted spatial symmetries of higher energy electron orbitals, and became especially vivid in Pauling's interpretation of what happens when electron orbitals, comprised of shared pairs of electrons, form hybrid orbitals.[19]

There is always the danger, as Coulson later points out, that this vision of hybridization—the example of this danger particularly relevant to Pauling's portrayal of the chemical bond—may prove to be an artifact of the mathematical technique for approximating the outcome in specific chemical systems. But Pauling's statement of the fundamental principle of quantum chemistry, in his text on the structure of molecules, makes it quite clear how significant was Lewis' insistence that theoretical chemistry must look beyond the structure of atoms (reflected in the periodic table) to the more difficult task of accounting for the consequent structure of molecules. Hybrid orbitals are just one particularly telling example (of many possible examples that flow from the nature of the chemical bond) of how quantum theory can provide the structural results chemistry requires.

Even Sidgwick, the major spokesman within chemistry to argue that it might be better to retain a more dynamic model of the atom (against Lewis's call for a more static model) and to rely upon Coulombic forces (against Lewis's call for shared pairing of electrons), acknowledges later in the 1920s:

> The relation between the nature of a valency group and its stability, and the connection of this with the structure of the atom as a whole, form the most fundamental problem in the theory of valency, because it is on this stability that the chemical properties of every element depend.[20]

Lewis may not have convinced Sidgwick of the priority of covalent bonding (any more than Pauling's theoretical constructs convinced Coulson), but he did succeed in resetting the criteria of the theoretical debate in terms of stability and structure. For a viable theory of chemistry, if not physics, one must account for the structure of molecules.

NOTES AND REFERENCES

1. COULSON, C.A. 1961. Valence. 2nd ed. Oxford University Press. Oxford. pp. 123, 132, 239–240; 109–110, 196ff; 233ff; 118.
2. HEITLER, W. & F. LONDON. 1927. Wechselwirkung neutraler Atome und homoopolare Bindung nach der Quantenmechanik. Z. Physik. **44**: 455.
3. COULSON, Valence, pp. 118–119.

4. PAULING, L. 1939. The Nature of the Chemical Bond and the Structure of Molecules and Crystals: An Introduction to Modern Structural Chemistry. Cornell University Press. Ithaca, NY. Already in 1931: The Nature of the Chemical Bond: Application of Results Obtained from Quantum Mechanics and from a Theory of Paramagnetic Susceptibility in the Structure of Molecules. J. Am. Chem. Soc. **53:** 1367; and earlier: 1928. Proc. Natl. Acad. Sci. USA **14:** 359; and with SLATER, J. 1931. Phys. Rev. **37:** 481 and **38:** 1109.
5. COULSON, Valence, p. 233.
6. Ibid., p. 171.
7. Ibid., p. 208.
8. Ibid., p. 113. Earlier, on p. 57, he had warned that the formulation of the wave function can be approximated "with great effectiveness, if we bring sufficient chemical intuition to bear on the choice."
9. PAULING, Nature of the Chemical Bond, p. 11.
10. Lewis's presentation was soon published: 1917. The static atom. Science **46:** 297–302. His original proposal for shared-pair bonding was in: 1916. The atom and the molecule. J. Am. Chem. Soc. **38:** 762–785.
11. LEWIS, "The static atom," p. 299.
12. Ibid., pp. 297–298.
13. Ibid., p. 302.
14. LEWIS, G.N. 1966. Valence and the Structure of Atoms and Molecules. Dover. New York. p. 57. [Originally published in 1923 by the Chemical Catalog Company, as one of a series of monographs sponsored by the American Chemical Society.]
15. Ibid., p. 146.
16. Ibid., p. 149.
17. Ibid., p. 81.
18. Ibid., See Chapter XIV.
19. Even the bold claim so often attributed to P. Dirac, that the "underlying physical laws necessary for the mathematical theory of a large part of physics and the whole of chemistry are completely known" (1929. Proc. R. Soc. **A123:** 714), although clearly an exaggeration, can be read as an acknowledgment on Dirac's part that quantum physics (before it can claim to be completely known) will have to accommodate itself to the needs of theoretical chemistry. Even then he went on to say: "the difficulty is only that the exact application of these laws leads to equations much too complicated to be solvable." And we only discern, as Coulson points out, how to approximate these equations from chemical experience.
20. SIDGWICK, N.V. The Electronic Theory of Valency. Oxford University Press. Oxford. p. 163.

The Superiority of "Chemical Thinking" for Understanding Free Human Society According to Hegel

MARK R. NOWACKI[a] AND WILFRIED VER EECKE[b]

[a]*Department of Philosophy, George Washington University, Washington, DC 20052, USA*
[b]*Department of Philosophy, Georgetown University, Washington, DC 20057, USA*

ABSTRACT: This paper examines the claim of G.W.F. Hegel (1770–1831) that "chemical thinking"—the method of thinking employed in chemistry—marks a significant advance upon (and hence is superior to) "mechanistic thinking"—the method of thinking characteristic of physics. This is done in the context of Mancur Olson's theory of collective action and public goods. The analogy between the efficiency of a catalyst in bringing about chemical transformation and the function of leaders in free human society in developing latent groups to provide public goods is explored.

KEYWORDS: Hegel; chemical thinking; mechanism; public goods; Mancur Olson; latent group; leadership; leader; free human society

G.W.F. Hegel (1770–1831) asserts that "chemical thinking"—the method of thinking employed in chemistry—marks a significant advance upon (and hence is superior to) "mechanistic thinking"—the method of thinking characteristic of physics. In particular, chemical thinking is superior to mechanistic thinking when we attempt to give an accurate account of human society. In this paper we summarize Hegel's argument for the superiority of chemical thinking, and then offer independent arguments that confirm the truth of Hegel's position.

With Hegel, we hold that the methods of thinking employed in chemistry are useful for coming to understand a wide variety of phenomena, perhaps the most important example of which is coming to an understanding of the true nature of the fundamental political organization of free human societies. As we demonstrate in what follows, the adoption of a chemical way of thinking makes it possible to explain several important aspects of human society that mechanistic thinking necessarily precludes. First, and most importantly, chemical thinking makes it possible for us to view society as something other than a simple aggregation of component human atoms. Human society is something far richer—a compound whose properties and whose intrinsic value add up to more than a sum of its elements. Second, careful application of distinctively chemical concepts allows us to appreciate and account for

---

Address for correspondence: Mark R. Nowacki, Department of Philosophy, George Washington University, Washington, DC 20052. Voice: 202-994-6265.
nowacki@gwu.edu

certain features we find in every human society. In relation to this second point, we develop the example that chemical thinking makes it possible to understand the necessary presence of leaders within society, and also their positive function.

Hegel raises the question of mechanistic versus chemical thinking in his Science of Logic.[1] He begins by noting that certain concepts are "independent"—concepts that do not depend on (or appeal to) other concepts to be understood. Examples of such independent concepts are "being" and "magnitude." On the other hand, there also are concepts that cannot be thought of (or understood) without immediately appealing to other concepts. Examples of such "dependent" concepts include "appearance" and "illusion." In thinking about an appearance or an illusion, one immediately calls to mind the particular ground that underlies the appearance or illusion.

Having distinguished these two sorts of concepts, Hegel proceeds to distinguish different ways of applying concepts. He argues that there are general ways (or patterns) of thinking that can be broadly characterized. One characteristic way of thinking that Hegel identifies he calls "mechanism." Mechanism is a way of thinking characteristic of physicists.[2] As Hegel writes,

> This is what constitutes the character of mechanism, namely, that whatever relation obtains between the things combined, this relation is one extraneous to them that does not concern their nature at all, and even if it is accompanied by a semblance of unity it remains nothing more than composition, mixture, aggregation and the like.[3]

To simplify Hegel's position somewhat, the basic picture mechanism presents us with is that of perfectly rigid bodies endlessly colliding with one another in the void. Mechanistically conceived objects are like idealized billiard balls caroming about on a billiard table: the balls collide with each other, and in colliding change their specific momentum and direction, but through it all the intrinsic structure of the balls remains unchanged.[4] In mechanism, cause and effect are reduced to changes in momentum and position, and the only differences one object brings about in another are extrinsic and come about through contact.[5] Interactions among mechanistic objects always retain an element of violence: changes come through the application of force, one object shoving another object it encounters along its path.[6]

The objects of mechanism may be aggregated, and in their aggregation be considered as a new object.[7] An aggregate mechanistic object is simply the sum of its component parts, like a heap. Hegel expressly extends this mechanistic concept of a mechanistic aggregate to human societies:

> The relations of pressure, thrust, attraction, and the like, as also aggregations or mixtures, belong to the relationship of externality which forms the basis of the third of this group of syllogisms [that is, the constellation of Hegelian arguments detailing the relations that obtain between citizens, the government, and the needs or external life of individual citizens].[8]

It may be noted that the unity of a heap can easily be disturbed, for there is no basic affinity that binds the constituent objects of a heap together into an essential unity. Applied to human society, the mechanistic way of thinking immediately suggests a Hobbesian account of human nature and political organization. In what follows, we will make the connection between mechanistic thinking and Thomas Hobbes' political philosophy explicit—once a few more pieces of Hegel's argument are put on the table.

In contrast to mechanism, Hegel distinguishes a second way of thinking, which he calls "chemism." Chemism, as the name suggests, is the way of thinking characteristic of chemists.[9] Chemical objects are intrinsically related and ordered to one another in ways that the externally related objects of mechanism are not. For our present purposes, the most salient feature of chemical thinking is that it recognizes modalities of interaction among chemical objects over and above the violent impact described in mechanism. Specifically, unlike mechanism, chemism allows for a coming together of two objects that results, not in a mere heap, but rather in a new, essential unity that possesses emergent properties not explainable as a mere sum of the properties of its parts.[10] For instance, the freezing point of water is not a simple sum or average of the freezing points of hydrogen and oxygen.[11]

As is the case with mechanistic thinking and mechanistic concepts, chemical thinking and chemical concepts find application outside the strictly defined borders of their original science. Chemical thinking, according to Hegel, can be used to shed light on such rich concepts as "love" and "friendship."[12] Chemical thinking, as we will argue shortly, can also lead us to a deeper understanding of free human society.

With these precisions in hand, we are in a position to appreciate the shortcomings of mechanistic thinking—and see how chemical thinking can go beyond these shortcomings. We begin by noting that Hegel uses his two levels of analysis, first the level of concepts and second the level of various ways of thinking, to argue that even if human beings are free, there are nonetheless necessary steps humans must take in their thinking. We must follow certain patterns of thought and use certain types of concepts, if we wish to truly understand the world as it is. This implies that freedom, and in particular the sort of freedom found in the thinking of free human beings, is not inherently chaotic but rather comes intrinsically ordered. In other words, the kind of freedom exercised by free human knowers is an essentially ordered freedom.

Within the context of this understanding of freedom, namely that human freedom is necessarily an ordered freedom, Hegel argues that chemical thinking is a superior way of thinking to mechanism when what we wish to understand is the inherent order found in free human societies. As mentioned above, mechanistic thinking characteristically employs a general conception of elements within an ordered domain wherein one element can influence another element only through contact. One mechanistic object influences another only by hitting it—the only differences found in hitting are whether an individual object hits hard or softly.

If one thinks about human beings in this mechanistic way, then thinking about human communities inevitably leads to a Hobbesian account of politics. As Hobbes has shown, a systematic application of mechanism implies an account of human society wherein the life of human beings in their natural state is "solitary, poor, nasty, brutish, and short."[13] Human beings, according to Hobbes, are atomistic individuals. They are naturally free, equal, and independent of one another.[14] In the presence of desirable objects, the fundamental equality and independence of human beings naturally causes them to develop a mutual diffidence. After all, since human A is the equal of human B, why should it be A that enjoys the consumption of scarce resources that B also desires?[15] From this mutual diffidence arises what Hobbes argues is the natural state of human beings: a general war of all against all.[16] Rephrased in explicitly mechanistic terms, human beings effect one another through colliding with each another. Collisions with other human beings tend to be harmful, as may be seen in the case where one human forcefully steals the fruit of another's labor.[17]

To put an end to their constant struggle and mitigate their natural impulse to constantly engage in the social equivalent of hitting one another, human beings collectively establish a social contract wherein they all agree to invest power in a common sovereign. This sovereign is called by Hobbes "the Leviathan." The Leviathan alone possesses the degree of force necessary to stop people from hitting one another. This is because the Leviathan carries the biggest stick of all—absolute political power—and with that stick the Leviathan imposes social order by force.[18] Thus, according to a mechanistic approach, free human society is little more than a projection of Boyle's account of an ideal gas. With an ideal gas we have equal, mutually independent gas atoms constantly colliding with the walls of their container and with one another. In Hobbesian political society, we have equal, independent, and mutually diffident human atoms competing for scarce resources in a war of all against all. Only the threat of a worse collision could cause a human atom to swerve from its natural impact with another human atom—providing that threat is the explicit function of Hobbes' Leviathan.

Chemical thinking presents a picture of human society that is not only less bleak but also truer to social reality. To display some of the advantages of adopting a chemical way of thinking when approaching the problem of free human societies, let us first consider the distinctively chemical notion of a catalyst.[19] For example, in the Haber process, iron oxide acts as a catalyst for the formation of ammonia from dinitrogen and dihydrogen. A chemical catalyst can speed up, or even absolutely enable, a particular chemical reaction.[20] The enabling activity of a chemical catalyst is essentially positive, and thus contrasts sharply with the kind of enabling typical of mechanism. Whereas the kind of enabling characteristic of mechanism is essentially negative—for instance, some impediment is removed, thus permitting a mechanistic rearrangement of elements—an enabling chemical catalyst positively brings about or positively moderates the becoming of something fundamentally new, a novel entity that cannot be adequately explained by the summation of its parts.

Once chemistry has introduced and clarified the notion of a catalyst, the notion of a catalyst can be used to clarify the connections between the elements of other ordered domains of study.[21] In particular, the chemical notion of a catalyst allows us to better understand the connection between order and freedom in human society.

How is this clarification brought about? We begin by noting that the role of a leader in human groups may be likened to that of a chemical catalyst. A leader organizes, gives focus to, and determines the character of group activity. The presence of a leader in a group, like the presence of a catalyst in a mixture of chemical substances, decisively influences the action and reaction of the reactants. The presence of a human catalyst influences or even absolutely enables certain outcomes not otherwise obtainable. As studies by Robert Bales, Edgar Borgatta, Arthur Couch, and others have shown, if two different leaders are placed in similar groups, the outcomes are different. Superior group performance is closely linked to superior leadership. Bales' studies suggest that it is the presence of a leader—and of a certain kind of leader at that—which makes the difference. A great leader catalyzes what would be an average group with average products into a superlative group with superlative achievements.[22]

Not only do leaders influence the outcomes of group activity, but the presence of a leader can be shown to be necessary for certain outcomes to occur. As Mancur Olson has demonstrated, the presence of a leader is necessary for the attainment of certain desirable public goods. Without a leader's catalyzing presence, these public goods would never be obtained and some desirable social benefit would be foregone.[23] As a matter of empirical fact, public goods such as Olson discusses have been pursued and obtained in human societies. That is, there are now, and in the past there have been, public goods obtained in free human societies that have the presence of a catalyzing leader as a necessary condition of their procurement. Since the catalyzing function of a leader can be explained by appealing to the concepts and ways of thinking appertaining to chemistry, it follows that an adequate explanation of free human society is aided by chemical thinking. And let us here remark that the concept of a catalyst is antithetical to mechanistic thinking: by definition, a catalyst moderates a different sort of causal interaction than that which mechanism, with its simple impact model, allows physicists to describe. Since the concept of a catalyst is foreign to mechanism but is native to chemism, the usefulness and superiority of chemical thinking to mechanistic thinking is demonstrated. If we were to remain at the level of mechanistic thinking, a crucial piece required for understanding the dynamics of free human society would be missing.

Olson proves that the presence of a catalyzing leader is necessary for the attainment of some public goods in the following fashion.[24] First, he notes that if certain types of public goods exist for large groups, for example, the good of workplace safety for factory workers, then the procurement of such goods will require legislation. Individuals cannot, on their own, secure passage of such legislation. Passage of the legislation requires lobbying.[25] Lobbying requires money, and lobbying money must come from interested parties willing to pay it. Since the naturally interested individuals (for example, individual factory workers) cannot achieve the desired outcome on their own, if the public good is to be obtained then the interested parties (called by Olson the "latent group") must be organized.[26] Organization of a latent group requires that some leader take the initiative.

What is it that the leader does, and what is it about a latent group that makes the presence of a leader necessary? The basic function of a leader organizing a large latent group is to furnish that crucial mix of selective private incentives that can be used to motivate group action.[27] As Olson argues, for large groups, the common good can only be obtained as a byproduct of the individual pursuit of private interest, by each group member.[28] In the particular case of worker safety, members of the latent group might be offered selective private incentives that make them interested in paying dues to belong to the group.[29] These group dues can then be used to pay for lobbying for the desired legislation. For example, unions might organize latently interested workers by offering health benefits to members at reduced rates.[30] Paying union dues then becomes desirable, because of the opportunities for private benefit that accrue from membership. The latent group of workers who would derive benefit from the passage of worker safety laws thus becomes mobilized through the activation of the group members' individual private interests.

Without procuring such private incentives, group dues would not be paid and the latent group would fail to organize.[31] Without organizing, the public good desired by the latent group would not be provided. As history has shown, laws mandating workplace safety do not get passed without the active involvement of mobilized worker organizations.[32] And, as history has also shown, workers do not organize without leaders who can identify the selective private incentives that motivate them to join and financially support their group.[33]

A leader who takes on the task of motivating a latent group for the attainment of some public good is clearly analogous to a catalyst in a chemical reaction. Without the presence of a catalyst certain chemical reactions either do not occur or occur at an exceedingly slow pace. Without the presence of an organizing leader who can identify and implement selective private incentives, the potential power of a latent group fails to become actual—and the public good aimed at by the latent group fails to materialize. A leader catalyzes a latent group into activity, transforming the elements of the group into a new product: a mobilized latent group.

Inasmuch as there do exist public goods that conform to Olson's analysis, we find that what is rationally desirable for free humans cannot, at least in some instances, be obtained without the catalyzing presence of a leader. The concept of a catalyst is a distinctively chemical concept, and the activity of a catalyst with respect to the elements catalyzed can be properly understood only if we adopt a chemical way of thinking. Since the concept of a catalyst captures and explains the function of a leader, the adoption of chemical thinking is superior to mechanistic thinking when we wish to explain the domain of ordered freedom characteristic of human societies.[34]

NOTES AND REFERENCES

1. Most of the relevant passages are to be found in Wissenschaft der Logik. Vol. 2, sect. 2, ch. 1–2. The following translation will be used for all quoted passages: HEGEL, G.W.F. 1989. Science of Logic. A.V. Miller, Trans. Humanities Press International. Atlantic Highlands, NJ. A valuable guide to understanding Hegel's thought that we have consulted is LÉONARD, A. 1974. Commentaire littéral de la logique de Hegel. J. Vrin. Paris. Éditions de l'Institut Supérieur de Philsophie. Louvain.
2. For Hegel's discussion of mechanism and mechanistic thinking, see Science of Logic. Vol. 2, sect. 2, ch. 1.
3. Ibid., "Mechanism."
4. As Léonard comments: "en raison de l'extériorité réciproque des objets, cette efficience (Wirksamkeit) de chaque objet sur les autres (ou sur l'autre en général) demeure une relation extérieure qui n'affecte en rien la structure interne des objects. Ce type de relations extrinsèques entre objets composant un simple agrégat sans unité interne est le MÉCANISME FORMEL..." (Commentaire littéral. p. 438–439.)
5. Thus Hegel writes: "The relationship in which the unessential single bodies stand to one another is one of mutual thrust and pressure ..." (Science of Logic. Vol. 2, sect. 2, ch. 1, subsect. C(a).)
6. Léonard brings out the violence implicit in mechanism quite clearly. Concerning the mechanistic object he observes: "il souffre VIOLENCE, car, dès lors qu'il est par ailleurs autosubsistant, la puissance efficiente qui agit sur lui est absolument étrangère à l'universe clos de l'objet et s'exerce donc sur lui aveuglément." (Commentaire littéral, 439.)
7. Mechanistic objects are "capable of mixing and aggregating and of becoming, as an aggregate, one object." (Science of Logic. Vol. 2, sect. 2, ch. 1, subsect. B.)
8. Ibid., Vol. 2, sect. 2, ch. 1, subsect. C(a).
9. For Hegel's discussion of chemism and chemical thinking see Science of Logic. Vol. 2, sect. 2, ch. 2. It is Hegel's recognition of the full generality of chemism, namely, that chemism employs a distinctive way of thinking applicable to diverse subject matters, that marks Hegel's account as a significant theoretical advance. (See Note 11 for a discussion of an important philosophical school that holds a position on chemical compounds similar to that of Hegel.)
10. Hegel advances this thought through a systematic development of the idea that chemical objects admit internal relations. He begins his discussion of chemism thus: "The chemical object is distinguished from the mechanical by the fact that the latter is a totality indifferent to determinateness, whereas in the case of the chemical object the determinateness, and consequently the relation to other and the kind and manner of this relation, belong to its nature." (Ibid., Vol. 2, sect. 2, ch. 2, subsect. A.)

11. Hegel is not the first philosopher to insist that certain combinations of elements result in essentially new compounds that cannot be reduced to the sum of their elemental parts: Aristotle adopted such an anti-reductionist stance more than two thousand years earlier. The importance of accounting for such compounds is recognized by both medieval and contemporary exponents of Aristotelian philosophy. No less a figure than Thomas Aquinas dedicated an entire treatise to the problem: see his De Mixtione Elementorum. Two contemporary representatives of the Aristotelian tradition who have written on the subject of chemical combination are Alan Wolter and Joseph Bobik. For an excellent introduction, see: WOLTER, A.B. 1960. Chemical substance. *In* Philosophy of Science. St. John's University Press. Jamaica, NY. pp. 87–130. See also BOBIK, J. 1998. Aquinas on Matter and Form and the Elements: A Translation and Interpretation of the De Principiis Naturae and the De Mixtione Elementorum of St. Thomas Aquinas. University of Notre Dame Press. Notre Dame, IN. Bobik's work is noteworthy in that he uses the conceptual tools developed by Aristotle and Aquinas to answer contemporary questions such as how quarks can be said to exist in protons. Summarizing the position of Aquinas, which he believes can be defended on its own merits, Bobik writes: "A mixing (mixis, mixtio)...results neither in something which is like a heap, e.g., of bricks and stones, in which its constituents are simply thrown together; nor does it result in something which is like a house, i.e., something which is not simply thrown together, but carefully put together out of constituents, e.g., wood and bricks and stones, arranged in an orderly way, and held together by certain joining materials, e.g., nails and mortar and glue. It results rather in something, i.e., a mixed body (like a molecule of water, or a proton), which differs in kind from any and all of its constituents, and in which the constituents, having undergone a mutual interactive alteration, remain in a special way, i.e., virtually, though not actually—virtually, meaning: with their powers, but these powers appropriately altered by means of their preceding interactive alteration (as well as retrievably, dispositionally, and instrumentally); though not actually, meaning: not with their substantial forms, because a substance (e.g., a molecule of water, or a proton) can have actually but one substantial form, its own." (Ibid., p. 272.)
12. Hegel writes: "With regard to the expression chemism for the relation of the difference of objectivity as it has presented itself, it may be further remarked that the expression must not be understood here as though this relation only exhibited itself in the form of elemental nature to which the name chemism so called is strictly applied. Even the meteorological relation must be regarded as a process whose parts have the nature more of physical than chemical elements. In the animate world, the sex relation comes under this schema and it also constitutes the formal basis for the spiritual relations of love, friendship, and the like." (Science of Logic, Vol. 2, sect. 2, ch. 2, subsect. A.)
13. HOBBES, T. 1958. Leviathan: Parts I and II. [Introduction by Herbert W. Schneider.] Ch. 13. Macmillan. New York. Collier Macmillan. London.
14. As Thomas Hobbes writes in Leviathan, ch. 13: "Nature has made men so equal in the faculties of the body and mind as that, though there be found one man sometimes manifestly stronger in body or of quicker mind than another, yet, when all is reckoned together, the difference between man and man is not so considerable as that one man can thereupon claim to himself any benefit to which another may not pretend as well as he."
15. "From this equality of ability arises equality of hope in the attaining of our ends. And therefore if any two men desire the same thing, which nevertheless they cannot both enjoy, they become enemies; and in the way to their end, which is principally their own conservation, and sometimes their delectation only, endeavor to destroy or subdue one another." (Ibid.)
16. "Hereby it is manifest that, during the time men live without a common power to keep them in awe, they are in that condition which is called war, and such a war as is of every man against every man." (Ibid.)
17. "Again, men have no pleasure, but on the contrary a great deal of grief, in keeping company where there is no power able to overawe them all. For every man looks that his companion should value him at the same rate he sets upon himself; and upon all

signs of contempt or undervaluing naturally endeavors, as far as he dare (which among them that have no common power to keep them in quiet is far enough to make them destroy each other), to extort a greater value from his contemners by damage and from others by the example." (Ibid.)

18. "The only way to erect such a common power as may be able to defend them from the invasion of foreigners and the injuries of one another, and thereby to secure them in such sort as that by their own industry and by the fruits of the earth they may nourish themselves and live contentedly, is to confer all their power and strength upon one man, or upon one assembly of men that may reduce all their wills, by plurality of voices, unto one will.... This is the generation of that great LEVIATHAN (or rather, to speak more reverently, of that mortal god) to which we owe, under the immortal God, our peace and defense. For by this authority, given him by every particular man in the commonwealth, he has the use of so much power and strength conferred on him that, by terror thereof, he is enabled to form the wills of them all to peace at home and mutual aid against their enemies abroad." (Leviathan. Ch. 17.)

19. Making allowances for the advances in scientific chemistry that have occurred in the two centuries since Hegel wrote, it is clear that Hegel has, at the very least, a conception of a chemical "enabler," that is, the concept of an agent within a chemical reaction whose presence makes possible a second chemical reaction: "Even ordinary chemistry shows examples of chemical alterations in which a body, for example, imparts a higher oxidation to one part of its mass and thereby reduces another part to a lower degree of oxidation, in which lower degree alone it can enter into a neutral combination with another [chemically] different body brought into contact with it, a combination for which it would not have been receptive in that first immediate degree." (Science of Logic. Vol. 2, sect. 2, ch. 2, subsect. C.) While it would be difficult to read this as a strict application of the modern notion of a catalyst, it is clearly a step in that direction.

20. The specifically chemical notion of a catalyst has received considerably less philosophical attention than it deserves. Notable exceptions include: CERRUTI, L. 1999. Historical and philosophical remarks on Ziegler-Natta catalysts: A discourse on industrial catalysis. Hyle **5:** 3–41; WITZEMANN, E.J. 1943. The role of catalysis in biological causation. Philos. Sci. **10:** 176–183; and MCCLOSKEY, H.J. 1964. Some concepts of cause. Rev. Metaphys. **17:** 586–607.

21. The usefulness of the concept of a catalyst has long been recognized in other intellectual domains. A brief survey reveals that the concept of a catalyst plays a pivotal role in such varied disciplines as social philosophy (see, for example, DONALDSON, T. 1990. Social contracts and corporations: A reply to Hodapp. J. Bus. Ethics **9,2:** 133–137), business and professional ethics (for example, REEVES, M.F. 1994. The gadfly business ethics project. J. Bus. Ethics **13,8:** 609–614), and philosophy of religion (for example, SCHALOW, F. 1989. Dread in a post-existentialist era: Kierkegaard reconsidered. Heythrop J. **30:** 160–167).

22. Summarizing the results of their study, we find the following claim: "Thus, it may be said that great men tend to make 'great groups' in the sense that both major factors of group performance—product and satisfaction of the members—are simultaneously increased." (BORGATTA, E.F., A.S. COUCH & R.F. BALES. 1966. Some findings relevant to the "great man theory of leadership" in Small Groups. *In* Studies in Social Interaction. A.P. Hare, E.F. Borgatta & R.F. Bales, Eds. Alfred A. Knopf. New York. pp. 700–706.)

23. For the material that follows, see OLSON, M. 1971. The Logic of Collective Action: Public Goods and the Theory of Groups. Harvard University Press. Cambridge, MA. The technical justification for the remarks on large groups is to be found in ch. 1.

24. Instead of "leaders," Olson prefers to couch his discussion in terms of "entrepreneurs." See Logic of Collective Action. pp. 175–177.

25. Olson discusses the necessity of lobbying Logic of Collective Action. pp. 10–11.

26. A "latent group" is characterized thus by Olson: "It is distinguished by the fact that, if one member does or does not help provide the collective good, no other one member will be significantly affected and therefore none has any reason to react. Thus an individual in a 'latent' group, by definition, cannot make a noticeable contribution to

any group effort, and since no one in the group will react if he makes no contribution, he has no incentive to contribute. Accordingly, large or 'latent' groups have no incentive to act to obtain a collective good because, however valuable the collective good might be to the group as a whole, it does not offer the individual any incentive to pay dues to any organization working in the latent group's interest, or to bear in any other way any of the costs of the necessary collective action." (Ibid., pp. 50–51.)

27. Olson describes the situation thus: "Only a separate and 'selective' incentive will stimulate a rational individual in a latent group to act in a group-oriented way. In such circumstances group action can be obtained only through an incentive that operates, not indiscriminately, like the collective good, upon the group as a whole, but rather selectively toward the individuals in the group. The incentive must be 'selective' so that those who do not join the organization working of the group's interest, can be treated differently from those who do. These 'selective incentives' can be either negative or positive, in that they can either coerce by punishing those who fail to bear an allocated share of the costs of the group action, or they can be positive inducements offered to those who act in the group interest." (Ibid., p. 51. Olson's footnotes omitted.)

28. "The common characteristic which distinguishes all of the large economic groups with significant lobbying organizations is that these groups are also organized for some other purpose. The large and powerful economic lobbies are in fact the by-products of organizations that obtain their strength and support because they perform some function in addition to lobbying for collective goods." (Ibid., p. 132.) See also Olson's remarks on p. 133.

29. "Though all of the members of the group ... have a common interest in obtaining this collective benefit, they have no common interest in paying the cost of providing that collective good. Each would prefer that the others pay the entire cost, and ordinarily would get any benefit provided whether he had borne part of the cost or not." (Ibid., p. 21.)

30. Railroad unions have used this strategy. The incidence of injury to railroad workers was once so high that established insurance companies would refuse to insure railroad workers. The railroad unions were thus, for a time, the only source of health insurance open to railroad workers. See OLSON, Logic of Collective Action, p. 72.

31. As Olson summarizes: "If the members of a large group rationally seek to maximize their personal welfare, they will not act to advance their common or group objectives unless there is coercion to force them to do so, or unless some separate incentive, distinct from the achievement of the common or group interest, is offered to the members of the group individually on the condition that they help bear the costs or burdens involved in the achievement of the group objectives. Nor will such large groups form organizations to further their common goals in the absence of the coercion or the separate incentives just mentioned. These points hold true even when there is unanimous agreement in a group about the common good and the methods of achieving it." (Ibid., p. 2.)

32. "[The] laboring, professional, and agricultural interests of the county make up large, latent groups that can organize and act effectively only when their latent power is crystallized by some organization which can provide political power as a by-product." (Ibid., p. 143.)

33. "Thus entrepreneurs [that is, leaders] will strive mightily to organize large groups. Many of the entrepreneurial efforts in this area, as in markets for private goods, will come to naught. But in some cases...imaginative entrepreneurs will be able to find or create selective incentives that can support a sizeable and stable organization providing a collective good to a large group. The successful entrepreneur in the large group case, then, is above all an innovator with selective incentives." (Ibid., p. 177.)

34. We would like to thank the reviewers for their helpful comments and suggestions.

An Organic Framework for a Philosophical Appreciation of Chemical Phenomena

RICHARD K. KHURI

Council for Research in Values and Philosophy, Falls Church, Virginia 22041-3736, USA

ABSTRACT: The aim of this paper is to show how chemistry gives rise to philosophical reflection. Initially, some general remarks on philosophy and science are offered in the light of recent reductive controversy surrounding them both. An appeal is made for a broader and deeper outlook. The philosophical outlook favored is one of openness towards the phenomena: rather than confine them to some "-ism" or other. The phenomena are seen as at least potentially "bathed" in more encompassing levels of being. The term "ontological suppleness" is later introduced in this spirit, to express how something at a given level, say electronegativity in chemistry, has significance at far higher levels of being—in this case, the emergence of life, and our planet's ecology as a whole. Similar consideration is given to the conversion of a carcinogen (benzene) into a "miracle drug" (aspirin). These illustrations are intended to highlight how science in general, including chemistry in particular, offers seamless transitions to what lies beyond the scope of science. What remains for us is to find a language adequate to deal with such transitions.

KEYWORDS: reductive dichotomies; subjective; objective; truth; metaphor and poetry in science; m/Mechanism; ontological suppleness; aspirin; benzene; quantitative; qualitative; water; electronegativity; polarity; membrane; ecology

FUNDAMENTALS

Philosophy

Any serious discussion of the philosophy of chemistry tacitly or explicitly introduces preconceived notions of what philosophy and science are or ought to be. The influence recently gained by the so-called postmodernists has introduced a new urgency to the need for us to be clear about what we are up to, from the ground up. It will often be stressed here that prolonged and sometimes bitter debate between two sides creates the illusion of starkly dichotomous choices—for instance, between absolutism and relativism, or objectivism and subjectivism. When it comes to philosophy itself, such polarization pressures many to choose between what is usually known as analytical philosophy and a more politicized approach announced in courses whose titles bespeak various forms of activism (feminism, for instance) and suggest that no inquiry is independent of prevalent power structures. An obvious ca-

---

Address for correspondence: Richard K. Khuri, 3713 South George Mason Drive, Apt. 1208-W, Falls Church, VA 22041-3736. Voice: 703-820-4726.
richardkhuri@yahoo.com

Ann. N.Y. Acad. Sci. 988: 322–334 (2003). © 2003 New York Academy of Sciences.

sualty of this turf war between technically minded objectivists and politically ambitious activists is "continental philosophy"—a loose and not terribly informative term linking together such diverse European approaches as phenomenology, existentialism, Marxism, and an amorphous area crossing the boundaries between philosophy, literature, and social science with ease.

While continental philosophy, as exemplified in the work of Patrick Heelan,[1] offers a richness and suppleness often lacking in the philosophical consideration of the sciences, and while it never yields to dichotomous choices—because it always keeps the layeredness, depth, and complexity of whatever it considers in mind—there is a further elemental dimension of philosophy that has almost been completely lost in the thickets of academic competition and argumentation. Quite simply put, philosophy involves both thinking and living. If it was more common to regard philosophy as a way of life in ancient Greece, Rome, or China than it is in our society, this does not mean that philosophy has, over time, ceased to be such. There are definite historical reasons why philosophy was relentlessly winnowed to its purely intellectual (and eventually technical) aspect, a process one can trace back to the Middle Ages.[2]

An immediate link between philosophy as a way of life and science is found when we consider the ancient practice that eventually became chemistry. Alchemy, after all, was not born from an avaricious drive to convert metals to gold, but from the desire to purify the human soul. Gold was merely a symbol for a purer substance thought to be within reach if the proper procedures were followed. Spiritual purification was the context for an art-science, in the course of which much knowledge about metals, their properties and transformations, was gained. The early core of chemistry was formed in the crucible of an attempt to live better, in a manner more befitting the human potential. If today chemistry has become entirely abstracted from its alchemical progenitor—as much philosophy has turned into recondite manipulations within purely abstract fields—this is not to say they no longer have anything to do with how we live or ought to live. The relation between chemistry and biology (including ecology), as it is found in nature as well as in areas such as pharmacology, opens new windows onto the relevance of chemistry for life. Even in the area of inorganic chemical invention, one can highlight the aesthetic, functional, or ecological appeal of various new substances—certain polymers come to mind—from the standpoint of how they improve our lives. A chemist at work on the creation of a new substance may have nothing but the technical aspects in mind, but within a working environment consciously encouraging the search for better ways to live—a cleaner environment, safer construction, building materials amenable to more elegant design—it becomes at least possible to echo the nobility of alchemy at its best. Meanwhile, the urgent call of life is never far away from philosophy, however trapped it may be in academic preciosity or political infighting. A very long tradition of asking fundamental questions, and the issues bound to arise in our own time, are always there to remind philosophers of their proper calling.

Philosophy, then, primarily involves thinking and living. More precisely, the fundamental aim of philosophy is to think better (or more truthfully) with a view towards living better (or more truthfully). It is not enough to say "better" without also saying "more truthfully," because to aim only at what is better without further specification implies relativism. One immediately asks: "Better in relation to what?" On the other hand, the traditional absolutist notion of truth is far too rigid. While it is applicable at the level of syllogisms and simple chemical reactions, as one probes the basic questions of life, and life itself, more deeply, one realizes how truth is no longer a matter of merely assigning a truth value to the assertions we make (or the propositions we state, to use the language of the logicians). How truth might be viewed with more openness is a very long story whose telling does not belong here. Nevertheless, two ways this might be appreciated are, first, in the notion of truth as something that is lived (rather than merely known), or truth as the dynamic ground of conviction, openness, and so on (so that it makes perfect sense to speak of growing into the truth); and, second, in how the truth embedded in a work of art discloses itself as we become more appreciative of and receptive to, say, a great poem or piece of music. To keep this more pregnant notion of truth in mind, one that restores what the word carried historically in any event—be it in the ancient Greek term "aletheia" or in pre-Elizabethan English[3]—is also to avoid the spurious conflict between rigid objectivists who are unable to see beyond "truth" in its simplistic absolutist sense, and equally shortsighted postmodernists who, with precisely this sense of truth in mind, find it to be oppressive and, with indignation, go on to deny that there is truth. Again and again, we witness such spurious conflicts (which have apparently spilled over into the inner circles of those responsible for chemical education[4]): in the first place, a key term such as "truth" is progressively narrowed down; then this narrowed down term becomes accepted even by those opposed to its implications, so that their only option is to question whether the term has any meaning in the first place.

In chemistry, the two concepts of "truth" are operative as well. Everything a chemist says about water, for instance, may well be true in the narrow, rigidly objective sense. But that the chemistry of water gives rise to a substance uniquely fit for life and at the center of a remarkable harmony between the four geospheres (the four geospheres are atmosphere, biosphere, hydrosphere, and lithosphere) provides a glimpse into another more elusive, but far more meaningful sense of "truth." As we shall see, the same chemical reality may be seen in two different ways: exactly for what it is; and as cause for meditation, to the extent that it points beyond itself, towards a domain no longer amenable to precise language, but which is at the heart of what makes our world as it is. In other words, we shall examine a major instance where chemistry seamlessly gives rise to philosophy.

Science

When we consider how science has evolved since Faraday shifted our attention from forces to fields early in the nineteenth century, it is surprising how little attention has been paid to the poetic dimension. Modern science is no doubt a practical enterprise. It was driven by the need to reach a concrete understanding of the natural world. The prevalence of epidemics in Europe made knowledge of human anatomy and physiology especially urgent. Nevertheless, even in the heyday of mechanism, those who were contributors to science rather than technicians who rode their coattails seemed more sensitive to the architectonic beauty of their visions of the natural world than the urge to understand nature so as to increase humanity's control over her. Furthermore, words like "gravity" remain metaphors more than three centuries after their introduction. We understand how gravity works, but we do not quite know what it is. In a different vein, the imagery of Faraday's notion of lines of force within fields such as the magnetic would inspire painters like Turner to render nature with a more potent and expressive dynamism than hitherto. Turner could hardly be called a more "scientific" painter than his predecessors, yet he was able to bring out the effect of powers otherwise hidden, the same powers grouped under the rubric of another metaphor, "energy" (Faraday intensified the elusiveness of the concept when he showed how the various forms of energy are interchangeable without ever being able to say quite what energy is—and neither are we, despite the readily available mathematical definition).

The theory of relativity then became the vehicle for a still more explicit role for what remains conceptually hidden. Physicists now routinely postulate a domain beyond physical (or any other natural) law, be it infinitely large or infinitely small (both the same for the Renaissance thinker Nicholas of Cusa), from which the known universe has emerged. They speak of an undefined substrate, sometimes called "radiant energy," that is ontologically and chronologically prior to all crystallizations found in nature, from elementary particles to clusters of galaxies. The substrate, or some other veiled stage of being, is also said to contain within itself the potential for ever more complex forms of organization, in a manner that vaguely recalls ancient theories that acknowledged the priority of form or archetype (as well as twentieth-century variants of those theories, notably that of Alfred North Whitehead). Quantum mechanics has contributed further to these developments, above all in recent experiments set up to decide the issue of non-locality or non-separability. The weight of the evidence increasingly points in the direction of some indivisible global reality locally manifest in the form of the elementary particles whose existence has been established by physicists. The global reality itself is inaccessible to any conceivable physical theory as things stand at present. The contemporary scientific picture of nature thus has a strikingly poetic aspect: Outwardly, it uses clear concepts and precise mathematical formulations; but it is underlain, underpinned, even intertwined with some indefinite, global, possibly unbounded substrate. This evokes the contrast between the poet's skill in the precise use of language and the ineffable depths embodied in the best poems.

When we turn from physics to chemistry, the poetic dimension may not be so apparent. It is usually said that chemistry is the most practical and down to earth of the "hard" sciences. Not so, argues Roald Hoffmann.[5] He makes a persuasive case in favor of the artistry, creativity, and—yes—poetry in the synthesis of compounds not to be found in nature. There is also the imaginative use of synthetic compounds in order to understand chemical processes we do find in nature, for instance, in living organisms. Again, the laboratory technician at work on some stage of chemical synthesis may not be sensitive to what Hoffmann highlights, but the process as a whole, whether contemplated by a chemist or not, shows a dimension invariably buried within the practical exigencies of experimental work.

Science needs and promises results. For instance, the exact formula for a certain medication must be known, as well as the nature of its intended effects and unintended side effects. However, to reduce science entirely to its practical dimension is to turn the whole enterprise into the relentless activity of an extraordinarily intelligent ant colony. Science would not be what it is without its poetic and metaphysical reverberations, ignited as it is by wonder and astonishment, these in turn fed by the strangeness, beauty, and elusiveness of what is encountered at every stage, however advanced. Who would have predicted that relativity and the quantum theory would lead us right back to the metaphysical speculations of the ancients, this time with the hindsight of vastly more detailed knowledge about the various workings of nature?

The poetic also happens to pay practical dividends. We should remember that some of the greatest contributors to science in the twentieth century—Einstein, Bohr, Schrödinger, Pauli, and Heisenberg come to mind—all were highly literate individuals, well formed in philosophy and the human sciences while being physicists of the highest order. It is precisely their flexibility—their facility with philosophy and poetic thinking—that enabled them to make the imaginative and conceptual breakthroughs that have radically changed the hard sciences. Their achievement can never be matched in depth and scope by a computationally driven science, which is what we may be in danger of falling into.

A strong affinity in spirit thus exists between philosophy and science. Just as philosophy is lived thought and thoughtful living before it becomes mere thought mindless of life, so is science a practice steeped in the poetry of nature and driven at its highest level by a poetic imagination—before it falls into the routine of a dry and workmanlike practice. Our philosophy of science, and of chemistry in particular, suffers the consequences if, right from the outset, we have in mind the merely practical aspect of science, and philosophy reduced to the temptation of pure abstraction that constantly stares it in the face. For one thing, those who intuitively feel restless within such limited confines, but lack the discursive and expressive means for a constructive exit owing to the prolonged negative conditioning prevalent at so many universities, are likely to angrily question the validity of both science and philosophy and throw the general culture into turmoil. This is not to brush aside the interesting dialogue between abstract philosophy and practical science, but simply an appeal to be mindful of the broader and deeper world in which both come into their own, and where their interaction becomes most fruitful.

PHILOSOPHY AND SCIENCE

Among all philosophical viewpoints found in science, perhaps the most bleak and dry is mechanism. Once one is enticed with the vision of the entire universe and everything within it as a finely tuned machine, it seems there is nothing left to do other than discover the various algorithms that describe how the machine works. There is nothing within mechanism itself to elicit wonder or contemplation. And yet, even if we should confine ourselves to mechanism, which has of course been well and truly left behind in the light of a continuous stream of scientific thought over the last two hundred years, there is more to consider philosophically than is first apparent. Most obviously, there is the architectonic beauty and elegance of the alleged machine. The precise manner of the ordering of the solar system, and the fit and form of the mathematical representation of its movements, had tremendous aesthetic appeal for Kepler, Newton, Leibniz, and their contemporaries. Some went further, and took nature's clockwork regularity and fine tuning to be signs of a divine intelligence at work. Kepler, Newton, and Leibniz were all favorably disposed towards such a view of a divine presence in universal order. Lesser thinkers, including men of religion throughout much of Europe, were so keen on identifying the divine intelligence with mechanical ordering that they willfully ignored all possible gaps in the theory. Not surprisingly, Newton and Leibniz were not quite so rash. In his letters to Father Bentley,[6] Newton made clear his ignorance of precisely what gravity may be, what sort of agency it might indicate, how just one set among infinitely many initial conditions for the solar system happened to obtain (namely those conditions according to which all the bodies within the solar system would have their exact size and weight and be at exactly the observed distances from one another, so that they orbit regularly and reliably without crashing into one another or falling into the sun, a problem he knew to be computationally insoluble, and which remains so to this day), and how the solar system happened to occupy the same plane when it might far more easily not have (Pluto had not been discovered yet). The "machine" brought to light by Newton raised questions from many different angles. It signified a level of intelligence beyond the mechanical at several points. As for Leibniz, since he was a philosopher in the full sense of the word (and not just a natural philosopher), he never ceased to emphasize the metaphysical aspect. His view of matter as being endlessly divisible, and of the relativity of space and time (because they were respectively nothing without the need for matter to be able to occupy different places at the same time, and the same place at different times), sound as though they belong a lot more to the twentieth century than the seventeenth.[7] He also intuited the irreducibility of perception and consciousness to the mechanisms we are able to discover and describe, however ingeniously and precisely.

Whenever we deal with a mechanism, however simple, philosophical issues arise. For instance, there is an order in and rate at which events within a mechanical system occur, a duration they have, and so on. When we are merely computing these values, we do not usually pause for thought. But an obvious set of questions arises: are terms like "order," "rate," and "duration" themselves conceived with a mechanical mindset? As found in nature, are they mechanically determined or do they determine mechanisms? What are they anyway? How do they enter the picture in the first place? These questions gain weight when we look at various cellular mechanisms and reflect on the degree to which they are ordered, finely tuned, show a purposiveness (as in the case of the behavior of slime molds in relation to available nutrients), and are context dependent (as in what an egg does upon fertilization over and above what it is genetically programmed to do, say with regard to differentiation and the specificity of acquired form).

Nature is full of mechanisms. Yet careful thought about them shows the various ways in which they cannot be used to justify a mechanistic philosophy, even before subsequent ideas, theories, experiments, and discoveries have eroded the tenability of Mechanism. The leap from mechanisms, however neat and many, to Mechanism cannot be philosophically justified. Mechanism (with a capital "M") requires the willful suppression of a whole line of questioning. If intelligent people throughout Europe were attracted to it for a long time, it was surely for reasons other than intellectual. By the middle of the seventeenth century, a combination of war and economic deprivation created a widespread yearning for order, political or otherwise—the more easily computable, the better.[8] Many still find comfort in the notion of a world susceptible to some kind of rational ordering, computable, predictable, and controllable. This is no way to truth.

There are two philosophical mindsets in our approach to nature. The first is narrow and dogmatic. It hastens to proclaim some doctrine such as Mechanism or genetic determinism. All phenomena are subsequently considered within such intellectual and spiritual confines. The second is broad and open. It has no need for a label, whether its suffix is an "ism" or otherwise. It always looks at nature and things in nature as potentially "bathed" in deeper levels of being. It sees every discovery, every established theory as simultaneously enclosing a certain aspect of the phenomena while calling forth questions pointing beyond the enclosures. The questions have both horizontal and vertical signification. Horizontally, they reaffirm the awareness shared by all great scientists that the methodical inquiry into nature really has no end. Vertically, they remind us of a level beyond the reach of methodical inquiry, a level where there is only poetry, contemplation, and inspiration, some of which is responsible for the creative leaps that enable the advancement of science in the first place.

PHILOSOPHY AND CHEMISTRY: THE SYNTHESIS OF ASPIRIN

The distinction between the broad and narrow approaches can be applied even for a specific example of a chemical process. Let us consider the industrial synthesis of aspirin from petroleum. The steps are simple enough to describe:

- Benzene is separated out from a petroleum fraction.
- One of the hydrogen atoms in the benzene molecule is replaced with SO_3H when the benzene molecule is reacted with sulfuric acid.
- The SO_3H group is then replaced with ONa by reacting the modified molecule with sodium hydroxide.
- By a further reaction with dry ice and water, ONa is replaced with an hydroxide radical, and another hydrogen atom in the benzene molecule is replaced with C(OH)O.
- The hydroxide radical is finally replaced with $CH_3C(O)O-$ by a reaction with acetic anhydride. The resultant molecule, of empirical formula $C_9H_8O_4$, is aspirin (acetylsalicylic acid).[9]

The foregoing process hardly seems to elicit philosophical questions. But this would only be the case if we closed our minds to everything that lies outside of the industrial synthesis of aspirin from petroleum, if we were single-minded about the practical aspect. In contrast, if we were to remember what many know about benzene and aspirin, a question immediately raises itself. All we have got to do is think of these in relation to human health. Benzene is a carcinogen. The hysteria over the seeping of some benzene molecules through a faulty filtration process into "Perrier" bottled water some years ago is a stark reminder of the health risks posed by benzene. Aspirin, on the other hand, is widely regarded as a "miracle drug." It not only helps in the relief of an assortment of mild ailments, from headache and fever to arthritis, but small daily doses are now known to be good for the heart. Is there nothing to be said when, in five steps, we are able to convert a carcinogen into a drug with many benefits for our health? We know precisely how to synthesize aspirin from benzene industrially. This is chemistry. But what about the conversion of a carcinogen into a miracle drug? Is this not some sort of alchemy?

If we think with greater philosophical openness about the industrial synthesis of aspirin, and about many other chemical processes, we may repeatedly notice a dimension other than the mere reporting of reactions and results. What chemistry is in our contemporary sense may have a surprising quasi-alchemical aspect. Or perhaps it is not so surprising after all, if we first remind ourselves of whatever in nature lies beyond systematic research and then realize how a relentless methodical approach will sooner or later collide with a level of reality we have been conditioned to ignore. One is free to insist on the hard facts: we begin with benzene, we end up with aspirin, benzene is a carcinogen, aspirin is generally good for our health. Nevertheless, the transformation of a substance with a given effect on human health to one with an effect dramatically different provides reason for wonder. To deny this is the same as denying that there is anything more to a spectacular sunset than a particular combination of optical effects, given the atmosphere's specific chemistry at that location and the shape, density, and distribution of clouds. It is a matter of whether we believe natural beauty to be a mere accident, if not some conventional socially conditioned reaction, or something rather more.

Other kinds of philosophical issues present themselves when we discuss a chemical process such as the industrial synthesis of aspirin. In general, what does it mean for some molecule to have a pronounced effect on human health one way or another? The problem arises because the levels of chemistry and human health are different. Just as it makes no sense to speak of acids or bases, or equilibrium constants, in physics, so is it nonsensical to speak of a headache or fever in chemistry. Health is what we attribute to an organism as a whole, or to an environment, or even to our whole planet. In the same way that we are unable to account scientifically for how the least possible change at the atomic level as we move from one element in the periodic table to its neighbor as governed by the Pauli Exclusion Principle shows up as a great difference at the macroscopic level (say, between gold and mercury), so is it difficult to understand the shift from the introduction of a given molecule into the human body to the effect on its overall health. We do understand that certain molecules or drugs have certain effects, often with great precision. But we have no understanding of the transformations whereby a chemical, in the context of something as complex as human physiology, predispositions, and perhaps emotional or psychological states, can have an effect far more global than its mere formula suggests. The problem is not only one of complexity, not only a matter of finding a more elaborate analytical account of what happens. This would describe only the horizontal aspect. There is also a vertical aspect. The effect of a chemical on human health implies an influence at a higher level than mere chemistry (in our contemporary sense) suggests. Defined in our strict sense, in other words, a chemical has no business affecting human health. Yet it does. How so?

One can only conclude that chemicals such as benzene and aspirin have what may be termed "ontological suppleness." In general what has ontological suppleness manifests itself at levels of being beyond that in which it has its initial appearance and definition. For instance, if it turns out that something at the subatomic level of metals accounts for the great differences in their macroscopic properties, then such a subatomic phenomenon has ontological suppleness. More radically, if, as some physicists suppose, there is a veiled substrate (veiled in the sense of being physically inexpressible) that manifests itself throughout the entire spectrum of natural phenomena, whatever their level or complexity, then such a substrate would have maximum ontological suppleness from a naturalist perspective. In the effects of something like electronegativity on the emergence of life and our planet's overall ecological stability, we shall presently see an illustration of ontological suppleness not too far from its natural limit (The use of the word "natural" is intended to at least suggest a more purely metaphysical dimension for ontological suppleness, which extends its possibilities further still). So a chemical with effects on human health also has ontological suppleness: a human being is far more than an aggregate of chemicals, yet human biology and psychology as a whole provide a context within which chemicals can have effects that their literal (objective and isolated) study would not justify, effects that are nevertheless nontrivially linked to the chemistry of the substances in question. Nothing exemplifies this better than mind-altering drugs.

Another case of ontological suppleness with its beginnings at the chemical level is when two molecules with exactly the same formula, different only with regard to their "left-" or "right-handedness" (chirality), may have pronouncedly different effects on the human body: large numbers of one may taste sweet, of the other sour; one may be harmful to human health, the other beneficial. The recognition of chemicals by the human body, or the manner in which they become integrated into its hierarchical structure, suggests that the simple chemical formula does not tell the entire story. Some aspect of the chemical's behavior enables it to be recognized one way or the other, to enhance or disturb the body, to please or displease our senses—something no formula can contain. Some combination of our physiology and consciousness becomes the context for the significance of chirality to show itself fully. The relevance of chirality therefore "stretches" from the given enantiomers to their biological and psychological significance, hence the usefulness of a neologism such as "ontological suppleness."

In some cases, we lack the language to express this elasticity. The problem is one of interface. Science is bound to encounter it more often as what is discovered and established by experiment becomes more and more intriguing. It arises when formal language is no longer adequate for the expression of what happens. How, for instance, can the language of chemistry ever express a changed mental attitude brought about by a given medication? There is a case to be made for a two-tier approach: on one hand, a precise quantitative description of what happens to the extent that this is possible; on the other, a qualitative account of effects that it would be nonsensical to attempt to quantify (combined with a qualitative account of the transformation from the quantitative to the qualitative). A new boundary zone is emerging where science flows seamlessly into a more holistic domain, where issues such as harmony and consciousness come to the fore. After all, an experience as simple as whether something tastes sweet or sour forces one to deal with the problem of consciousness, even were one a dogmatic materialist.

PHILOSOPHY AND CHEMISTRY: A CLOSER LOOK

In the foregoing spirit, we shall consider an extremely common chemical substance, namely water. To understand what is extraordinary about water, another chemical term without physical sense needs to be introduced. This is electronegativity, defined as "a measure of the tendency of an atom in a stable molecule to attract electrons within bonds." The value of electronegativity is calculated with the help of two functions to which it is simultaneously related. The first function relates electronegativity to bond dissociation energy, which is the energy required to break a particular bond. The second relates it to ionization energy, which is "the energy required to remove an electron from a free atom or ion in the gaseous state," and the electron affinity, which is "the energy released when an electron is added to a neutral atom in the gaseous state."[10]

It so happens that oxygen has an unusually high electronegativity (another sentence without sense in physics). This has the following immediate implications: The two electrons contributed by the two hydrogen atoms in the two OH bonds in a water molecule tend to spend more time in the vicinity of the oxygen nucleus than either hydrogen nucleus. A polarity (asymmetry) is thereby created within the water molecule: near each hydrogen nucleus, the charge has a positive bias, near the oxygen nucleus, a negative bias. Whenever one of the two hydrogen atoms in a water molecule is near the oxygen atom in its neighbor, the two molecules attract one another. This is known as the hydrogen bond. It gives water internal "strength" (Chemists will prefer "structure" to "strength.")

The polarity and internal strength (structure) of water have a series of consequences, among which the following are noted:

- The latent heat of melting and vaporization for water are high. This means that compared with other substances, a lot of energy is required for water to change phase from solid to liquid, and from liquid to gas. The stability of the oceans is thereby guaranteed. Too much energy would be needed for them to evaporate away.

- The heat capacity for water is unusually high. To raise the temperature of water, a lot of energy is also required. Thus an organism like a human being, which mostly consists of water, is protected from large or rapid changes in the temperature of the surroundings. Similarly, the oceans protect the temperature of the earth from such changes.

- The polarity of water is crucial in the formation of membranes. Here, it should be mentioned that polar molecules interact with one another, but not with non-polar molecules, and vice versa. So a non-polar substance, if it can form into sheets that close in upon themselves, can separate two regions filled with a polar substance like water, for instance, the watery interior of a cell and its aqueous surroundings. Thanks to the strict differentiation between polarity and non-polarity, there is the possibility of an "inside" and an "outside." "Ambiphiles" present us with a more complex case. These molecules are polar at one end, non-polar at the other. An example would be sheets of polar lipids with an oily (non-polar) end, with the polar end immersed in water. Dishwashing liquid has these properties, so it can cut through grease and yet dissolve in water. Ambiphiles allow interaction with an aqueous environment and protection from it at the same time.[12]

The crucial role of the polarity of water in the formation of membranes, without which life as we know it would not exist, and in our whole planet's ecology, highlights "ontological suppleness," the main theme of this discussion, quite dramatically. A phenomenon such as electronegativity, which causes water to be polar, influences a far higher level of nature, namely, the only way that we know life to have become embodied in practice, whatever the primordial possibilities may have been, as well as overall ecological stability. With this in mind, let us review the situation in a philosophical spirit.

(1) We have a measure of enormous consequence (electronegativity), yet its definition depends on concepts such as bonding and ionization that make sense only at the chemical (and not the physical) level. In general, we have countless situations in science where terms only have meaning at one or more levels removed from the physical (it makes no sense to speak of cells in chemistry, or of society in biochemistry). This means that the languages of different scientific disciplines are not within the same "plane" of meaning, but occur within different levels of a "hierarchy" of meaning. There is the further point that a phenomenon encountered "lower" within the hierarchy, such as electronegativity, has consequences at a much "higher" level. This suggests that the phenomenon in question is not exhaustively described at the "lower" level, but has further aspects that become manifest at "higher" levels for which the language of the "lower" level is obviously inadequate. The language of physics does not adequately express electronegativity, yet oxygen has a high electronegativity with nothing else manifest to us other than its physical structure. Either we need a more open understanding of the physics of oxygen, or there is something about oxygen that is "veiled" from physics, and which later becomes manifest as a high electronegativity. Similarly, the language of chemistry does not express membrane formation, yet the polarity of water, which is a chemical phenomenon, causes conditions uniquely suitable for the formation of membranes. Again, there is something about water that either demands a more open understanding of chemistry or remains veiled from it. The same can be said of the ecological fitness of water, which issues from its chemistry, but for which the language of biochemistry and biology (in the ordinary sense) is inadequate. We see here an occasion for a strong claim made on behalf of ontological suppleness. Beginning with the chemistry of water (and even the physics of hydrogen and oxygen, if not the primordial physics of radiant energy), we note phenomena that have consequences all the way up to the ecological level. It would be strange, then, to close water off within a strictly chemical definition. Something in our language and philosophical attitude ought to be mindful of the deep linkages between a chemical aspect and its ecological consequences (and all those in between). Once again, we find ourselves faced with a choice. As has been stated earlier, there are two philosophical mindsets in our approach to nature, one narrow and dogmatic, the other broad and open. The latter, we have said, always looks at nature and things in nature as potentially "bathed" in deeper levels of being. So here too, we may resort to the reductive approach and simply brush aside the issue, stripping it of all that is philosophically interesting about it. Or we may adopt a more fluid and sophisticated approach, one that acknowledges the need for a precise language at the center of chemical activity and description, together with some kind of open boundary to show appreciation of the continuum of effects (of something like the high electronegativity of oxygen), from chemistry to ecology. A science like chemistry needs to be metaphysically "loose" at its edges (physics even more so).

(2) A related point to the foregoing concerns a sense in which the reductive enterprise is an exercise in futility. It is all well and good to try to reduce

nature to the subatomic level. But how useful is this when already at the chemical level, there are emergent phenomena such as electronegativity, which differ sufficiently from one element to the next to bring about radically divergent effects? These divergent effects can in turn sometimes work together to bring about further effects on an even larger scale, often harmonious and stunningly creative. In other words, the reductive approach fails on two counts: taken at face value, it has nothing to say about emergent phenomena of the greatest significance as we move on to the chemical, biochemical, biological, and ecological levels; if, on the other hand, we wish to include the emergent potential within concepts operative at the reductive level, then these concepts will have to be so rich and open as to be poetic and metaphysical rather than merely physical. Again, the use of precise language, with well-defined terms, means, given the cascading and broadening effects as we examine nature at higher and higher levels, an actual hierarchy of languages presenting us with real problems of mutual translatability. Moreover, an open boundary is required beyond the core of each precise language to maintain expressive suppleness with regard to the full scope of emergent phenomena and how these suggest continuity with smaller scale indicators such as oxygen's high electronegativity (the same problem would present itself if we took the macroscopic features of metals into account in relation to their atomic structure).

(3) There is also cause to reflect on just what it means for the electronegativity differential between hydrogen and oxygen to be such that water is extraordinarily fit for life both within organisms and in their environment as a whole, as well as at the boundary or interface between the two. Philosophically, the point is no different from that raised by astrophysicists who ponder the meaning of the "fine tuning" of fundamental universal properties such as temperature and density. What interests us here is that chemistry also gives rise to such reflection. The idea is not to hint at an argument from design. The temptation to arguments from design seem linked to Mechanism. It is easy to imagine a finely tuned machine requiring some master mechanic to design it. The contemporary scientific view of nature is rather more open, more organic. One way to imagine a metaphysical setting for fine tuning may begin with the thermodynamic view of some energy flow between a source and a sink. The energy between the two then becomes available for many levels of self-organization.

At the cosmic level, sources and sinks are put forward, and between them our planet, the solar system, our galaxy, or a cluster of galaxies may form. There would be nothing unusual about such thinking among scientists who are able to apply the assumptions and laws of thermodynamics on a very large scale. But as philosophers, we can perhaps appeal to the tradition as we contemplate what it might further mean to posit a Source and a Sink on such a large scale as to allow the self-organization of the entire universe. For "Source" and "Sink," the philosophical tradition might substitute "Being" and "non-Being." The infinite intensity of the contrast between absolute plenitude, and utter nothingness, generates, among other kinds of flow, the energy necessary for the successive transformation of radiant energy into subatomic particles, atoms, molecules, macromolecules, living organisms, society, and culture, when Mind enters the picture among what exists. The emergence of large natural bodies, such as stars and planets in their galactic context, is another aspect of this flow. Note, however, that the sequence from atoms to Mind does not only entail greater complexity, but some kind of vertical ascent, so that thought, values, and consciousness enter the picture. Some physicists have recently linked these, however loosely, with the subatomic and prior levels, only no longer atomized, but envisaged as some seamless all-encompassing whole beyond the reach of physical law (David Bohm)[12] or a level of reality at any rate veiled from physics for which poetic or mythical language might be more appropriate (Bernard d'Espagnat).[13]

In the wake of experiments performed by Aspect and Gisin, the tendency to regard mind and matter as somehow inseparable is growing.[14] This harks back to the days when Aristotle and Plotinus also saw nature as infused with intelligence at every level, and regarded Mind in all of its forms as essentially one. The remarkable fitness of water for life, among many, many cases of uncanny fit that we know of, can then be seen not as the product of design, but as the expression of a more integral presence of intelligence in natural phenomena. This is not to give an unqualified endorsement of such speculation, but simply to draw attention to the proximity of contemporary science to ancient thought should one have the openness and motivation to pursue the implications of some of the more striking recent findings to their full philosophical potential. And chemistry, for all its obvious practical bias, has a crucial contribution to make in the fashioning of a new dynamic, scientifically informed, metaphysics. The philosophy of chemistry, in common with the philosophy of any other major natural science, will at some point show how some of what we know through chemistry becomes a part of philosophy, not as mere argument and discussion, but philosophy in the grand sense, namely the fashioning of a worldview, in part contemporary, in part timeless.

NOTES AND REFERENCES

1. HEELAN, P. 2003. Paradoxes of Measurement. Ann. N.Y. Acad. Sci. This volume.
2. HADOT, P. 1995. Philosophy as a Way of Life. Blackwell. Oxford, UK.
3. "Troth." Ed.
4. SCERRI, E. 2003. Ann. N.Y. Acad. Sci. This volume.
5. HOFFMANN, R. 1995. The Same and Not the Same. Columbia University Press. New York.
6. NEWTON, I. 1958. Paper and Letters on Natural Philosophy. I.B. Cohen, Ed. Harvard University Press. Cambridge, MA.
7. LEIBNIZ, G.W. 1951. On Newton's mathematical principles of philosophy (Letters to Samuel Clarke, 1715–6). In Selections. P.P. Wiener, Ed. Charles Scribner's Sons. New York.
8. TOULMIN, S. 1990. Cosmopolis. University of Chicago Press. Chicago.
9. HOFFMANN, The Same and Not the Same.
10. SHARP, D.W.A., Ed. Dictionary of Chemistry. 1990. Penguin Books. London.
11. MOROWITZ, H. 1987. Cosmic Joy and Local Pain. Charles Scribner's Sons. New York.
12. BOHM, D. 1995. Wholeness and the Implicate Order. Routeledge. London.
13. D'ESPAGNAT, B. 1983. In Search of Reality. Springer-Verlag. New York.
14. NADEAU, R. & M. KAFATOS. 1999. The Non-Local Universe. Oxford University Press. Oxford, UK.

Chemistry Beyond Positivism

WERNER W. BRANDT

*Department of Chemistry, University of Wisconsin-Milwaukee,
Milwaukee, Wisconsin 53201, USA*

ABSTRACT: Chemistry is often thought to be quite factual, and therefore might be considered close to the "positivist" ideal of a value-free science. A closer look, however, reveals that the field is coupled to the invisible realm of values, meanings, and purpose in various ways, and chemists interact with that realm loosely and unevenly. Tacit knowledge is one important locus of such interactions. We are concerned in this essay with two questions. What is the nature of the knowledge when we are in the early stages of discovery? and In what ways does the hidden reality we are seeking affect our search for an understanding of it? The first question is partly answered by Polanyi's theory of tacit knowledge, while the second one leads us to realize the limitations of our language when discussing "reality"—or certain chemical experimental results. A strictly positivist approach is of little use, but so is the opposite, the complete disregard of facts. The contrast between positivism and non-formulable aspects of scientific reasoning amounts to a paradox that needs to be analyzed and can lead to a "connected" chemistry. This in turn resembles networks described by Schweber and is more concerned than the chemistry "as it is" with aspects such as the image of chemistry, the challenges chemists face as citizens, and chemistry in liberal education.

KEYWORDS: positivism; metaphors; tacit knowledge; epistemology; Michael Polanyi; language games

INTRODUCTION

Chemists generally leave to others the search for contacts with the literary world, with humanism, with art, and perhaps with other "non-scientific" pursuits. Exceptions to this sweeping statement are important to us; here are some examples:

- The Nobel Laureate chemist Roald Hoffmann mentions the "links of the spiritual and material world," how "the processes of nature connect with the interior world of our emotions," and the effects of "forces deep within our psyche."[1] Hoffmann feels that the dualities he identified, such as chemistry/philosophy, lead to some healthy tension and in this way add interest to the entire field of chemistry. Hoffmann is also a poet of considerable stature, who has thought about the role of the poet, living and working at his own interesting interface of chemistry and psychology.

Address for correspondence: Werner W. Brandt, Department of Chemistry, University of Wisconsin-Milwaukee, Post Office Box 413, Milwaukee, WI 53201. Voice: 414-332-9078.
wwbrandt@uwm.edu

- The Nobel Laureate chemist Dudley Herschbach is very interested in the use of metaphors, in connection with his search for novel pedagogic methods.[2] He considers an impressionist painting as a metaphor for chemistry, in that one needs to see the entire painting from an appropriate distance, not just the individual paint spots that one would see if one stood too close. Similarly, chemists need to use, in quick succession, different tools at their disposal (thermodynamics, molecular geometry, electronic structure theories, etc.) to solve a chemical problem they face. In the humanities as well, the context of a problem is usually very important.

- The eminent chemist David W. Oxtoby, obviously motivated by his Christian faith, connects, in a metaphorical sense, the chemistry and physics of phase transitions, his research interests, to God's action in the world.[3]

Metaphors can greatly help chemists: they give hints of the connections to the wider context, ultimately of "the way the world is," which we seek to describe, the philosopher's goal in a nutshell. It has been said that metaphors often say something better (more conducive to understanding), but unfortunately they do not say more than the corresponding literal expression.[4] Literal and metaphoric languages frequently overlap,[5] and metaphors, and even hints at metaphors, are often the "currency of the teacher-student transaction in teaching,"[5] but they do not automatically go beyond "positivism," taken to be "the view, originally advanced by Comte, that 'positive facts' concerning observable phenomena and their relations are all that can be known, and that inquiry into causes, origins, and purposes should be abandoned."[6]

In contrast to this, there are some other aspects of a chemist's work that can serve as stepping stones into that which lies beyond positivism. We are therefore talking here about the realm of values, meanings, and purpose. Some of these facets are as follows: ethical constraints on proliferation in print,[7] the ideal of service,[8] the possible future migration of chemical subject matter into other academic disciplines,[9] and the attempted epistemological[10] revolution of Michael Polanyi (1891–1976)[11–15] concerning the study of the nature, origins, objects, and limitations of knowledge.

In this essay we are concentrating on the latter topic, because its tentacles seem to run deep and are ubiquitous in the bloodstream of our general culture, and we hope they are of interest to many.

Some of these tentacles have grown well in the recent past, and will be discussed in some detail, because they illustrate that chemistry can in fact contribute a lot to the wider context—that is, to our culture beyond positivism. We think it is important that Michael Polanyi, a great person and chemist-turned-philosopher, is slowly getting the recognition he deserves.

LIFE AND WORK OF MICHAEL POLANYI—AN OVERVIEW

It helps to understand Polanyi's work and to appreciate the driving force for his philosophical studies if one knows something about his life and his education. Polanyi was born in Budapest and raised in a very cultured family. He studied medicine and received his medical degree in 1913. From 1914 to 1918 he served as a medical officer in the Austro-Hungarian army. While recuperating from a war related illness, he produced a thesis on the adsorption of gases on solids that earned him a Ph.D. in physical chemistry, in 1917. For many years he worked as a very productive physical chemist at the Kaiser Wilhelm Institute for Fiber Research in Berlin, doing both theoretical and experimental work. He became the mentor of two future Nobel laureates, Eugene Wigner and Melvin Calvin. In 1933 he was appointed chair of Physical Chemistry at Manchester University in England. Polanyi switched to the field of philosophy in 1946, and worked on a theory of tacit knowledge (TTK), which is an attempt to revolutionize the epistemology of science. Unfortunately Polanyi did not find the resonance he was hoping for amongst his colleagues at Oxford University where he had moved in 1958. These matters will be at the center of the discussion that follows.[16,17]

Polanyi was a gentle and conciliatory person[18] and had many friends at Berlin. He was persistent. For example, it was more than a decade before his theory of adsorption was accepted by the scientific community; it is in use even today. On the other hand, he was not stubborn when confronted with good experimental results, as his friend and co-worker Wigner testified.[19]

One reason why we are interested in Polanyi's work is his outstanding reputation in chemistry: at the height of his scientific career, from 1933 and 1937, he had 37 publications, and many of these were of great importance. In his career as chemist he worked in three areas: adsorption of gases on solids, X-ray structure analysis of the properties of solids, and chemical reaction rates. He published 217 items in these fields. Later, he wrote 138 publications in other fields, notably philosophy, not counting many newspaper articles and reviews.[20]

Polanyi gave us his own description of the deeper qualities that enable a scientist to pursue his vision, and even to disobey external authority. The example relates to a context familiar to chemists, that is, to the decision on the merits of a theory. He starts by reminding us that *simplicity*, as well as *symmetry* and *economy*, contribute to the excellence of a theory:

> [These properties] can only account for its [the theory's] merits if the meanings of these terms are stretched far beyond their usual scope, so as to include much deeper qualities which make the scientist rejoice in a vision such as that of relativity. They [the meanings] must stand for those peculiar intellectual harmonies which reveal, more profoundly and permanently than any sense-experience, the presence of objective truth.[21]

Here, Polanyi clearly relates an inner experience, *intellectual harmony*, to *objectivity*, which most of us think we understand as something far more down-to-earth, and, as Polanyi might say, as something more superficial.

The inner strength of Polanyi is also seen in his ability to survive a professional near-catastrophy.[22,23] Polanyi had developed his own theory of adsorption, which was in reasonable agreement with experimental results. At one point he had to defend his theory vis-à-vis Albert Einstein and Fritz Haber, and he fared badly; he survived professionally, as he put it, only "by the skin of his teeth." The amazing aspect is that he knew that his theory must be essentially correct, and in fact later on he collaborated with Fritz London on the development of the theory of dispersion forces, which was instrumental in putting his adsorption theory on a solid footing.

It is not surprising that a person of strong convictions and clear visions like Polanyi also at times made mistakes: "*Wo viel Licht ist, ist starker Schatten*" (Where there is much light, there is deep shade [Goethe]).[22]

As an important example, Polanyi tended to use terms from the common language, but give them a different meaning, such as the word "body" or "universal intent."[24,25] Another of Polanyi's weaknesses is the liberal creation of neologisms, such as "from-to structures." We plan to bypass, in this short essay, the burden of presenting lengthy definitions of such technical terms, by rendering their meaning in simple common language, thus following the example given by the aged Wittgenstein in *Philosophical Investigations*.[26] Other problems in Polanyi's writings are that he mixed psychology, scientific, and technical knowledge with epistemology,[27] and used a somewhat cumbersome style.[28]

Polanyi was very sincere even when he used such unfortunate techniques. By contrast, the dean of the philosophers of science during the last century, Karl Popper, who also used potent technical language, has been accused of wholesale deceit of his readers, based on detailed analysis of his writings.[29] Nevertheless, Popper is much better known than Polanyi!

POLANYI'S "THEORY OF TACIT KNOWLEDGE"—I

The overall goal of Polanyi's philosophical efforts was to establish the personal, passionate, committed, and communitarian components of knowledge. He even wanted to assimilate scientific knowledge to moral knowledge. This is a considerable task, right on a collision course with positivism, but, if successful, it

> ... illuminates both the nature of science and the nature of morals both of them distorted by arbitrary modern separations of descriptive from evaluative expressions.[30]

In the end, then, in Polanyi's own words:

> The freedom of the subjective person to do as he pleases is overruled by the freedom of the responsible person to act as he must.[31]

We will return to this saying at the end of this essay.

It seems best to organize the following outline of Polanyi's philosophy by referring to two questions, formulated by Wigner; the first one is: "What is the nature of the knowledge when we are in the early stages of a discovery?"[32]

Polanyi disagreed very much with Descartes' method of systematic doubt; instead, he regarded

> ... a person's whole being, all his experience, as the ground from which rational, articulate thinking grows.[33]

This conviction explains the emphasis he placed on two important tenets of his theory:

> We know more than we can tell. ... All knowledge is either tacit (implicit, unconscious) or is rooted in tacit knowledge.[34]

We can become aware of this tacit knowledge only in a "subsidiary" (supplementary, indirect, or secondary) fashion. According to gestalt psychological thought, subsidiary knowledge often contains clues that somehow on "integration" (combination, blending)[35] can yield explicit knowledge concerning the whole of a problem. Polanyi is thinking here of "bodily"[24] clues that we "indwell."[36] This process cannot be focally (clearly) observed by us. There will always be a tacit dimension in our knowledge that is held together by the person. The explicit knowledge resulting from the integration, finally, is quite accessible and can be manipulated.[37–39]

We believe that Polanyi's medical background was important when he formulated the TTK. Physicians cannot simply be technicians, handling only the explicit evidence, but must intuitively seek to understand their own and the patient's psyche. True objectivity, according to Polanyi, involves more than impersonal technical knowledge, it must include tacit aspects, which are likely to be filtered out and set aside by a radical rational approach. Polanyi and his followers considered this a loss, an error of Enlightenment thought. Examples of the importance of tacit knowledge arise in the work of an art connoisseur, or of a trial judge, people who have to arrive at a decision, somehow.[40] Polanyi thinks of an appraisal where the skills, as acts of knowing, involve a *personal* coefficient, which shapes all factual knowledge. In doing so, it bridges the disjunction between subjectivity and objectivity. It implies the claim that man can transcend his own subjectivity by striving passionately to fulfill his personal obligation to "universal standards."[41]

It is of particular interest to us that two chemists, Grosholz and Hoffmann, have used their chemical experience to add to our understanding of TTK in important ways. After carefully reading an experimental research paper by Hamuro et al.,[42] which dealt with the synthesis of an antibody mimic for recognition of protein surfaces,[43] they comment on it, saying:

> [T]he intuitive may be analyzed, the tacit may be articulated, but never completely and all at once: certain indeterminacies and logical gaps always remain, even as scientists achieve a consensual understanding of reality.[44]

The word "never" is intriguing here. The authors seem to say there is some tacit knowledge that is accessible while some of it may be entirely out of reach. This contrasts with Enlightenment confidence in the unlimited powers of the human mind—which Polanyi considered to be in error. The delineation of accessible and inaccessible is often unknown, and may in fact change with time, as methods and insights change. In any case, if "consensual understanding" does not overcome logical gaps encountered, objectivity, as usually defined, seems unattainable.

Recently, many of Polanyi's speculations concerning the TTK have been found to be correct. Fishman has considered *hunches* as alternate ways of knowing.[45] Hunches differ from scientific beliefs in that sufficient grounds for explicit persuasion are absent, but they may lead to action, if that is needed, for example, in medicine, or in chemical research. A scholar thus proceeds in a hybrid mode: on the one hand, there is an incomplete framework, based on facts and theories that are accepted and brought into focus for the purpose. On the other hand, there may be intuitions that are not related to this framework, but, perhaps in despair, and as physicians might say *"ut aliquid fiat"* (Latin 'let something be done'),[46] they are still taken seriously, because inaction is not advisable. As Fishman puts it,

> [T]here are cerebral phenomena here, at times nonlinguistic, at times not self-conscious; but not on that ground, necessarily irrational.[47]

Work on artificial neural networks (ANNs) has shown that these systems can lead to the discovery of correlations that were not suspected by those who supplied the data to the computer, and can yield unexpected predictions. These results have nothing to do with any rules or algorithms supplied to the computer earlier. Using this as an analogy for the intuitions or hunches mentioned above, one finds that these hunches are no longer subjective; instead, they are now properties of an ANN system as a whole. Fishman concludes his essay by naming the phenomenon considered *"intuitive rationality,"* and says that while its mechanisms remain mysterious, the term is not an oxymoron.[45]

In Reber's recent work with artificial grammars, subjects were shown statements produced by well-defined rules of grammar of an artificial language; the rules were unknown to the person tested. The subjects acquired some facility in the use of this language, without ever becoming aware of its rules.[48]

As another contribution from chemistry, Hoffmann reported on an intriguing laboratory situation.[49] During a complicated synthesis, a student tried a number of things to improve the yield of a particular synthesis step and, finally, in desperation, did something for which there was no precedent, and it worked. Just like the physician mentioned above, the student had to act, and he needed to use all he had, even an intuition, or a "jump of imagination," as Hoffmann put it. The student simply had no basis in fact to expect a favorable result.

The nature of this hunch is interesting. Perhaps, in the student's previous experience there was some correlation of earlier events, of good and bad results, things that were cerebrally stored but were subconscious to the student, inaccessible to his reasoning, or to his recollection. His desperate action thus was based on tacit knowledge, and was certainly a *personal* matter, which interacted with the explicit, factual knowledge he had. Only later, when this decision led to successful action, and was therefore used by others, did the new knowledge lose its subjective character. The disjunction between the subjective and objective components thus arose quite late.

POLANYI'S "THEORY OF TACIT KNOWLEDGE"—II

The second question Wigner formulated is:

> In what ways does the hidden reality we are seeking affect our search for an understanding of it?[50]

The complications we mentioned in the last section would clearly affect whatever idea we might have concerning reality. Since we know very little about these shifts, and since reality surpasses our speech, we must conclude that reality is, in effect, "unspecifiable."[51] Torrance spoke of a "legislative authority of reality," and Jaspers discussed ciphers (hints) one might receive concerning that which transcends our experience,[52] but we will think about two ideas that are closer to chemistry. One is that we

> understand the reality whose independence we honor as requiring scientific methods that are not univocal and reductionist precisely because reality is multifarious, surprising, and infinitely rich.[53]

The second one relates to languages used in chemistry:

> ... there is no single correct analysis of the complex entities of chemistry expressed in a single adequate language, as various reductionist scripts require, and yet the multiplicity and multivocality of the sciences, and their complex "horizontal interrelations" do not preclude, but in many ways enhance their reasonableness and success.[53]

The "horizontal interrelations" are the multiple connections of ideas and theories, of experimental results and whatever else is needed for the creation of a conceptual network as a stepping-stone for future progress. Like a ship at sea, that network can, and sometimes must be, repaired while in use.[54] Hoffmann has discussed elsewhere "horizontal understanding," a term closely related to the above, which is in stark contrast to reductionism. Chemistry, due to its positivist heritage, has been said to live in a suburb, separate from the central city, where "suburb" and "city" are intended to denote linguistic communities.[55] By contrast, positivism aims at using a clear and unambiguous language to deal with those "positive (certain, manifest) facts" that are to be trusted—facts that hopefully will never change.

We recall at this point that Wittgenstein, in his later stages of thinking, argued that every word or sentence we use is in effect modified by the connotations, or the shades of meaning, it is exposed to and is meant to express.[26] In fact, so Wittgenstein held, language is doing its work only if it interacts with all comers, and thus is not "idling"; he calls these interactions "language games." Positivists, however, tend to freeze language in place, and thereby stymie intellectual life in general. Many scholars interested in Polanyi's work regret deeply that Polanyi apparently mistook Wittgenstein's concepts and as a result did not recognize him as a potentially valuable ally.[55]

In summary, we have little to add to the perennial philosophical discussion of "reality." The debate between realism and instrumentalism appears to remain at an impasse.[56] On the other hand, we find that the linguistic aspects of experimental chemistry have much import, even at the deeper level. Grosholz and Hoffmann even spoke of "transcendental" elements.[57] Regrettably, philosophers often ignore this aspect.

A PHILOSOPHY OF CHEMISTRY "UNDER CONSTRUCTION"

At this point we wish to arrive at the principal conclusion from the last three sections. Ten years before Polanyi became a full-time philosopher, he wrote a letter to a journal editor[58] in which he first expounded, with youthful vigor and with poetic flair, ideas central to what later became his theory of tacit knowledge with its opposition to positivism. The "value of the inexact" he discussed in that letter seemed to be paradoxical when one remembers that Polanyi was a highly successful physical chemist at that time. The general situation has been subjected to a Hegelian analysis by John Wisdom, who speaks

> ... of philosophy as progressing through paradox to discovery. Antithetical theses, each affirming in turn what the other had denied, and denying what the other had affirmed, bring cumulatively into view opposed but complementary aspects of reality.[59]

Wisdom also maintains that philosophy succeeds precisely by being paradoxical and provocative. In a similar way, Daly suggested that

> Polanyi's thought, having triumphantly reinstated against positivism the non-logically-formulable aspects of scientific reasoning might be developed in the direction of showing how the non-logical in science is reconciled with the logical, the creative with the inductive, the subjective with the objectifiable, the 'personal' with the 'critical.' Having shown the non-truth in positivism and the non-positivistic in truth, he might complete his service by helping us to see in synthesis the truth in positivism and the non-positivistic in truth. ... [If] it is necessary to show how like non-scientific forms of knowing science is, it is also necessary to show how unlike them it is, and why.[59]

We conclude that, in order to correct the hubris of Enlightenment thought, one should not categorically deny the importance of objectivity, without concern for the possible adverse side effects, such as a wrong sense of accomplishment. The problem with positivism is not that it is entirely mistaken but that it all too often neglects to see important aspects of reality, and in particular the function of languages.[60]

We suggest that to arrive at a more complete philosophy of chemistry we need to accept an unsettled, dynamic state of chemistry so it reflects "that is how the world is," to consider this a likely permanent feature, and recall that at least some of the tacit knowledge is not decipherable, either for a long time to come, or forever. Going beyond positivism then means connecting to the wider cultural context, the realm of values, meanings, and purpose, and being concerned, more than before, for example about the image of chemistry, the challenges chemists face as citizens, and the problems and opportunities chemists may find in liberal education. One might think in terms of a "connected chemistry" that conforms to the "multifarious reality" Grosholz and Hoffmann speak about—a chemistry that no longer needs to rescue or reanimate reductionism to define its philosophy.

This "connected chemistry" offers yet another interesting advantage: it does not automatically force us to enter other difficult discussions, as physics and biology would do, and it may thus be considered more of a "normal" science,[61] and perhaps even as a model, in certain respects.

CONCLUDING REMARKS

One could very well wish to expand on the above, by looking at some of the "tentacles" alluded to earlier. For example, the "unity of existence" mentioned by Grosholz and Hoffmann may indicate that these authors are reaching beyond chemistry "as is."[43] Secondly, chemists, poets, or others, might wish to discuss the "objects" hidden in poems, that is, values and meanings unnoticed by many a reader, ineffable even to the poet himself, and in certain ways similar to Polanyi's tacit knowledge.[62] The second tentacle reaches much farther than the first, and intermediate cases may also be of interest, whenever they bear on our discussions. An example would be the efforts to arrive at a "new humanism" of a few decades ago.[63] Next, there is the important historical dimension of this work, starting with Heraclitus (about 500 B.C.), who taught that "everything flows" and thus might have alerted us to such changes as the abatement of the recent "science wars"[64] and the separation of logical empiricism from (logical) positivism,[65] changes continuing even at the present.

Finally, given such tentacles, we are painfully aware that often the motivation is missing to pursue such thoughts—not only in chemistry.[66] Polanyi speaks to this problem when he discusses freedom, a basic or primary value in our society: personal freedom means we can build on the work of our intellectual forefathers, but we are also free to go further, from modern skepticism to Polanyi's theory of knowledge discussed above. In addition it also means we are free not to let the saying "freedom entails responsibility" become a pale stereotype, but to follow Polanyi's ideal saying that the "responsible person" is one whose freedom is being "overruled" by another, higher freedom, such that he will act "as he must." This higher concept is beckoning all of us who teach, or explain, or publish. It is a sign of the greatness of Polanyi that he, in his own way, posited this, hoping to see many, including chemists, become attuned to the wider culture surrounding them, and thus leave behind their former semi-voluntary confinement and isolation.

NOTES AND REFERENCES

1. HOFFMANN, R. 1995. The Same and Not the Same. Columbia University Press. New York. pp. 83–84. There are many more interesting metaphors in this book. See pp. xiii, xiv, xvi, 6, 123, 250, and 256.
2. HERSCHBACH, D. 1996. "Liberal Education." Fall. p. 89.
3. OXTOBY, D.W. 1994. Zygon **29**: 547.
4. MONTUSCHI, E. 2000. *In* A Companion to the Philosophy of Science. W.H. Newton-Smith, Ed. Blackwell. Malden, MA. p. 277.
5. BHUSHAN, N. & S. ROSENFELD. 1995. J. Chem. Educ. **72**: 578.
6. BOYD, R., P. GASPER & J.D. TROUT, Eds. 1993. Philosophy of Science. MIT Press. Cambridge, MA. p. 779.
7. LASZLO, P. 2001. Hyle **7**: 125.
8. An ideal of service is that which "goes beyond the requirements of ordinary morality and law." See KOVAC, J. Paper presented at the Third Summer Symposium on the Philosophy of Chemistry and Biochemistry. University of South Carolina. July 28–August 1, 1999. p. 1.
9. ZARE, R.N. 1998. Quoted in GETTYS, N.S. 1998. J. Chem. Educ. **75**: 665.
10. BOYD *et al.*, Eds. Philosophy of Science, p. 777.
11. POLANYI, M. 1958. Personal Knowledge: Towards a Post-Critical Philosophy. Harper & Row. New York. [Corrected edition, 1962; paperback edition, 1974.]
12. GELWICK, R. 1977. The Way of Discovery. Oxford University Press. New York. pp. 31–42.
13. SANDERS, A. 1988. Michael Polanyi's Post-Critical Epistemology. Rodopi. Amsterdam. Chapter 1.
14. NYE, M.J. 2002. Hyle **8**: 123.
15. GILL, J.H. 2000. The Tacit Mode: Michael Polanyi's Postmodern Philosophy. State University of New York Press. Albany, NY.
16. WIGNER, E.P. 1977. Biograph. Mem. Fellows Royal Soc. **23**: 413.
17. GELWICK, The Way of Discovery, pp. 29–53.
18. ARON, R. 1968. *In* Intellect and Hope. T.A. Langford & W.H. Poteat, Eds. Duke University Press. Durham, NC. p. 341.
19. WIGNER, Biograph, p. 417.
20. Ibid., p. 424.
21. POLANYI, Personal Knowledge, p. 16.
22. WIGNER, Biograph, pp. 417–418.
23. GELWICK, The Way of Discovery, p. 33.
24. "Body" is taken to be "the complex of instruments of subsidiary awareness and performance which serves as the starting point for all extraverted attention and inquiry." See WEBB, E. 1988. Philosophers of Consciousness. University of Washington Press. Seattle, WA. p. 43. "Bodily" means "pertaining to the 'body.'" "Universal intent" is described as closely related to a "prototype of intellectual commitment," or as that which applies a stranglehold to a judge's discretion and thus narrows it down to zero. See POLANYI, Personal Knowledge, pp. 65, 309.
25. We seek to render the meaning of several technical terms using common language to conserve space.
26. WEISCHEDEL, W. 1973 Die Philosophische Hintertreppe. DTV. Munich. P.295; WITTGENSTEIN, L. 1978. Philosophical Investigations. Blackwell. Oxford, UK.
27. See WIGNER, Biograph, p. 431.
28. Samples of Polanyi's style are also seen in Ref. 13. (SANDERS, Michael Polanyi's Post-Critical Epistemology.)
29. STOVE, D.C. 1982. Popper and After. Pergamon Press. New York. p. 52. Popper is here compared to the frustrated fox in Aesop's well-known fable, but this Popper-fox, frustrated in trying to show how rational he is, in the end winds up writing heavy books on viticulture!
30. DALY, C. B. *In* Intellect and Hope, p. 136. (And especially p. 150.)
31. POLANYI, Personal Knowledge, p. 309.
32. WIGNER, Biograph, p. 430.
33. Ibid., p. 428.

34. SANDERS. Michael Polanyi's Post-Critical Epistemology, p. 3.
35. "Integration" is an accomplishment of the whole person, done within our "body," defined in Note 24.
36. "Indwelling" is a term borrowed from Dilthey, meaning a deep relationship and understanding of that which we "dwell in" and which we seek to penetrate (German: "Verstehen").
37. WIGNER, Biograph, p. 431.
38. GELWICK, The Way of Discovery, pp. 55–82.
39. SANDERS, Michael Polanyi's Post-Critical Epistemology, Chapter 1.
40. POLANYI, Personal Knowledge, pp. 54, 309.
41. Ibid., pp. 16–17.
42. HAMURO, Y., M. CRECO CALAMA, H.S. PARK & A.D. HAMILTON. 1997. Angew. Chemie **36:** 2680.
43. GROSHOLZ, E.R. & R. HOFFMANN. 2000. *In* Of Minds and Molecules: New Philosophical Perspectives on Chemistry. N. Bhushan & S. Rosenfeld, Eds. Oxford University Press. Oxford, UK. pp. 230–247.
44. Ibid., p. 244.
45. FISHMAN, L. *In* P.R. GROSS, N. LEVITT & N.W. LEWIS, Eds. 1996. The Flight from Science and Reason. Vol. 775: 87–95. New York Academy of Sciences. New York.
46. *"Ut aliquid fiat"* (Latin: "Let something be done"), traditional saying of physicians to their colleagues when in a quandary, to impart to the patient the feeling of being cared for.
47. FISHMAN. *In* The Flight from Science and Reason, p. 93. Polanyi would agree, definitely.
48. REBER, A.S. 1993. Polanyiana **3:** 97.
49. HOFFMANN, R. 1988. Am. Sci. **76:** 182.
50. Torrance's statement, quoted by Wigner. See WIGNER, Biograph, p. 433.
51. GELWICK, The Way of Discovery, p. 108.
52. JASPERS, K. 1970. Chiffren der Transzendenz. Serie Piper. Munich.
53. GROSHOLZ & HOFFMANN, Of Minds and Molecules, p. 231.
54. SCHWEBER, S.S. 1997. *In* Experimental Metaphysics. R.S. Cohen *et al.*, Eds. Kluwer Academic Publishers. Netherlands. p. 171.
55. DALY. *In* Intellect and Hope, pp.147–148.
56. LEPHIN, J. A Companion to the Philosophy of Science, p. 393.
57. GROSHOLZ & HOFFMANN, Of Minds and Molecules, p. 237, note 5.
58. POLANYI, M. 1936. The value of the inexact. Philos. Sci. **13:** 233.
59. DALY, *In* Intellect and Hope, p. 166.
60. Unfortunately, Polanyi did not create a philosophy of chemistry, even though he was in a much better position to do so than many others. Jha suggested that if, for example, he had taken the Prigogine theory of irreversible thermodynamics as a starting point, he might have been able to create a complete philosophy of chemistry. Jha's suggestion, however, only deals with the bridging of disparate discourses, and discusses important concepts only as metaphors. See JHA, S.R. 1998. Science in Context. **11:** 89. Also see PRIGOGINE, I. 1980. From Being to Becoming. W.H. Freeman. New York.
61. EISVOGEL, M. 1996. *In* Philosophie der Chemie: Bestandsaufnahme und Ausblick. Koenigshausen & Neumann, GmbH. Wuerzburg. p. 95.
62. Eliseo Vivas has discussed the objects of a poem, that is, the ineffable values and meanings a poet embedded in it without being (fully) aware of it. The parallelism of such hidden objects to Polanyi's tacit knowledge is noteworthy. See VIVAS, E. Creation and Discovery. Chicago. [Copyright by the author, Eliseo Vivas. Library of Congress Catalogue Number 54-11731. pp. 137–142. Quoted in HOFFMAN, The Same and Not the Same.]
63. GREEN, M. 1964. Science and the Shabby Curate of Poetry. Longmans, Green & Co. London. Also, see excerpts in KREUZER, H., Ed. 1969. Literarische und naturwissenschaftliche Intelligenz. Klett Verlag. Stuttgart.
64. LABINGER, J.A. 1995. Soc. Stud. Sci. **25:** 285.
65. SALMON, W.C. 2000. *In* A Companion to the Philosophy of Science. W.H. Newton-Smith, Ed. Blackwell. Malden, MA. p. 233.
66. MERMIN, N.D. 1985. *In* BOYD *et al.*, Eds., Philosophy of Science, p. 501.

Chemical Self-Organization, Complexification, and Process Metaphysics

JAMES F. SALMON

Chemistry Department, Loyola College in Maryland, Baltimore, Maryland 21210, USA

Woodstock Center, Washington, DC, 20057, USA

> ABSTRACT: Recognizing the self-activity of matter through complexification can lead to seeing evolution as a process of union. This process approach emphasizes a metaphysics of becoming rather than of being. Advances in the science of thermodynamics lead to understanding chemical self-organization as a stage in this process of complexification.
>
> KEYWORDS: complexity; Teilhard de Chardin; union; chemical thermodynamics; self organization

This paper analyzes "complexification"—a term suggested by the French geologist and paleontologist Teilhard de Chardin (1881–1955)—as a potential model for development of a metaphysics of becoming. The analysis manifests the unique and important role of developments in theoretical and experimental chemical thermodynamics in understanding the evolutionary process.

COMPLEXIFICATION

In the mid 1960s, Ian Barbour argued:

> Most of [Teilhard's] interpreters have neglected the function of process metaphysics as a "middle term" between his evolutionary and biblical concepts. Even though Teilhard was not professionally a philosopher, I will suggest that it is in his informal metaphysics that one must seek the unity of his thought, not only within *The Phenomenon of Man* but among his writings.[1]

A literature search indicated there has been little evaluation of Teilhard's metaphysics since Barbour's argument.[2]

Teilhard's essay, "Man's Place in Nature," offers his brief and mature interpretation of the general stages of evolution: matter, life, thought, and society. Of course, there are different and more popular interpretations of evolution by scientists,[3] such as recent formulations of neo-Darwinism. The intention here is to discuss Teilhard's interpretation within an empirical perspective of matter in the process of transformation.

Address for correspondence: James F. Salmon, S.J., Chemistry Department, Loyola College in Maryland, Baltimore, MD 21210-2699. Voice: 410-617-2350; fax: 410-617-2125.

jsalmon@loyola.edu

To begin his essay, Teilhard writes:

> We must try, if possible, to build a bridge, or at any rate the skeleton of a bridge, between physics and philosophy.[4]

Although he had taught physics in a high school in Egypt early in his Jesuit career, and continued to keep up with scientific advances until the end of his life, his professional reputation as a scientist is based on his contributions to field and theoretical geology and paleontology. Therefore his vision is based primarily on that experience.

Teilhard denies that life is an epiphenomenon of matter and thought an epiphenomenon of life. He argues:

> ...for a whole group of solid reasons that come to light as soon as we begin to realize the intimate structural link that connects the "accident of life" to the vast universal phenomenon (so obvious and yet so little understood) of *complexification of matter.*[5] [italics in original]

For purposes of this study, matter is *Weltstoff*, the term that is ordinarily opposed to energy in scientific literature. Moreover, he notes:

> One of the most curious intellectual phenomena to be produced in the field of scientific thought during the last fifty years, is without doubt the gradual irresistible invasion of physics and chemistry by history: with the prime elements of matter exchanging their quasi-absolute mathematical position for that of contingent, concrete, reality, and with physics and chemistry, formerly branches of calculation, beginning to appear as preliminary chapters to a "natural history of the world." A strange reversal, indeed, of our picture of the universe.[6]

Complexity, the result of complexification, is neither a simple aggregation like a heap of sand, nor a repetition of units as in crystallization. Rather, it is

> ...*combination*, i.e. that higher form of grouping whose property it is to knit together upon themselves a certain fixed number—whether great or small matters little—of elements, with or without the secondary addition of aggregation or repetition—within a closed whole of determined radius: such as the atom, the molecule, the cell, the metazoan, etc.[7]

A combination is a type of grouping that, by becoming unified, becomes structurally completed around itself at each moment, although it is indefinitely extensible from within. Combinations manifest autonomous phenomena at certain higher levels of internal complexity.

Teilhard was fond of drawing pictures. In FIGURE 1, he represents the three principal phases of the evolutionary process that engender consciousness and reflection.[8] The three zones of concentric iso-spheres represent three levels of evolution: pre-living Matter, Life (the Biosphere), and Thought (the Noosphere—mind). The left sketch outlines the general process, and the right sketch explains each stage of the process in more detail.

The top sketch on the right in FIGURE 1 illustrates fragments of pre-living matter. The fragments are open at each end, offering a sort of curvature. At this degree of disjunction of the fragments, there is only a disposition to come together and to fit in with one another—not by intention—but through the play of chance. At this phase, practically the whole of time and space is quantitatively evaluated. The reason is that "the play of large numbers requires a more extensive laboratory for its experiments."[9] Although the accuracy of measurements is by 1950 standards, the principle remains:

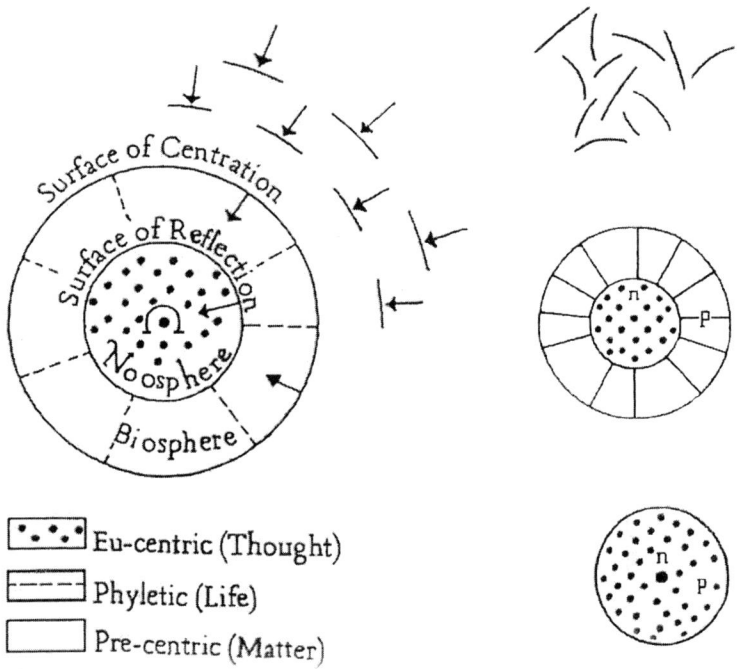

FIGURE 1. Complexificaion process.

> While it [consciousness] is completely imperceptible to our observational methods below an atomic complexity of 10^5 (the virus), it can be plainly detected when we reach that of the cell (10^{10}); but it enters into its major developments only in the brains of large mammals (10^{20}), in other words when we have atomic groupings astronomic in order.[10]

Particles of low complexity appear inanimate, just as other fundamental properties of matter, such as relativistic variation of mass and curvature of space, only become perceptible in the very small or very large. The universe is represented as a system in process—from dispersed heterogeneous fragments lacking unity to organized heterogeneous fragments. Each fragment of matter is an infinitesimal psychic center—like one of Leibniz's monads (although Leibniz was ignorant of evolution). Evolution does not correspond, as Spencer maintained (in his *First Principles*), to transition from a relatively indefinite, incoherent homogeneity to a relatively definite coherent heterogeneity.[11]

By passing from pre-life into life through the complexification process, the fragments cross a critical surface of "centration"—not of spatial symmetry, but of action. This is conceived as the general process where matter folds on itself in a process of becoming more conscious and unified. Centration and "centricity," the degree of centration, are important concepts in the Teilhardian approach. He calls the process of centration the "law of complexity-consciousness." Two essays that develop these themes, "Centrology" in 1944 and "The Analysis of Life" in 1945, analyze the development of the internal properties of interiorization and psychism in matter. He notes that the approach is

> ...not an *a priori* geometric synthesis starting from some definition of "being," but an experiential *law of recurrence* which can be checked in the phenomenal field and can appropriately be extrapolated into the totality of space and time.[12]

He writes:

> The living... has long been regarded as an accidental peculiarity of terrestrial matter. As a result, the whole of biology has been left out on its own with no intelligible connection with the rest of physics. This is corrected if... life is, in scientific experience, no other than a specific effect (the specific effect) of complexified matter: a property in itself co-extensive with the whole stuff of the cosmos, but perceptible to us only where (after stepping over a number of thresholds...) complexity exceeds a certain critical value—below that value we cannot perceive it at all.[13]

He offers an analogy:

> The speed of a body must approach that of light for the variation in its mass to be apparent to us. Its temperature must reach 500 °C for its radiation to be visible to us. Is it not then reasonable to expect that through just the same mechanism matter, until it begins to approach a complexity of a million or half a million, should appear "dead" (though pre-living would be the better term) while beyond that figure it begins to show the red glow of life?[14]

The middle sketch in FIGURE 1 proposes the joining up of fragments—as a result of the repeated play of large numbers—to form the oldest and outermost of the biosphere. It is proposed that, at this stage, chemical self-organization participates in closing fragments to form the "surface of centration." Except for cases of spontaneous nuclear disintegration, pre-living matter is indefinitely stable by human standards. Elements in the biosphere are different. Every living entity contains two aspects:

- "p" represents the "peripheral" aspect, that is divisible and transmissible; it accounts for biological continuity.

- "n" represents a "nuclear" aspect that is incommunicable.

Within this sphere, beginning with mono-cellular stages, each entity is capable of producing only an infinitesimal advance along the phylum to which it belongs; after that it disappears, though not without undergoing multiplication.

The bottom sketch in FIGURE 1 proposes the final known transition of complexity-consciousness. Although there is not sufficient space available here to amplify on the stages in the complexification process, this iso-sphere is included for the purpose of closure. The process has passed through a critical point on the "surface of reflection" to form a unified nuclear portion that permits reflection: awareness of self, personality, and highly organized social structures—even a global community.

A PROPOSED PROCESS METAPHYSICS OF UNION

While at the front during World War I, Teilhard first proposed "Creative Union" as a way to see the cosmos in evolution. Between 1918 and 1955, as he tried to clarify his thought, we find 155 entries, into his private journals that are available, on the subject of union. As a Christian, he gradually closed the "cleavage with the physical world" in his vision regarding "the Divine," although, of course, never perfectly. (To include mention of the Divine here is only to indicate that there is room in Teilhard's approach to extend the vision.)

The basis of the proposal may be enunciated in the following two statements[15]:

- *Plus esse* = *plus a pluribus uniri.* (More **to be** = to be more united from many.)
- *Plus esse* = *plus plura unire.* (More **to be** = more to unite the many.)

The first statement realizes the passive aspect of being, and the second statement realizes the active aspect. Being becomes realized and maintained union. These two statements reveal the dynamics of complexification in evolution as an ongoing process of bringing forth new entities that emerge from, but are not identical with, fragments that unite. This vision can be absent in more traditional, static philosophies of being.[16]

The approach is related to classical metaphysics in the philosophies of both Aristotle and Plato. Like Aristotle, Teilhard built his metaphysics on the physics of his time. Separated from the reality each of them knew, their respective approaches become a set of abstract concepts. Also, one of the universals rooted in the Platonic tradition may be applied because "every being is one" becomes "every being is united." Obviously the proposal requires further exploration. As a provisional metaphysics, however, the analogy of union that explains the dynamics of complexification in the ladder of evolution permits the expansion of a philosophy of being to a philosophy of becoming. Being becomes a temporary aspect of becoming. This is significant for chemistry because, as Ilya Prigogine writes: "Becoming is the *sine qua non* of science, and indeed of knowledge itself."[17]

SELF-ORGANIZATION IN CHEMICAL SYSTEMS

In his essay, "What is Life," Erwin Schrödinger emphasized the Boltzmann equation,

$$\text{entropy} = k \log D$$

to discuss the thermodynamics of entropy of natural processes, where k is the Boltzmann constant and D is a quantitative measure of disorder in a system. Therefore Schrödinger regarded $1/D$ as a direct measure of "negative entropy," that is, order in the system.[18] Although Schrödinger's application of negative entropy to the environment might be judged a bit primitive to contemporary mature environmentalists, the principle holds:

> Thus the device by which an organism maintains itself stationary at a fairly high level of orderliness really consists in continually sucking orderliness from its environment. Indeed, in the case of higher animals we know the kind of orderliness they feed upon well enough, *viz*, the well-ordered state of matter in more or less complex organic compounds which serve as food stuffs. After utilizing that food they return it to the environment in a much degraded form—not entirely degraded however—for plants can still make use of it. (These, of course, have their most powerful supply of negative entropy in sunlight).[19]

It is clear that Schrödinger is referring to the biosphere (in FIG. 1). In thermodynamic terms, living organisms are open systems, where matter and energy can be exchanged with the environment in irreversible processes.

What is the role of chemical self-organization in natural processes? Prigogine notes:

> It is now generally admitted that biological evolution is the combined result of Darwin's natural selection as well as of self-organization which results from irreversible processes.[20]

Complexification through union in evolution is not an *a priori* geometric synthesis; rather Teilhard calls it a "law of recurrence in nature"—detected by paleontologists in the field, compatible with contemporary Big Bang cosmology, and with the evolution of complex matter. Does the metaphysics of union that explains complexification, along with chemical self-organization, contribute to a possible interpretation of the irreversibility of evolution?

The significance of chemical self-organization in natural processes may be understood if one briefly reviews three stages of the development of thermodynamics. Chemists are familiar with stage 1—the developments in science and technology that are based on reversible processes, with the application of Gibbs Phase relations, Carnot's cycle, and Clausius' two laws: the energy of the universe is constant, and the entropy of the universe approaches a maximum.

Many philosophers and some chemists are not familiar with stage 2. In the early 1930s, Lars Onsager, using linear equations, analyzed irreversible microscopic fluctuations around chemical equilibrium. Onsager's reciprocal relations showed for the first time that nonequilibrium thermodynamics, like reversible thermodynamics, could lead to general results independent of particular molecular models.

> Equilibrium thermodynamics was an achievement of the nineteenth century, nonequilibrium thermodynamics was developed in the twentieth century, and Onsager's relations mark a crucial point in the shift of interest away from equilibrium toward nonequilibrium.[21]

This discovery stimulated theoreticians to investigate the irreversibility that occurs in most natural processes. Linear near-equilibrium systems and the second law of Clausius can be reconciled with the existence of biological systems. Such systems can maintain steady, nonequilibrium states by absorption of negative entropy, $1/D$, from the surroundings—as noted by Schrödinger in 1944. But how could far-from-equilibrium complex systems, such as biological organisms, come into existence? How can such complexification take place?

The third and present period of development of thermodynamics, led by Prigogine and the "Brussels school," has studied situations that occur far from equilibrium. Nonlinear differential equations may be applied to the study of these far-from-equilibrium physical, chemical, and biological systems. The interest here is chemical and biochemical complexification processes in open systems, that is, where a system can exchange matter and energy with the surroundings. It is shown experimentally that new types of ordered systems—called "dissipative structures"—can come into existence through self-organization by millions of ionic and molecular reactants. Open systems can be maintained in ordered states in space and time by dissipating positive entropy to the surroundings, as heat and degraded chemical substances—as predicted by Schrödinger. Such order can persist, provided the system reaches a steady state in which the total entropy remains unchanged—that is, $d_e S = d_i S < 0$, where the subscripts represent entropy changes in the surroundings ($d_e S$) and the system ($d_i S$), respectively. Such a nonequilibrium state is said to be a "far-from-equilibrium self-organizing steady state." It is striking that such complex self-organized systems, comprising a large number of chemical entities, or "fragments" as Teilhard called them, can act in a unified fashion—as organized, coherent wholes.

CONCLUSIONS

(1) Evidence based on data from fields as disparate as field paleontology and cosmology can lead to seeing evolution as a process of complexification. Matter can become more complex in its self-activity, as unified wholes come into existence through union. These entities manifest emerging properties whose existence depends on the process of their union. Traditional philosophy knew only a static universe and developed a metaphysics based on being. An empirical process philosophy of becoming that brings about union through complexification seems more compatible with evolution.

(2) In open systems, nature also exhibits certain steady state complex (in time and space) chemical structures that come into existence through self-organization. Processes in these chemical systems exhibit a form of ordered union that is not found at or near equilibrium. The existence of order is verified by entropy values of structures that form. The phenomena can be understood by application of far-from-equilibrium thermodynamic theory for irreversible processes. The theory is also applicable to living systems. But the form of union and communication in nonliving structures does not exhibit the integrality of union that exists in living structures. Structures are formed, but not with the functional order found in a living union. These nonliving complex systems, however, seem to be an example of a stage in the overall process of complexification that supports further investigation of this process approach to metaphysics.

NOTES AND REFERENCES

1. BARBOUR, I. 1968. Five ways of reading Teilhard. Sounding **2**: 115–145. For an explanation of process categories, *cf.* RESCHER, N. 1996. Process Metaphysics: An Introduction to Process Philosophy. State University of New York Press. Albany, NY.
2. SALMON, J.F. & T.M. KING. 1994. Work on Teilhard, 1980–1994: An annotated bibliography. Zygon: J. Relig. Sci. **30**: 131–142.
3. An example of another important interpretation is that of Stephen Jay Gould. A recent book discusses that interpretation in terms of the overall empirical evidence concerning cosmic evolution. *Cf.* SCHMITZ-MOORMANN, K. [In collaboration with J.F. Salmon] 1997. Theology of Creation in an Evolutionary World. Pilgrim Press. Cleveland, OH. pp. 33–36.
4. TEILHARD DE CHARDIN, P. 1966. Man's Place in Nature. Harper & Row. New York. p. 17.
5. Ibid., p. 19.
6. Ibid., p. 27.
7. Ibid., p. 20.
8. TEILHARD DE CHARDIN, P. 1970. Activation of Energy. Harcourt Brace Jovanovich. New York. p. 100.
9. Ibid., p. 106.
10. Ibid., p. 102. Teilhard writes: "By atomic complexity, I mean here the number of atoms contained in the particle in question."
11. SPENCER, H. 1900. First Principles. 6th ed. Werner. Akron, OH.
12. TEILHARD DE CHARDIN, P. 1970. Activation of Energy. Harcourt Brace Jovanovich. New York. p. 99.

13. TEILHARD DE CHARDIN, Man's Place in Nature, p. 23.
14. Ibid., p. 24.
15. SALMON, J.F. & N. SCHMITZ-MOORMANN. 2002. Evolution as revelation of a triune god. Zygon: J. Relig. Sci. **37:** 853–871.
16. EARLEY, J. 1998. Modes of chemical becoming. Hyle **4:** 105–115. [See http://www.hyle.org/journal/issues/4/earley.htm.]
17. PRIGOGINE, I. 1997. The End of Certainty. The Free Press. New York. p. 16.
18. Actually Schrödinger's interpretation of D as a direct measure of entropy has since been shown to be inaccurate. *Cf.* JAYNES, E.T. 1957. Informational theory and statistical mechanics. Phys. Rev. **106:** 620–630.
19. SCHRÖDINGER, E. 1956. What Is Life? & Other Scientific Essays. Doubleday. Garden City, NY. p. 73.
20. KONDEPUDI, D. & I. PRIGOGINE. 1998. Modern Thermodynamics. John Wiley & Sons. New York. p. xii.
21. PRIGOGINE, I. & I. STENGERS. 1984. Order out of Chaos. Bantam. New York. p. 138.

The Philosophy of Kitaro Nishida and Current Concepts of the Origin of Life

KO HOJO

81-7, Okitsu Kuboyamadai, Katsuura-shi, Chiba, 299-5246, Japan

ABSTRACT: The philosophy of Kitaro Nishida is a basic metaphysics that intends to have a perspective beyond both Eastern and Western traditions. Nishida holds that the fundamental reality of the world appears from "the place of the absolute nothingness." Nishida's thought can be seen to have striking relationships to certain aspects of contemporary science and philosophy.

KEYWORDS: science and spirituality; origin of life; nothingness; action-intuition; contradictory identity; Enlightenment; East and West

Science and spirituality are the two different ways of looking at the reality of the world. (Bas van Fraassen has discussed this point in a recent book.[1]) Recent interdisciplinary discussions of relationships between these two approaches focus on physics and biology and employ Christian perspectives.[2] Considerations of science and spirituality should deal with chemical concepts and also involve non-Western traditions of thought.

Kitaro Nishida (1870–1945) is a premier philosopher of Japan (his collected works[3] are hereafter designated as CVN). He grew to intellectual maturity in the final decades of the Meiji period (1868–1912), during which Japan opened to the world—and began modernization by rapidly absorbing Western culture. In his forties and fifties, he achieved recognition as Japan's leading establishment philosopher as Professor of Philosophy at Kyoto University. Nishida, and his younger colleagues such as Hajime Tanabe (1885–1962), Keiji Nishitani (1900–1990), and their students of the next generation, came to be known as philosophers of "the Kyoto School."[4] Contributions of Nishida's intimate lifelong friend, Daisetsu Suzuki (1870-1966)—a modern Buddhist philosopher—were important to his work. Suzuki calls spirituality "discrimination that transcends discrimination."[5] Nishida takes this to mean that religious spirituality does not identify the self as distinct from the rest of the world (CVN 11-348). (It is interesting to note that Albert Einstein said that the true value of a human being is determined primarily by the measure, and the sense, in which he has attained liberation from the self.[6])

Address for correspondence: 81-7, Okitsu Kuboyamadai, Katsuura-shi, Chiba, 299-5246, Japan. Voice/fax: 81-470-76-2580.

kohojo@mb.infoweb.ne.jp

Nishida was a Buddhist, practicing Zen. Buddhism is the religion in which one strives to attain Nirvana (perfect tranquility). The final goal for Buddhists is to become "a Buddha"—enlightened (or awakened) one. Originally, Gautama Siddhartha (563?–483? B.C.E.), the founder of Buddhism, was called by this name. In the course of Buddhist history, two main streams of thought have appeared—Mahayana and Theravada. Buddhism of the Mahayana type spread to China, Tibet, Nepal, Vietnam, Korea, and Japan, while Theravada Buddhism became established in Burma, Sri Lanka, and Thailand.

The most important idea in Mahayana Buddhism is *Sunyata* (nonsustainability)—the concept that no thing has either substance or permanence, since everything that appears or disappears depends on causation (emptiness). Whereas every whole consists of individuals, each individual contains the whole. The one who realizes this can comprehend the true nature of things—this is called "Enlightenment."[7]

Nishida rejected the idea that "spirit" and "nature" exist as two independent entities (CVN 1–178). These are two elements, phases, or functions of a single reality. For Nishida, God is the base of reality and the foundation of the universe. The relation between the universe and God is the same as the connection between phenomena of our consciousness and their unity. Our consciousness is part of God's consciousness—its unity comes from God's unity. The universe is not a *creation* of God but a *manifestation* of God. Nishida had the idea that true reality must be actuality just as it is, and that the so-called "material world" is something conceptualized and abstracted out of it (CVN 1–7). He rejected the "objective" viewpoint. He says that "we are born in the world, act in the world, and die in the world. It is impossible that we should stand outside the world." Nishida also rejected the distinction between individual and universal, many and one, ideal and reality, and theory and practice.

Nishida's fundamental philosophical question, throughout thirty five years of continual effort, was solely to explain the ultimate reality of the living world—beyond Eastern and Western traditions. This stance was manifested in his first and most celebrated book *An Inquiry into the Good*, published in 1911 (CVN 1–3),[8] and lasted until his last papers in 1945 (CVN 11–371).[9] Nishida's standpoints, and analyses of the ultimate reality of the world, were continuously explored and accumulated in new forms by a stream of original publications. "Pure experience" was his first standpoint from which to seek reality.[10] This is the unity of the consciousness, before any thought or discrimination begins to work—where there is not yet a subject or an object. In Nishida's second major work, *Intuition and Reflection in Self-Consciousness* (1917), he developed the standpoint of "absolute will" (CVN 2–3), and then in *From the Actor to the Seer* (1927), he shifted significantly from existential experience to logic—through mediation on the idea of "place" in Greek philosophy (CVN 4–3). For him, place is not only metaphysical but also logical, and here he developed the concepts of "individual" and "world," by confrontation with the ideas of Plato, Aristotle, Descartes, Spinoza, Leibniz, and Kant. He then concretized the idea of place as a "dialectical universe," in a series of papers that were collected into two volumes on *The Fundamental Problem of Philosophy* (1933–1934). There he gave that standpoint a direct expression in terms of "action-intuition" (CVN 7–173). Action-intuition is not a passive state but a formative activity, such as is seen in the work of great artists. "Self Identity of Absolute Contradiction" (1939) is a logical structure of the world. This was achieved by confrontation of Hegel, Marx, Husserl, Bergson, Jaspers, and Heidegger. In particular, Nishida owes many of his ideas to the thought of Nagarjuna (2–3 century) and also to Hegel's theory of dialectic (CVN 9-147).

Nishida holds that "the place of absolute nothingness" is the source from which everything originates. This is the ground that actively produces all beings—including the "true self" that exists deep inside every person. "Contradictory identity" is the formulation of this process. Nishida held that everything in the world maintains its identity by self negation of two opponents. Nishida's "nothingness" means the self identity of the place where both opposites exist—and therefore there is no form. To see oneself from the side of the world, to think by becoming things, and to act by becoming things, this is the reciprocal way of looking at the same formulation. *The Logic of the Place of Nothingness and the Religious Worldview*, completed in the last months before his death (1945), is a summation of Nishida's philosophy that develops this view (CVN 11–371). Here he maintains that:

> [A]ll life arises from the fact that it transforms itself by containing its own-self expression within itself. It is first biological, and instinctive in a spatially predominant way. It becomes historical life as it becomes concrete in a temporal dimension. In a historical life, there is always a dialectic of affirmative and negative: the former is the material world, the latter the world of consciousness, in the transpositional structure of the contradictory identity of matter and form.
>
> [E]veryday human existence has to be seen eschatologically. It is awareness of the absolute present. It is from that standpoint that we can conceive of a beginningless past and an endless future. It is because there is always this eschatological quality in ordinary experience that we can speak of the contradictory identity of time and space, of the immanent and transcendent that constitutes the creative process of history.

Nishida applied his philosophy to practical problems in various fields, including physics, mathematics, biology, arts, and religion (1943–1945). These papers were posthumously published in 1945–1946 and collected in edited volumes in 1949 (CVN 11 3–485).

The notion of "individual" is a basic concept that was explored by Nishida. This differs from Aristotle's notion of "substance" and is rather close to Leibniz's idea of "monad," but with different characteristics. The "absolute present" is another important concept. It is the eternal "now"—which encompasses infinite past and infinite future. The concept was explored by a synthesis of religious philosophy with quantum and relativity theory (CVN 6–188). In Nishida's philosophy, the individual is an independent entity—but each has close relationships. Each stands against the "other"—but also works together with others. When many individuals are in the state of contradictory identity with one whole, a "world" arises, moves, and works. By working together, individuals produce history, and history produces individuals (CVN 13–147). Form appears when individuals are seen spatially. In the physical world, an individual could be: quantum, quark, electron, atom, molecule, and so on—these are predominantly spatial and universal, whether continuous or discontinuous. The world becomes a dialectic of one and many. As the result of contradictory identity of many and one by negation, creative production of the world occurs—from the world created to the creative world (CVN 11–5, 11–289).

In Nishida's philosophy, negation is a logic of explaining changes in a historical world. We recognize changes in ourselves every moment in many respects. This is referred to as "contradictory identity" of the past and present of ourselves: our past is negated, and our present is created, while we remain identical. We can interpret chemical changes as similar to transitions that we ourselves undergo. The change of reagents to products can be seen as the negation of prior identities and creation of new identities, as happens continually in every human life. The reagents, products, and we are all individuals, which work together while retaining independence. The scientific story of the birth of the universe can be understood in terms of the contradictory identity of the world that arises from absolute nothingness.

For example: at the moment of Big Bang, the world was assumed to be seething with the activity of the primordial vacuum (without form, a oneness properly called absolute nothingness). Matter, antimatter, and photons were continuously forming and dissolving in eternal cycles of creation and destruction. The first particles (quarks, electrons, and neutrinos) that appeared were awash with photons. (This was the negation of one, the creation of many). Quarks clumped together to produce protons and neutrons, and these combined to form hydrogen and helium nuclei—by successive negations and creations. While every whole consists of individuals, each individual contains the whole. This feature is involved in every stage of cosmic evolution, in the origin of life, and in the evolution of DNA. An infinite number of such processes involving physical entities eventually produced a biological world, a human world, and a historical world. When an individual goes historically backward by self negation, through innumerable nothingness, one ultimately reaches absolute nothingness, from whence everything originates. This would correspond to the primordial vacuum of the Big Bang theory. We may say that we have come from absolute nothingness, a light ball of unimaginable high energy.

This insight—closely related to traditional Buddhist thought—has been given a logical basis by Nishida. Two statements that appear strange, but are reasonable in a spiritual sense, also express this insight. These statements (one from the West and one from the East) can be contrasted to the scientific, objective way of looking at reality.

Etienne Condillac (1715–1780), a French philosopher who had a great influence on the thought of Lavoisier, once said: "…when we first see light, it is not so much that we see it, but rather we are the light itself (CVN I–171)."[11]

In a lecture at an international meeting, Daisetsu Suzuki asked the audience:

> In the Bible, God said, "Let there be light": and there was light. But who in the world was watching it?

No one answered. Then he said:

> I watched it. I am the witness. Our minds themselves have a potentiality to notice it. We are practicing them for this every moment.[12]

We can understand both these statements to mean that human consciousness is produced by things that trace back to that first instant. We, in the world at the absolute present, see the world from the view point of an element of the world. This is the stance of Nishida's action-intuition (CVN 10–402).

Contemporary science has generated impressively coherent accounts of how the present material universe emerged from the primordial vacuum through "the Big Bang"—including how the chemical elements are produced in stars. There even are somewhat reasonable stories that can be told about the origin of life. With respect to the origin of life, the chemist and philosopher Michael Polanyi concludes that[13]:

> [T]he ordering principle which originated life is the potentiality of a stable open system; while the inanimate matter on which life feeds is merely a condition which sustains life, and the accidental configuration from which life had started had merely released the operations of life.

On the basis of results of computer experiments, Stuart Kauffman (a nonpracticing physician) developed a concept of "order for free"—growth in complexity leads eventually and ineluctably to a situation where coherence on a new level emerges, through closure of a network of catalytic and autocatalytic processes.[14] Kauffman utilizes Manfred Eigen's discussion of autocatalysis,[15] and Ilya Prigogine's concept of "dissipative structure."[16] Both of these concepts are closely related to chemistry, and less closely connected with physics and biology.

The grand sweep of the accounts of the origin of the material universe, and of life on earth, which have emerged from recent scientific research present a challenge for philosophers. For Alfred North Whitehead:

> God is involved in each event, in the concresence of every actual entity. God is the source of the "form of definiteness," the "subjective aim" of each occasion.[17]

Robert Neville (a philosopher and theologian who is quite conversant with both Western and Eastern thought) holds[20]:

> God is creator of everything determinate, creator of things actual as well as of things possible. Apart from the relative nature the divinity gives itself as creator in creating the world, God is utterly transcendent.... God is the immediate creator of the novel values or patterns by which an event is constituted as the harmonizing of a multiplicity. Since the real being of an occasion is the becoming of a harmonized integration of the multiplicity, its components stem either immediately from God or from what it prehends; since what it prehends are other occasions, themselves analyzable into novel and prehended features, it can be suggested that every feature at some in the present or past is or was a spontaneous novel pattern or value immediately created by God. Thus God is the creator of every determinate thing, each in its own occasion of spontaneous appearance.

This approach, in which Neville describes as "creation *ex nihilo*," seems to correspond to Nishida's philosophy (CVN 11–371):

> That which stands in relation to the absolute as its self-negation must itself be self-expressive through its own self-negation. This principle cannot be grasped in either mechanistic or teleological terms. We exist as the absolute affirmation through self negation of God, and this is the true meaning of creation. Mankind is scientific from the standpoint where it possesses itself in self-negation. In religious language, it is the fact that God sees himself through his own negation. In this sense the world of science may be said to be religious....[T]he historical world, as the self-determination of the absolute, exists in the form of the absolute present. Therefore, as the contradictory identity of that which expresses and that which is expressed, the historical world, as the expression of the self-expressive absolute, contains its own self-expression within itself and transforms itself. For this reason the historical world is religious, and is metaphysical, in essence. Every race of people, as a formulation of the historical world, is its own expression of God.

The thought of Kitaro Nishida, now called a world philosophy, should be explored in the context of the philosophy of chemistry, taking recent advances of science into account.

NOTES AND REFERENCES

1. VAN FRAASSEN, B. 2002. The Empirical Stance. Yale University Press. New Haven, CT.
2. For a comprehensive textbook in science and spirituality see: SOUTHGATE, C., *et al.* 1999. God, Humanity and the Cosmos. Trinity Press International. Harrisburg, PA.
3. NISHIDA, K. 1947–1953. Collected Volume of Kitaro Nishida (hereafter designated as CVN), 19 volumes (edited by T. Shimomura and others). Iwanami. Tokyo.
4. FUJITA, M. Philosophy of Kyoto School. 2001. Showado. Kyoto; HEISIG, J.W. 2001. Philosophers of Nothingness. University of Hawaii Press. Honolulu.

5. SUZUKI, D.T. 1972. Japanese Spirituality (translated by N. Waddel, originally written in Japanese in 1944, Iwanami. Tokyo). Japan Society for the Promotion of Science. Tokyo.
6. EINSTEIN, A. 1994 (originally written in 1934). Ideas and Opinions. The Modern Library. New York.
7. 1988. The Teaching of Buddha. Bukkyo Dendo Kyokai. Tokyo. See also CVN 7–419.
8. NISHIDA, K. 1990. An Inquiry into the Good (translated with introduction by Masao Abe & Christopher Ives; originally written in Japanese in 1921 as Zen No Kenkyu). Yale University Press. New Haven, CT.
9. NISHIDA, K. 1987. Last Writings: The Logic of the Place of Nothingness and the Religious Worldview. (translated with introduction by David A. Dilworth; originally written in Japanese in 1945). University of Hawaii Press. Honolulu.
10. Pure experience was a current contemporary thought when Nishida started his carrier. Philosophers like Gustav Fechner (1801–1887), Wilhelm Wundt (1832–1920), Ernst Mach (1838–1916), and William James (1842–1910) were quoted in his book (CVN 1–9).
11. Quoted by Nishida (CVN 1–171). Antoine Laurent Lavoisier (1743–1794) quoted and praised Condillac's philosophy in the preface of his famous books: Tratite Elémentaire de chemie (1789), and Méthode de Nomenclature Chimique (1787).
12. SUZUKI, D. 1967. Quoted by RIUMIN AKIZUKI in Words and Thoughts of Daisetsu Suzuki. Kodansha. Tokyo. p. 67.
13. POLANYI, M. 1974. Personal Knowledge. The University of Chicago Press. Chicago. Cited in EARLEY, J.E. Sr. 1998. Naturalism, Theism, and the Origin of Life. Process Studies, **27(3-4):** 275. I am grateful to Professor Earley for providing me with copies of related literature.
14. KAUFFMAN, S. 1993. The Origin of Order. Oxford University Press. New York and Oxford, cited in EARLEY, Naturalism, Theism, p. 275.
15. EIGEN, M. 1992. Steps towards Life. Oxford University Press. Oxford; New York. Cited in EARLEY, Naturalism, Theism, p. 275.
16. In an open, nonequilibrium system like a living cell, an order is generated by persistently dissipating matter and energy. See PRIGOGINE, I. & I. STENGERS. 1984. Order Out of Chaos. Bantam Books. New York.
17. Cited in EARLEY, Naturalism, Theism. Comparative studies of the philosophy of Nishida and Whitehead have been ongoing.[18] There is no evidence, however, to indicate that communication existed between these two philosophers.[19] Both developed their philosophy independently in the period between two World Wars, moving beyond their own traditions of Buddhist and Christian beliefs.
18. HANAOKA, E. 1998. Fundamental Studies of Religious Philosophy. Hokuju Shuppann. Tokyo; NOBEHARA, T. 2001. Between Whitehead and Nishida-Tetsugaku. Hozokan. Kyoto.
19. YAMAMOTO, S. 1978. Kitaro Nishida and Whitehead, Supplement 5. Collected Volume of Kitaro Nishida. Iwanami. Tokyo.
20. NEVILLE, R.C. 1995. Creativity and God. State University of New York Press. Albany. p. 8. Cited by EARLEY, Naturalism, Theism, p. 275.

Constructivism, Relativism, and Chemical Education

ERIC SCERRI

Department of Chemistry and Biochemistry, University of California, Los Angeles, California 90095, USA

ABSTRACT: Whereas most scientists are highly critical of constructivism and relativism in the context of scientific knowledge acquisition, the dominant school of chemical education researchers appears to support a variety of such positions. By reference to the views of Herron, Spencer, and Bodner, I claim that these authors are philosophically confused, and that they are presenting a damaging and anti-scientific message to other unsuspecting educators. Part of the problem, as I argue, is a failure to distinguish between pedagogical constructivism regarding students' understanding of science, and constructivism about the way that scientific knowledge is acquired by expert scientists.

KEYWORDS: constructivism; relativism; chemical education; Science Wars; realism

INTRODUCTION

For some years, the academic world has been in the midst of a fierce debate that shows little sign of abating. I am referring to what is popularly knows as the Science Wars, which began following the publication of Gross and Levitt's book entitled "Higher Superstition."[1] The charge made by these authors was that many who have written on the nature of science are seriously mistaken and are having a damaging influence upon scholarly work, the public image of science, and last but not least, on science education.

Briefly put, defenders of the traditional understanding of science (such as Gross and Levitt) complain that some sociologists, anthropologists, literary critics, and others have supported relativistic views, which threaten to undermine the fabric of scientific knowledge. The opposing side, which includes many of those belonging to the discipline that calls itself Science Studies, has defended itself in equally strident terms, although not as convincingly, to my mind. Many of the members of this opposing faction support constructivist views about scientific knowledge and about the learning of science. They draw their inspiration from a variety of sources ranging from Thomas Kuhn, in history and philosophy of science, to Jean Piaget, in psychology. There is much variety regarding the meaning of terms such as "constructivism"

Address for correspondence: Eric Scerri, Department of Chemistry and Biochemistry, UCLA, Los Angeles, CA 90095. Voice: 310-206-7443; fax: 310-206-2061.
scerri@chem.ucla.edu

among authors, and this has added to the general confusion among Science Wars adversaries and even allies.

More recently, the Science Wars reached something of a crescendo following the publication of Alan Sokal's article in the journal *Social Text*.[2] Sokal, a theoretical physicist who believes that the postmodern commentators on science are mistaken, wrote a paper in which he imitated the style of these scholars by drawing analogies between research in modern physics and mathematics. Sokal's article was accepted by the journal in question and promptly published. At the same time, the author revealed, in another journal, that the article had been a prank intended to expose the sloppiness of the review process among postmodern commentators on science. His prank seemed to show that complete nonsense could apparently be made to pass for scholarly work in these circles.[3] Not only did Sokal's ingenious mischief inflame passions within the already divided academic community, it also attracted the attention of lay readers. The fallout of the Sokal affair has been examined in many commentaries, editorials, and debates appearing in newspapers and public forums of various kinds.

WHAT ROLE FOR CHEMISTRY IN THE SCIENCE WARS?

It appears that, in keeping with the low profile they display in philosophy of science, chemists have been almost completely invisible in discussions on the Science Wars, with just a few exceptions.[4] But I would like to suggest that some chemical educators (I will cite a few below) are also actors in the unfolding drama, in a way that has not been generally acknowledged. In addition, I advance the more startling notion that, unwittingly, these chemical educators are fighting on the wrong side of the battle. If one looks closely at the philosophical positions offered by these chemical educators, one sees many radical themes that confirm that many of them have indeed defected to what the science lobby would regard as "the opposition."

As a recent article pointed out, there are now a number of U.S. institutions that award Ph.D.'s in chemical education research.[5] However, the field continues to be viewed by the majority of mainstream chemists with suspicion, and sometimes even with hostility. It is not uncommon to hear of junior tenure-track faculty who are under undue pressure to perform according to unrealistic criteria set by departments that do not understand, or value, the nature of research in chemical education. Indeed, one hears from some full professors, in the institutions that do have specialists in chemical education, that these are marginalized and misunderstood by their traditional chemical colleagues. It is frequently said that research in chemical education represents a soft option, suited for those who are not capable of succeeding in "real chemistry."

I believe that part of the blame for the current state of affairs lies, not with the majority of mainstream chemists, but with the field of chemical education itself. One has only to attend a chemical education session at an American Chemical Society meeting to see that the field has become somewhat inward looking and self-congratulatory. One of the biggest failings, as I see it, is a lack of engagement in issues of chemical content. Instead, chemical education research frequently withdraws into producing better visualizations, and developing multi-media projects, in the hope of improving the teaching of chemistry. Such innovations often leave the subject of chemical content as a mysterious black box that is supposed to look after itself. Mainstream chemists understandably view such activities as superficial busywork.

In this article, my aim is to concentrate on another aspect of research in chemical education, one that I believe to be more harmful to the reputation of the field. I refer especially to some dubious and abstract theoretical issues revolving around the themes of constructivism, relativism, and other philosophical "-isms."

My focus in what follows will be on the work of some chemical educators who call themselves "constructivists." Of course, mere adherence to a constructivist perspective need not be taken to mean any form of radical constructivism, of a social or individual kind, such as that which has recently angered the scientific community. But if one looks closely at the philosophical positions offered by some contemporary chemical constructivists, one sees many radical themes that are not only open to serious question but can also be construed as being anti-scientific. In other cases, I will suggest that chemical educators who call themselves constructivists are unwittingly supporting a very traditional conception of scientific knowledge that sits rather uncomfortably with constructivism as generally understood. In the following cases, I will be more concerned with philosophical motivations and commitments, as far as these may be discerned, than with detailed chemical examples, although some of the latter will also be touched on.

ORIGINS OF CHEMICAL CONSTRUCTIVISM

In a much-cited article that is regarded as a manifesto for chemical constructivism, Dudley Herron drew on Piaget's stages of psychological development and especially the transition between concrete and formal thinking.[6] Herron has become the undisputed leader in the movement to further what I will term "chemical constructivism." He is widely quoted in this context by authors who then proceed to offer what they regard as experimental support for the use of constructivism in chemical education. From a philosophical perspective, this tendency seems rather inexplicable. I can only surmise that the term constructivism is being used in a quite different sense of psychological or pedagogical constructivism rather than the philosophical or social constructivism often associated with Thomas Kuhn and others. But presumably there should be a connection between these two different forms of constructivism since a society of scientists comprises a collection of psychological individuals. If constructivism operates at the social level it might presumably be due to its also operating at the individual level. Perhaps some of the confusion philosophers experience on hearing the views that are voiced in chemical constructivism, and science education generally, is due to the gap between these two levels, the psychological and the social.

But to return to Herron, I believe that he is at heart an empiricist and that he makes no secret of this fact in many of his writings. Why he or his followers should label such views with constructivism is something that I propose to explore a little in this article. Herron has argued, as did Piaget before him, that many high school and beginning college students may not have effected the transition to the stage of formal reasoning. Herron's response is that we should take account of this fact in the way in which chemical education is approached. For example, in discussing the topic of acid-base chemistry, Herron adopts what seems to be an essentially empiricist stance.

> I have suggested that the concept of an acid as anything that will turn litmus red is a concrete concept. The meaning of the concept is easily apprehended from sensory observation and requires simple classification skills. But I have also suggested that the concept of an acid as anything that will produce hydrogen ions in water solution (Arrhenius), as a proton donor (Brønsted-Lowry) or as an electron-pair acceptor (Lewis) is formal. The meanings of acid cannot be made clear through the senses directly since there is no way to sense protons or electron pairs. Rather this concept of acid can have meaning only through imagination or through logical thought about the nature of molecules which interact.[7]

It appears that Herron is interpreting Piaget's sense of the concrete in a narrowly empiricist fashion. Clearly Herron regards only things that can be seen, or sensed directly, as being concrete.

But I think that Herron has introduced something of an inconsistency, since the kind of empiricism to which he appeals—namely, the demand that scientific knowledge should have its foundation in sense perception—stands in direct opposition to virtually all forms of constructivism. Constructivism instead upholds that scientific knowledge is not so much discovered but negotiated or "constructed" by social factors or in the mind of the scientist or the learner.

But in all fairness to Herron, a close inspection of his much-cited article, as well as a subsequent one entitled "Piaget for Chemists," reveals absolutely no reference to constructivism, either psychological or social. What these early articles show is that Herron advises chemical educators to make chemistry instruction more concrete, since so many students have apparently not reached the more formal or more abstract stage of reasoning. Herron does finally acknowledge that it might also be an idea to find ways of accelerating the student's entry into the formal level of operation.

> Chemistry, and most of science, is formal by its very nature. Recognizing this we cannot continue to duck our responsibility for the development of formal thought.[8]

But he immediately reverts to the concern shown in his entire paper, namely, the need to make chemical issues more concrete.

> There are some studies which show that education can lead to improvement in formal thinking. We are in the exploratory stage of research in this area but there are consistencies that seem to be emerging. First, the inclusion of concrete experience—i.e. opportunities to actually touch, smell, see, and manipulate materials that would lead to the concept—appears to be important.[9]

No attempt to connect these Paigetian views with any form of constructivism whatsoever has been conducted by Herron in any of his articles in the *Journal of Chemical Education*. Indeed, in all his publications in that journal, I don't believe he has used the word "constructivism" on a single occasion!

The first, and perhaps the only, article in that journal that has attempted to connect the work of Paiget and Herron with constructivism of a psychological kind is one written by George Bodner.[10] In this article, Bodner claims that constructivism is the accepted view among psychologists. Of course, this may be so. It is not for me to comment on this claim. But Bodner also makes a number of rather dismissive remarks on the subject of realism. These claims by Bodner show that there is indeed a gulf between psychological constructivism and philosophical constructivism, for the simple reason that constructivism is by no means the predominant view among philosophers of science. In addition, far from being an abandoned position in philosophy of science, scientific realism continues to flourish, and indeed appears to be the predominant view—opposed only by van Fraassen and his supporters.[11] I also note that Herron himself gives a brief discussion of how his views are supposed to constitute a form of constructivism, but this discussion appeared in a book that was published only in 1995.

The appeal to a nonspecific "constructivism" in the chemical education literature is somewhat ambivalent and continues to cause confusion. The only attempt to express disagreement with full-blown philosophical constructivism that has been made by any chemical educator, that I am aware of, was made by Herron in his book on chemical education, in which he cites another author approvingly as saying

> [even] though in some "ultimate" sense there is no way to determine whether one paradigm is a better approximation to the "real" laws on nature than another, the exclusion of nature and the empirical world from our model of how scientific knowledge grows makes it difficult to understand why some knowledge enters the core and most does not. Thus it is on practical sociological grounds that I select my realist perspective.
>
> Nature poses some limits on what the content of a solution adopted by the scientific community can be. By leaving nature out, the social constructivists make it difficult to understand the way in which the external world and social processes interact in the development of scientific knowledge.[12]

Herron then adds that

> [if] we are to understand learning, the only viable position to take is that an external reality exists, even though the understanding of it may differ from one person to another and from one point in time to the next.[13]

Although this word of caution represents a welcome improvement on the writings of other chemical constructivists, it does not go nearly far enough in moderating radical constructivist claims. It addition, it fails to distinguish clearly between philosophical and pedagogical constructivism. The author unfortunately also adds a footnote to tell readers that they can safely skip this entire section since it deals with an "obtuse point." As I see it, this section is absolutely essential to anyone involved in chemical education that might be drawn to constructivism, and should be made required, rather than optional, reading.

It is also unfortunate that Herron's followers in chemical education research, some of whom have been cited in the present article, have not seen the need to specify the precise sense in which they are using such terms as constructivism and relativism. Bodner and colleagues in particular appear to support an unqualified form of relativism, as I argue below, and which I maintain is anti-scientific in spirit.

THE "BEFORE AND AFTER" TREATMENT

Meanwhile, another chemical constructivist gives what can only be described as a simplistic comparison between what he terms "objectivism" and "constructivism" (TABLE 1).[14] Unfortunately, this tendency to present constructivism as though it were a form of weight reduction treatment, complete with "before and after" snapshots, is only too common in chemical education research.

The first of the three statements in TABLE 1 is difficult to interpret as it stands, since the author does not feel the need to qualify what is intended any further. Given

TABLE 1. Distinctions between objectivism and constructivism proposed by Spencer, a chemical constructivist

Objectivism	Constructivism
Truths are independent of the context in which they are observed.	Knowledge is constructed.
Learner observes the order inherent in the world. Aim is to transmit knowledge experts have acquired.	Group work promotes the negotiation of and develops a mutually shared meaning of knowledge. Individual learner is important.
Exam questions have one correct answer.	The ability to answer with only one answer does not demonstrate student understanding.

the scope of the article, namely chemistry and chemical education, I can only presume that the author is referring to scientific truths. The claim appears to be that objectivity is a myth regarding scientific findings whereas, according to the entry in the right-hand column, knowledge of scientific facts is constructed rather than objectively discovered. Needless to say, there may be ways of arguing for the importance of the context of scientific discoveries. After all, the growth of the Science Studies movement attests to such interests among historians and philosophers, but to reduce any such argument to the form of a one-line statement can only be described as an irresponsible move. This is especially so since such articles are intended for consumption by chemical educators who are generally not familiar with the detailed arguments that have been presented in the historical and philosophical literature. It is from chemical education researchers that chemical educators obtain their philosophical education, since they do not generally have the time or inclination to engage with the primary literature in history and philosophy of science.

In fact, to adopt a somewhat naïve view, the statement that "truths are independent of the context in which they are observed" is essentially correct, contrary to what the author implies. Indeed, it is a central belief for anyone either practicing or teaching science. If one were to believe the contents of the TABLE 1, one might conclude that a scientific truth would differ according to whether it was obtained at different geographical locations or at different times of the year, which is patent nonsense.

Similarly, if the author does not give any further qualification, the statement that "knowledge is constructed" is either plainly incorrect or so uncontroversial as to be superfluous. If the author implies that human preference dictates whether the magnitude of the speed of light is either 3 or 6 or 9×10^8 m/sec, approximately, this is simply untrue. If, on the other hand, the author is referring to the fact that all scientific knowledge is devised by human beings rather than being given to us directly by Nature itself, then, of course everyone, even the most rabid "objectivist," would probably concur.

The third entry in TABLE 1 is also a gross oversimplification. Unless the author is prepared to qualify the statement that "exam questions have one answer," which he implies to be mistaken, I don't believe he is expressing any position whatsoever. If the exam question is something along the lines of "What is the velocity of light in a vacuum?" then even a radical constructivist would have to concede that there is only one correct answer. One exception might be the possibility of quoting the velocity to varying degrees of accuracy, but this does not seem to be the kind of thing that the author intends. Indeed, in the particular case of the velocity of light, there is absolutely no possibility of there being more than one response to the question, given the peculiar nature of light.

Alternatively, if the author is thinking of an open-ended question, such as whether Bohr's theory resolved the question of the collapse of the Rutherford atom, then many might respond that there may be more than one answer. As in the previously considered case, one does not need to be a constructivist to accept the entries in the right-hand column under certain circumstances. But to claim that knowledge is constructed in general, or that the majority of exam questions have more than one answer is, I think, the height of folly.

It is not mature scientific knowledge that is constructed, but only the student's *understanding* of mature science, a theme that I return to below.

RELATIVISM WITH A VENGEANCE

One of the worst confusions set loose among chemical educators has been the notion of relativism. In an unpublished but widely distributed article, as well as a published one, George Bodner and colleagues leave the reader in no doubt about their own stance on this question.[15] Bodner and colleagues appear to have latched onto a rather idiosyncratic interpretation of relativism that they claim to support. This is what they write:

> The difference between the traditional and constructivist theories of knowledge mirrors the difference between the philosophy of science known as realist, objectivist, or positivist, and the philosophy of science known as relativist. ... Realists believe that logical analysis applied to objective observations can be used to discover the truth about the world in which we live. They view knowledge in science as cumulative; it builds upon existing knowledge as science progresses. They believe we can separate objective truth from our "means of knowing it." In other words the identity of the researcher and the choice of research methodologies will have no effect on the truth that comes out of the research. ... Relativists accept the existence of the world but question whether the world is "knowable." They note that observations, and the choice of observations to be made, are influenced by [the] beliefs, theories, hypotheses, and background of the individual who makes them. Statements about these observations are then expressed in a language whose words are embedded in a particular theoretical framework. Relativists therefore question whether a truly unbiased, objective observer can exist.[15]

I think this is a simply a misrepresentation of realism as well as relativism. To lump together realism, objectivism, and positivism is misleading, as is the implication that these positions are necessarily outmoded and inappropriate. Objectivism and realism, among the three positions grouped together, remain perfectly viable and are supported by the majority of scientists and philosophers of science. One does not need to be a relativist to accept that observations are influenced by the beliefs and background theories held by the observer. Most objectivists or realists would happily concede these uncontroversial claims regarding scientific knowledge.

Contrary to what Bodner and colleagues are claiming, the central idea in relativism is precisely that all knowledge is relative. This implies that the forms of knowledge derived from chemistry, black magic, or voodoo, to take three random examples, are all equally valid. I maintain that anyone who believes that science is worth teaching, in preference to these other pursuits, would not claim allegiance to this form of relativism. As far as I know, the only person to ever propose such an outrageous view was the self-proclaimed anarchist of science, Paul Feyerabend,[16] who did so in very similar terms. But even Feyerabend, unlike political anarchists, conceded that he did not intend others to take him seriously.

In the world of analytical philosophy, to be accused of being a relativist is tantamount to being accused of violating rationality itself. If all forms of knowledge are relative, why should one accept relativism as a worthwhile view to adopt? Relativism is simply a self-defeating position. I cannot believe that any scientist would seriously contemplate relativism as a viable philosophical position regarding the nature of scientific knowledge, or that science educators would be prepared to accept such a view. And yet this is precisely what Bodner and colleagues are recommending, in the mistaken belief that it represents a more enlightened and up-to-date philosophical approach to science.

But even the more extreme philosophers and sociologists of science who claim to be relativists have been forced to moderate their position in the light of criticism. It appears to have escaped the attention of the chemical constructivists that leading relativists like Harry Collins are now advocating what they term "methodological relativism" as opposed to full-blown, or philosophical, relativism. Collins now holds that

> Methodological relativism says nothing directly about reality or the justification of knowledge. Methodological relativism is an attitude of mind recommended to the social-scientist investigator: the sociologist or historian should act as though the beliefs about reality of any competing groups being investigated are not caused by reality itself.[17]

It appears that even the most extreme relativists are trying to distance themselves from full-blown relativism. Meanwhile the chemical educators quoted above still cling to an extreme version of relativism in the belief that it represents an improvement on "objectivism, positivism, and realism."

WHY THROW OUT THE BABY WITH THE BATHWATER?

To do full justice to the question of constructivism in science education would require a discussion of how this term is used by philosophers, sociologists, and anthropologists on one hand, and science educators on the other hand. It is important to distinguish the radical claims of the constructivists, who maintain that scientific knowledge itself is obtained by a process of negotiation and social forces, from the claims of constructivists in science education.

The first group of authors opposes the traditional belief that scientific knowledge results from investigating the way the world actually is. Meanwhile, the claims made by most constructivists in the educational sphere are more modest. They claim that students develop their understanding of science in a constructivist manner. This process is supposed to involve issues such as the preconceptions and misconceptions that students might bring to chemistry classes. One can, of course, accept such views about learning science while at the same time rejecting the more radical philosophical constructivism that claims that scientific knowledge itself is arrived at by a process of social negotiation.

Fully mature scientific knowledge, of the form that commands widespread consent by the community of scientists, does not differ according to the pedagogical evolution of the particular scientist concerned. Of course, the views of mature scientists may well have begun as "constructions" that might have been influenced by all manner of social factors, but mature science is largely free of personal idiosyncrasies.

If, on the other hand, some chemical educators do wish to support the more radical claim, that mature science itself shares constructivist elements, they should make this more explicit in their writings. But one suspects that only a small minority of chemical educators—most were trained as chemists—would want to go quite so far. Most educators are understandably attracted to educational constructivism, but overstate their case by drawing support from the more extreme and often anti-scientific writings of philosophical constructivists.

Of course, each individual developing student may have a slightly different initial conception of any particular phenomenon. One might also grant that this conception may be relative to the educational and even sociological background of the individual. But the process of learning science, perhaps more than any other field, involves reaching a position where the student has understood enough of the shared store of knowledge so that he or she can communicate with others, and even make contributions to the general scientific consensus.

I applaud chemical constructivists for encouraging teachers to be more conscious of the fact that students come to the study of chemical topics from a great variety of directions. But with respect to concepts such as constructivism and relativism, ideas borrowed from philosophy, chemical constructivists need to make it clear that they are not supporting the same brand of constructivism or relativism in the context of pedagogy. Unfortunately, the present appeal to a nonspecific "constructivism" continues to cause confusion.

SO WHAT?

Some readers may be asking whether any of these philosophical concerns have any real importance in chemical education. I believe that they have great importance, and that chemical education oversimplifies its philosophical content, as I have tried to suggest above. The current approach is sloppy and not conducive to the growth or wider acceptance of chemical education research. It is high time for chemical educators to become more philosophically informed and to begin to address the kinds of issues raised here. Otherwise, they will be providing further ammunition to what scientists generally regard as the "wrong side" of the Science Wars debate.

WHAT SHOULD BE DONE?

What I am recommending is not less use of philosophy in educational issues but more careful use. The obvious remedy is for chemical education researchers to become better acquainted with the philosophical positions to which they appeal in their writings. Secondly, philosophers of science have largely forsaken the search for an all-encompassing account of the scientific method and have concentrated instead on developing philosophical understandings of each separate natural science. Gone are the days of "heroic philosophy of science," when Popper, Kuhn, or Lakatos would try to pronounce on the nature of the whole of science.[18] It may be because these philosophers attempted to cast their nets too widely that they failed to obtain any lasting criteria to describe the nature of the scientific method.

And yet chemical constructivists continue to base a large part of their work on the views of a Kuhn or Feyerabend, to cite the most popular choices among science educators. Chemistry, like any science, has its own philosophical peculiarities that have been the focus of much investigation since the rebirth of philosophy of chemistry in the early 1990s. But whereas philosophy of chemistry is presently the fastest growing subfield in philosophy of science, it has been almost completely ignored by chemical education researchers, with a few exceptions.[19] Many resources are now available in philosophy of chemistry. All that is required is for chemical educators to begin to draw upon them.[20]

Chemistry is partly a liberal art, and is as much about thinking as it is about synthesis, experimentation, and computation. It is unfortunate that philosophy, which provides the most systematic analysis of ways of thinking, has been traditionally neglected by chemists. Even if chemical educators ignore recommendations that they should take an interest in philosophy, they should at least strive to obtain a good understanding of those philosophical concepts that have already crept into chemical education. Now that the situation has begun to change, and philosophy of chemistry is becoming an established discipline, there is no excuse for shoddy philosophical thinking on the part of chemical educators.

Just as scientists tend to be suspicious of the anti-science lobby in the Science Wars debate, they are also correctly suspicious of chemical or other educators who openly support relativistic views about science. The view that individual students may bring a variety of preconceptions to the study of chemistry is a valuable one, but this should not commit educators to relativistic views about the nature of mature science.

ACKNOWLEDGMENTS

An earlier version of this article appeared in the *Journal of Chemical Education*, volume 80, no. 5, 2003.

NOTES AND REFERENCES

1. GROSS, P.R. & N. LEVITT. 1994. Higher Superstition. Johns Hopkins University Press. Baltimore, MD.
2. SOKAL, A. 1996. Transgressing the boundaries: towards a hermeneutics of quantum gravity. Soc. Text **46–47:** 217–252.
3. SOKAL, A. 1996. Sokal's response. Lingua Franca **6:** 62–64.
4. LABINGER, J. 2001. Awakening a sleeping giant. *In* The One Culture? J.A. Labinger & H. Collins, Eds. University of Chicago Press. Chicago, IL. pp. 167–176; BAUER, H.H. 2000. Antiscience in Current Science & Technology Studies. *In* Beyond Science Wars. U. Segerstrale, Ed. State University of New York Press. Albany, NY. pp. 41–61; HERSCHBACH, D.R. 1996. *In* The Flight From Science and Reason. P.R. Gross, N. Levitt & M.W. Lewis, Eds. Vol. 775: 11–30. New York Academy of Sciences. New York; BARD, A. Chem. Eng. News. 22 April 1996. p. 5.

5. MASON, D. 2001. A survey of doctoral programs in chemical education in the United States. J. Chem. Educ. **78:** 158–160.
6. HERRON, J.D. 1975. Piaget for chemists. J. Chem. Educ. **52:** 146–150.
7. Ibid., p. 149.
8. Ibid., p. 150.
9. Ibid., p. 150.
10. BODNER, G.M. 1986. Constructivism: a theory of knowledge. J. Chem. Educ. **63:** 873–878.
11. VAN FRAASSEN, B.C. 1980. The Scientific Image. Oxford University Press. Oxford, UK.
12. HERRON, J.D. 1996. The Chemistry Classroom. American Chemical Society. Washington, DC.
13. Ibid., p. 47.
14. SPENCER, J.N. 1999. New directions in teaching chemistry: A philosophical and pedagogical basis. J. Chem. Educ. **76:** 566–569.
15. BODNER, G., M. KLOBUCHAR & D. GEELAN. 2001. The many forms of constructivism. J. Chem. Edu. **78:** 1107–1134.
16. FEYERABEND, P. 1975. Against Method. Verso. London.
17. COLLINS, H. 2001. One more round with relativeism. *In* The One Culture? J.A. Labinger & H. Collins, Eds. University of Chicago Press. Chicago, IL. pp. 184–195.
18. KUHN, T.S. 1970. The Structure of Scientific Revolutions. University of Chicago Press. Chicago, IL.
19. ERDURAN, S. 2000. A missing component of the curriculum? Educ. Chem. **9:** 168–168; Erduran, S. 2001. Philosophy of chemistry: an emerging field with implications for chemistry education. Sci. Educ. **10:** 581–593.
20. Chemical educators may be interested in *Foundations of Chemistry* (a journal published since 1999 by Kluwer Academic Press) and *Hyle* (a journal published since 1997 by the University of Karlsruhe Press—see especially volume 3 and subsequent volumes). Other writings include the following: BHUSHAN, N. & S. ROSENFELD, Eds. 2000. Of Minds and Molecules. Oxford University Press. Oxford, UK; VAN BRAKEL, J. 2000. Philosophy of Chemistry. University of Leuven Press. Leuven. Belgium; SCERRI, E.R. & L. MCINYRE. 1997. The case for the philosophy of chemistry. Synthese **111:** 213–232; SCERRI, E.R. 2000. Philosophy of chemistry—a new interdisciplinary field? J. Chem. Educ. **77:** 522–526; VAN BRAKEL, J. 1999. On the neglect of philosophy of chemistry. Found. Chem. **1:** 111–174.

The National Science Foundation and the Philosophy of Chemistry

BRUCE E. SEELY[a]

Department of Social Sciences, Michigan Technological University, Houghton, Michigan 49931-1295, USA

> ABSTRACT: Since its founding in 1950, the National Science Foundation has provided support for a variety of studies in history, philosophy, and social studies of science. The fact that a relatively small number of projects dealing with the philosophy of chemistry have received NSF support is due to the small number of such proposals that have been submitted. The NSF Science and Technology Studies Program (STS) welcomes proposals dealing with philosophy of chemistry.
>
> KEYWORDS: philosophy of chemistry; National Science Foundation (NSF); social studies of science; history of science; history of the NSF

Almost from the time that the Congress of the United States created the National Science Foundation (NSF) in 1950, NSF funds have supported scholarship in the history and philosophy of science. In recent decades, this support has expanded significantly, not only measured in terms of dollars but also in terms of support extended beyond history and philosophy to newer disciplines, such as the social studies of science, and different approaches to the study of science. Yet this expansion has not really touched the philosophy of chemistry. After briefly examining broad patterns of changes in the study of science by philosophers, historians, and social scientists, this essay will focus on available opportunities for supporting scholarly studies in the philosophy of chemistry. The intent is to encourage scholars interested in the history and philosophy of chemistry to take full advantage of the funding opportunities at the National Science Foundation's Science and Technology Studies (STS) Program.

The National Science Foundation's mandate has emphasized supporting basic research in the sciences. The lengthy fight for congressional and presidential approval of the organization in the late 1940s resulted in an initial budget that was quite limited. Program officers always had more good proposals than they had money.[1] Even so, the Foundation has almost always provided a measure of support for scholarship

---

Address for correspondence: Department of Social Sciences, Michigan Technological University, Houghton, MI 49931-1295. Voice: 906-487-2113; fax: 906-487-2468.

bseely@mtu.edu

[a]When an earlier draft of this paper was read at the Sixth Summer Symposium of the International Society for the Philosophy of Chemistry, the author was serving as Program Director for the National Science Foundation's Science and Technology Studies Program.

in the history and philosophy of science. Initial awards, apparently from the office of the director, rested upon the belief that a well-educated scientist ought to know the history and philosophy of his/her profession.[2] Earlier, this logic had helped justify the development of an academic journal in the philosophy of science in 1934, and further justified the establishment of the first academic departments in the history of science at Wisconsin and Cornell in the 1940s, as well as the expansion of the History of Science Society in the 1950s.[3] By the late 1960s, the NSF had established a regular program to support research in the History and Philosophy of Science (HPS), a step that coincided with the steady expansion of formal academic programs in these subdisciplines.

In the 1970s, the NSF's program was changing in several ways. The initial impetus came from the emergence of new fields of inquiry. The history of technology had emerged as a viable subdiscipline with the formation of the Society for the History of Technology (SHOT) in 1958, but formal recognition by the NSF program became evident only in the late 1960s—perhaps paralleling the slow acceptance of engineering itself within the NSF.[4] After the formation of the Society for the Social Studies of Science (4S) in 1975, NSF support for scholarship from sociologists of science and others interested in science studies soon followed. Over time, the NSF program's name changed to reflect this expanding domain: first to History and Philosophy of Science and Technology (HPST), and eventually to Science and Technology Studies (STS)—its current name—in the 1990s. The program's annual budget also grew with these changes, and in 2002 stood above $3.85 million.

The other important changes that marked the HPS program in this period followed from this expansion of the field—for researchers from new disciplinary groups brought with them new approaches to scholarship. The shifts were similar to those occurring in almost all realms of the social sciences and humanities at this time, as conceptions from the sociology of knowledge, social theory, literary theory, and other sources, worked their way to the consciousness of scholars. Within the community studying science and technology, Thomas Kuhn's path-breaking study, *The Structure of Scientific Revolutions*, buttressed a social approach to science visible two decades earlier in Robert Merton's *Science, Technology and Society in Seventeenth Century England*.[5] Kuhn is widely seen as making a pivotal contribution that steered historians and philosophers of science towards a new approach to their work. Among historians, the emergence, in the 1970s, of a contextual style of history that focused on social settings, culture, and people as much as on the genesis of hardware, was one indication of the shift.[6]

As scholars adopted theoretical approaches such as social construction in their work, additional significant changes in scholarly approaches to science followed during the 1980s and 1990s. Sociologists of science from Europe—especially those associated with the University of Edinburgh—were highly influential in introducing this outlook and approach, which emphasized science as a social activity. Their outlook reinforced the view that science was more than an intellectual activity, largely divorced from the rest of society and culture. Instead of envisioning science as a process in which ideas passed from great mind to great mind, scholars began to emphasize the manner in which a range of social, political, cultural, and economic factors influenced the development and growth of science.[7] Scholars such as Bruno Latour focused on scientific work inside laboratories, highlighting the highly subjective human interactions and motivations that were part of all scientific activities. Latour and others introduced concepts such as actor network theory, which have been widely adopted by scholars working in science studies.[8] The inevitable—indeed, the intended—result was to diminish the sense that science was a special or privileged form of knowledge, objective in nature. For philosophers, a key element in the resulting debates concerned scientific realism.

These scholarly stances inevitably aroused opposition in many quarters, but the "conversations" turned nasty in the early 1990s, during the so-called "Science Wars." Traditionalists dismissed these new approaches as anti-scientific, and the resulting controversy occasionally got very nasty indeed—especially after physicist Alan Sokal published a supposed parody of a postmodern essay on quantum gravity in the journal *Social Text* in 1996, touching off a furor of immense proportions.[9] Yet hidden beneath the turmoil was that fact that the intellectual challenge posed by new theoretical frameworks in science studies led to dynamic and exciting scholarship. Many scholars cannot completely accept constructivist approaches to science, as shown by Elliot Scerri's essay (in this volume) decrying the simplistic—if not simple-minded—manner in which some writers have adopted social construction. A similar controversy followed the publication of Steve Fuller's recent book on Thomas Kuhn.[10] But few scholars working today within the community of the history, philosophy, or the social studies of science advocate studying science strictly as a branch of intellectual history

From the perspective of the NSF's STS Program, this attention to science as a social enterprise has not been the only change affecting the philosophy of science and the broader field of science studies. Another recent trend has been the increasing attention by philosophers to several areas of their discipline, generally paralleling the unfolding of new fields of science. During the 1990s, the philosophy of biology has become an especially vibrant area of scholarship, reflecting the significant scientific developments in terms of genomics. And at the end of the decade, scholarship in the philosophy of psychology showed a similar tendency, as philosophers of science and philosophers of mind found points of departure in new understandings brought forward in the cognitive sciences and through new research tools such as brain-scan technology. The emergence of another working group, the History of the Philosophy of Science (HOPOS), marked another target area for philosophers of science. The NSF's STS Program has attempted to support all of these developments.

Despite this expansion across so many areas of science studies, scholarship in the philosophy of chemistry has not been a recipient of support from the NSF's STS Program. This is not a product of deliberate exclusion. The real problem is that few, if any, proposals for such work have been submitted for consideration. The STS Program has certainly paid attention to chemistry, but most all of the proposals supported have been historical examinations. The chemistry history projects funded by the STS program within the past ten years include a range of topics across time: John Robison and natural philosophy in the Scottish enlightenment (David Wilson, Iowa State); early modern atomism (William Newman, Indiana); the rise of synthetic chemistry (Alan Rocke, Case Western Reserve); Andreas Libavius and chemistry and culture in early modern Germany (Bruce Moran, Nevada-Reno); the American organic chemical industry from 1910–1930 (Kathryn Steen, Drexel); the interplay of commerce and science in the 19th century American chemicals industry (Paul Lucier, RPI); and a study of career patterns of African-American chemists (Willie Pearson, Wake Forest).

The program also supported a conference on chemists and World War I (Jeffrey Johnson, Villanova), and a CAREER award whose research component focused on the Paris Academy in the early 18th century and the role of Wilhelm Homberg in developing the "new chemistry" (Lawrence Principe, Johns Hopkins). Additionally, the program has supported a number of doctoral dissertations, on such topics as Athanasius Kircher and Occult Traditions (Daniel Stolzenberg, Stanford); Herman Boerhaave and chemical education in the 18th century (John Powers, Indiana); chemistry and culture in 18th century France (John Dettloff, Princeton); Humphrey Davy and modern chemistry (Julian Tuttle, Indiana); and color and chemistry (Sarah Lowengard, SUNY-Stony Brook). All of these projects are strongly historical in nature, although several of them deal with philosophical issues. Thus Mi Gyung Kim's (North Carolina State) current study of several French chemists at the *Jardin du Roi* and the *Academie Royal des Sciences* deals with chemical philosophy in understanding the tension between theory and practice during the "chemical revolution." In the end, the only participant in the Sixth Summer Symposium of the International Society for the Philosophy of Chemistry who has received support for the STS Program at NSF is Nathan Brooks (New Mexico State), who is preparing a biography of Dmitrii Mendeleev, another historical project.[11]

The STS Program is more than interested in changing this state of affairs. The philosophy of chemistry clearly falls within the purview of the program, and the program strongly encourages proposals from interested scholars. Awards are made for a range of purposes, as the above list suggests. The most common grants are made as **STS Scholars Awards**, which provide support for individual investigators for an academic year, with the possibility of partial support for up to another two years (three years total). Research projects may be in any of the disciplines supported by the program.[12] Similar awards are made for small groups of scholars working on the same project through **Grants for Collaborative Research**. With solid justification, it is possible to seek such awards to support projects yielding educational infrastructure, including the writing and editing of reference works, editions of scientific and personal papers, or databases and digital libraries. Funding limits apply, and only a few projects can be supported, so applicants should talk to STS Program officers early in the planning stages of such proposals. Indeed, this is good advice in developing any proposal.

The STS Program also supports two types of fellowships. **Classic Postdoctoral Fellowships** are available for researchers within five years of receipt of their Ph.D. All applicants must be U.S. citizens or permanent residents. Please note that the STS program requires that these proposals include both a training and a research component; in addition, the site for the fellowship must be different from the institution where the applicant received their doctorate. These fellowships, which may be for two years, carry an annual stipend of up to $36,000 (including fringe benefits), as well as a research and travel expenses allowance of up to $3,000/year and an institutional allowance of $3,000/year (in lieu of indirect costs). The STS Program also offers **Professional Development Fellowships** to allow scholars trained in history, philosophy, or the social studies of science and technology to expand their knowledge of science or engineering; conversely physical and natural scientists and engineers may apply for training in STS disciplines. Normally, this training is accomplished by working with an individual scholar at another institution to learn the appropriate research methods, analytical tools, and current scholarly understandings. These fellowship applications also must have training and research components. Stipends depend upon the applicant's current salary and work history.

The STS Program supports an extensive **Doctoral Dissertation Research Improvement Grants** program. The program awards an average of 20 such grants each year as its primary means of assisting graduate study in the science studies area. These awards do not carry a stipend, and are limited to research expenses not normally available through the student's university. The usual limit on a dissertation award is $8,000 for research in North America and $12,000 for work abroad. The only other training program within the STS Program is its **Small Grants for Training and Research (SGTR)**. These awards are intended to support program development by providing up to three years of support for a group of graduate students and postdoctoral fellows conducting research on a targeted research theme within the history, philosophy, or social studies of science. Awards are for up to $100,000 per year for three years, and provide no support for faculty. SGTR proposals are considered each fall.

Finally, the STS Program makes small awards for **Conferences and Workshops**. Such grants are to support national or international conferences, symposia, and research workshops. Ideally, the program prefers to support smaller gatherings that bring together groups of scholars who have not previously been in contact, in order to advance research agenda-setting goals. Participants should be as diverse as possible. Unless a very strong justification is provided, the program normally limits its support for conferences to $10,000.

CONCLUSION

The STS Program at the NSF invites scholars working in the philosophy of chemistry to develop proposals that respond to any of these funding opportunities. This avenue has not been well utilized by philosophers of chemistry up to this point in time, unlike their historian colleagues. The roster of presenters at the Sixth Annual Summer Symposium on the Philosophy of Chemistry is one indication of this situation. But there is no reason for this situation to continue. Indeed, exciting work is already taking place in this area of the philosophy of science that could become the subjects of competitive grant proposals. An indicator of this circumstance can be found in the call of papers for *Hyle,* the international journal for philosophy of chemistry, which circulated at the symposium. The call sought papers for a special issue devoted to aesthetics and visualization in chemistry, a set of exciting topics that clearly fall within the mainstream of the work currently supported by the STS program. Among the specific topics the editors identified as being of special interest within those general headings, were the aesthetics of the laboratory and the aesthetics of materials, with one possible question under the latter heading being scholarship on the "sensual qualities of materials." Intriguingly, both of these categories resonate well with contemporary scholarship elsewhere within the community of scholars who explore science. The focus on the laboratory has been a major subject of attention for almost two decades. Research on the sensual aspect of chemistry, on the other hand, matches an area that is just beginning to unfold in science studies, as scholars begin exploring how a range of stimuli (sound, touch, smell—the last an especially appropriate issue of chemistry!) are connected to science and technology studies. The issue of visualization identified by the editors of *Hyle* fits right into this emerging area of scholarship. These are areas, then, in which the STS program would be as interested as the editors of the journal. Perhaps some grant proposals can begin to move to Washington as a result. For more information on the program or for guidance about contacts, please go to the STS Program's Web site: http://www.nsf.gov/sbe/ses/sts/start.htm.

NOTES AND REFERENCES

1. Proposed by Vannevar Bush in his classic report issued at war's end, legislation creating the NSF was vetoed by President Truman in 1947. The key issue, executive control over the organization, delayed the NSF's emergence until 1950. BUSH, V. 1960. Science, The Endless Frontier. National Science Foundation. Washington, DC. The early history of the NSF is treated in a number of historical summaries. See, for example, KEVLES, D. 1979. The Physicists: The History of a Scientific Community in America. Vintage Books. New York. pp. 342–366; also MAZUZAN, G. 1988. The National Science Foundation: A Brief History. NSF Publication 8816, http://www.nsf.gov/pubs/stis1994/nsf8816/nsf8816.txt.
2. Expressions of this logic can be found in HERSHBERG, J.G. 1993. James B. Conant: Harvard to Hiroshima and the Making of the Nuclear Age. Alfred A. Knopf. New York. pp. 407–412; NASH, L.K. An historical approach to the teaching of the history of science. J. Chem. Educ. **28:** 145–151; DE MILT, C. 1952. The value of the history and philosophy of science in the training of graduate students in chemistry. J. Chem. Educ. **29:** 340–344; and EDGE, D. 1995. Reinventing the wheel. *In* Handbook of Science and Technology Studies. S. Jasanoff *et al.*, Eds.: 7–8. Sage Publications. Thousand Oaks, CA.
3. The History of Science Society had taken shape in 1924, following by twelve years the formation of what became its journal—*Isis*. Both were very much a product of their founder, George Sarton, but a strong professional structure was a product of the postwar period. See THACKRAY, A. & R.K. MERTON. 1972. On discipline building: the paradoxes of George Sarton. Isis **63:** 473–495; CLAGETT, M. 1951. Teaching the history of science at the University of Wisconsin. Isis **43:** 51.
4. Engineering was initially deemed too much an applied science, so research in engineering was not at first supported by the NSF. See BELANGER, D.O. 1998. Enabling American Innovation: Engineering and the National Science Foundation. Purdue University Press. West Lafayette, IN.
5. KUHN, THOMAS S. 1962. The Structure of Scientific Revolutions. University of Chicago Press. Chicago, IL; MERTON, R.K. 1938. Science, Technology and Society in Seventeenth England. Bruges. See also KUHN, T.S. 2000. The Road since Structure: Philosophical Essays, 1970–1993 [with an autobiographical interview]. University of Chicago Press. Chicago, IL; and EDGE, D. Reinventing the wheel. *In* Handbook of Science and Technology Studies. S. Jasanoff *et al.*, Eds. Sage Publications. Thousand Oaks, CA. pp. 5–11.
6. On the impact of Kuhn, see such works as HORWICH, P., Ed. 1993. World Changes: Thomas Kuhn and the Nature of Science. MIT Press. Cambridge, MA; BARNES, B. 1981. T.S. Kuhn and Social Science. Columbia University Press. New York; FULLER, S. 2000. Thomas Kuhn: A Philosophical History for Our Time. University of Chicago Press. Chicago, IL; BRUSH, S.G. 2000. Thomas Kuhn as a historian of science. Sci. Educ. **9:** 39–58; BABER, Z. 1998. "Deconstruction Gone Mad?": The Social Construction of the Kuhnian Revolution and Its Consequences for the Sociology of Science. Centre for Advanced Studies. University of Singapore. Singapore; and STOVE, D.C. 2001. Scientific Irrationalism: Origins of a Postmodern Cult. Transaction. New Brunswick, NJ.
7. For examples of these developments, see EDGE, D.O. & J.N. WOLFE, Eds. 1973. Meaning and Control: Essays in Social Aspects of Science and Technology. [Based on the Seminar on Social Aspects of Science and Technology, University of Edinburgh,

1970.] Tavistock Publications. London; BARNES, B. 1974. Scientific Knowledge and Sociological Theory. Routledge & K. Paul. London; BARNES B. & S. SHAPIN. 1979. Natural Order: Historical Studies of Scientific Culture. Sage Publications. Beverly Hills, CA.

8. LATOUR, B. 1979. Laboratory Life: The Social Construction of Scientific Facts. Sage Publications. Beverly Hills, CA; LATOUR, B. 1987. Science in Action: How to Follow Scientists and Engineers Through Society. Harvard University Press. Cambridge, MA; SHAPIN, S. 1985. Leviathan and the Air-Pump: Hobbes, Boyle, and the Experimental Life. Princeton University Press. Princeton, NJ; LAW, J. 1984. Science for Social Scientists. Macmillan Press. London; LAW, J. & J. HASSARD, Eds. 1999. Actor Network Theory and After. Blackwell Publishers/Sociological Review. Oxford; WOOLGAR, S. 1988. Knowledge and Reflexivity: New Frontiers in the Sociology of Knowledge. Sage. London; and CALLON, M. 1986. Mapping the Dynamics of Science and Technology: Sociology of Science in the Real World. Macmillan Press. Houndmills, Basingstoke, Hampshire. In the realm of technology, the key works in this vein are BIJKER, W., T. P. HUGHES & T.J. PINCH, Eds. 1987. The Social Construction of Technological Systems: New Directions in the Sociology and History of Technology. MIT Press. Cambridge, MA; BIJKER, W.E. & J. LAW, Eds. 2000. Shaping Technology/Building Society: Studies in Sociotechnical Change. MIT Press. Cambridge, MA; and BIJKER, W.E. 1995. Of Bicycles, Bakelites, and Bulbs: Toward a Theory of Sociotechnical Change. MIT Press. Cambridge, MA.

9. See comment by EDGE, D. 1995. Reinventing the wheel. *In* Handbook of Science and Technology Studies. S. Jasanoff *et al.*, Eds. Sage Publications. Thousand Oaks, CA. p. 15. Edge's comment provides an early instance of the events that figure prominently in GROSS, P.R. & N. LEVITT. 1994. Higher Superstition: The Academic Left and its Quarrels with Science. Johns Hopkins University Press. Baltimore, MD; GROSS, P.R., N. LEVITT & M.W. LEWIS, Eds. 1996. The Flight from Science and Reason. New York Academy of Sciences. New York; ROSS, A., Ed. 1996. Science Wars. Duke University Press. Durham, NC; KOERTGE, N., Ed. 1998. A House Built on Sand: Exposing Postmodernist Myths about Science. Oxford University Press. Oxford. On the Sokal hoax, see Editors of *Lingua Franca*. 2002. The Sokal Hoax: The Sham That Shook the Academy. University of Nebraska Press. Lincoln, NE; and SOKAL, A. 1998. Fashionable Nonsense: Postmodern Intellectuals' Abuse of Science. Picador USA. New York. A complete bibliography of articles, comments, reviews, and other information on this event can be found at http://www.physics.nyu.edu/faculty/sokal/

10. See HOLLINGER, D.A. May 28, 2000. Paradigms Lost. The New York Times Book Review. **105:** 23; RAYMO, C. 2000. Review: Thomas Kuhn. Sci. Am. **283:** 104–105; ANDERSEN, H. 2001. Critical Notice: Kuhn, Conant and Everything—A Full or Fuller Account. Phil. Sci. **68:** 258–263; MCOUAT, G. 2001. The mistaken gestalt of science studies: Steve Fuller takes on Kuhn. Can. J. Hist. **36:** 523–529.

11. The NSF's Division of Chemical Education (DUE) has funded some "course development" projects that relate, more or less directly, to the philosophy of chemistry—including two grants for which the editor of this volume was principal investigator. —Ed.

12. Please see http://www.nsf.gov/pubs/2001/nsf01159/nsf01159.html for detailed information about this and other awards, including budgetary guidelines.

Index of Contributors

Akeroyd, M., 293–301

Bogaard, P.A., 307–312
Brandt, W.W., 335–344
Brown, T.L., 209–216
Burewicz, A., 244–249

Del Re, G., 133–140

Earley, J.E., Sr., xi–xii, 80–89

Fisher, G., 16–21

Goodwin, W., 141–153

Harré, R., ix–x, 1–15
Heelan, P.A., 114–127
Hendry, R.F., 44–58
Hojo, K., 353–358

Jacob, C., 239–243
Job, G., 171–181

Kaji, M., 302–306
Khuri, R.K., 322–334
King, R.B., 158–170
Kohout, L., 250–256
Kovac, J., 233–238
Kühl, E., 203–208

Lankau, T., 171–181, 203–208

Mattingly, J., 193–202
Miranowicz, N., 244–249

Needham, P., 99–113
Nowacki, M.R., 313–321

Ostrovsky, V.N., 182–192

Prigogine, I., 128–132

Roberts, J.A., 257–268
Rothbart, D., 250–256

Salmon, J.F., 345–352
Scerri, E., 359–369
Schummer, J., 269–281
Seely, B.E., 370–376
Spector, T.I., 227–232
Stemwedel, J.D., 217–226

Ulanowicz, R.E., 154–157

Van Brakel, J., 30–43
Vemulapalli, G.K., 90–98
Ver Eecke, W., 313–321
Vihalemm, R., 59–70
Vollmer, S.H., 71–79

Weisberg, M., 282–292
Wood, R., 282–292
Woody, A.I., 22–29

OHIO UNIVERSITY LIBRARY

Please return this book as soon as you have finished with it. In order to avoid a fine it must be returned by the latest date stamped below. All books are subject to recall after two weeks or immediately if needed for reserve.

CF